Springer Series in Reliability Engineering

For further volumes:
http://www.springer.com/series/6917

Hoang Pham
Editor

Safety and Risk Modeling and Its Applications

Springer

Prof. Hoang Pham
Department of Industrial and Systems Engineering
Rutgers University
96 Frelinghuysen Road
Piscataway
New Jersey 08854-8018
USA
e-mail: hopham@rci.rutgers.edu

ISSN 1614-7839
ISBN 978-0-85729-469-2 e-ISBN 978-0-85729-470-8
DOI 10.1007/978-0-85729-470-8
Springer London Dordrecht Heidelberg New York

British Library Cataloguing in Publication Data
A catalogue record for this book is available from the British library

Cover design: eStudio Calamar, Berlin/Figueres

Printed on acid-free paper

Springer is part of Springer Science+Business Media (www.springer.com)

Dedicated to

Michelle, Hoang Jr. and David

Preface

Today's products and safety critical-systems of most applications have become increasingly complex to build while the demand for safety and cost effective development continues. The interest in safety and risk modeling and assessment has been growing in many years to come.

This book, consisting of 14 chapters, aims to present latest methods and techniques for quantifying the safety and risk with emphasis on safety, reliability and risk modeling and their applications in several areas including Aviation Systems and Security, Sensor Detection and Management Decision-Making, and Systems of Systems and Human Integration. The subjects covered include safety engineering, system maintenance, safety in design, failure mode analysis, risk concept and modeling, software safety, human safety, product safety in operating environments, human-hand safety modeling, sensor management, safety decision-making, nuclear safety detection, sensor data modeling, uncertainty modeling, aircraft system safety, aviation security and safety, passenger screening models, checked baggage screening models, security inspection, risk-based analysis, economic reliability-safety, quantile risk approach, risk importance measures, probability risk assessment, safety analysis, complex system reliability modeling, risk-informed decision-making, in-service inspection, Markov modeling, hybrid uncertainty system safety assessment, risk prediction, warranty concepts, maintenance policies, and risk management.

The book consists of three parts. Part I—Safety and Reliability Modeling—contains four papers, focuses on the modeling and prediction of complex systems and finance management with respect to safety, reliability and maintenance, and economic aspects.

The first chapter by Kołowrocki is concerned with the application of limit reliability functions to the reliability evaluation of complex systems. Reliability evaluation of a port grain transportation system and a rope elevator are also discussed. Chapter 2 by Tapiero discusses an economic approach to reliability design and safety based on economic considerations its safety consequences which depend on both system reliability and their safety-consequential effects. Chapter 3 by Chakraborty and Sam deals with the safety evaluation of complex systems.

The chapter discusses various developments of reliability evaluation of structures with emphasis on the safety evaluation of hybrid uncertain system. An application of reinforced concrete beam is given to illustrate the proposed safety evaluation procedure of hybrid uncertain system. Chapter 4 by Park and Pham discusses on the developments of warranty concepts with various maintenance policies and mathematical methods that used to formulate the mathematical warranty modeling. The concepts of warranty and the overall information about the warranty policy such as warranty's role, concept, and different types will also be discussed.

Part II—Risk Modeling—contains four papers, focuses on the risk modeling, methods, and methodologies. Chapter 5 by Aven provides basic principles and methods in risk analysis. Some key challenges related to the treatment of uncertainties and the incorporation of human and organizational factors are discussed. Chapter 6 by Zio focuses on the use of various classical importance measures such as Birnbaum's measure, Criticality importance, Fussell-Vesely importance measure in risk-informed applications relates to the ranking, or categorization of components or, more generally, basic events, with respect to their risk-significance. Several applications of risk importance measures are also discussed. Chapter 7 by Wang and Pham discusses recent modeling the reliability of complex degradation systems with competing risks as well as random shocks. Further research related to this subject on the condition-based maintenance and the maintenance policies for multi-component systems embedded within the framework of dependent competing risk of degradation wear and random shocks are also discussed. Chapter 8 by Otsuka discusses the framework of the system design review based on failure modes (DRBFM) based on the design concept and its procedure to identify both the latent problems and misunderstood problems during the system evaluation process. It can be used to help design management team to predict number of failure modes which can improve reliability and safety of products in use.

Part III—Applications—contains six papers, aims to discuss various applications in three areas that are related to safety and risk such as: Aviation Systems and Security, Sensor Detection and Management Decision-Making, and System of Systems and Human Integration.

In the application of aviation systems and security, Chap. 9 by McLay discusses various analytical modeling approaches including checked baggage screening models and passenger screening models for managing risk in aviation security screening systems based on probabilistic methods. It focuses on passenger screening problems, an important and highly visible aspect of aviation security. Chapter 10 by Oztekin and Luxhoj discusses a generalized hybrid Fuzzy-Bayesian methodology for modeling the risk and uncertainty associated with complex real-world systems and the emergent Unmanned Aircraft Systems (UAS) operations in the National Airspace System (NAS) in particular.

The application of sensor detection and management decision-making, Chap. 11 by Carpenter, Cheng, Roberts, and Xie describes a variety of approaches to sensor management problems of nuclear detection. Their approaches in the project include: the methods to exploit data from radiation sensors and shipping manifests for classification and decision-making; ways to optimize sequential decisions in layered inspection processes; detection using a fleet of mobile radiation sensors; and data sampling strategies for nuclear detection. Chapter 12 by Verma, Srividya, Gopika, and Rao focuses on the risk-informed decision-making based on probabilistic safety assessment (PSA) methodology with applications in nuclear power plant safety such as technical specification optimization and risk informed in-service inspection (ISI). The goal of risk informed ISI is to allow the use of risk assessment, understanding of component specific degradation mechanisms, and to establish an effective plant integrity management program, which maintains plant safety.

The application of system in systems and human integration, Chap. 13 by Schneidewind describes an application case study that illustrates how software risk and safety reliability analysis can be used to not only reduce the risk of software failure but also to improve the reliability of the entire software product using sequential testing approach. Chapter 14 by Marler et al. focuses on the predictive modeling from a biomechanical perspective between human body performance, hand modeling capabilities and cognitive modeling. A three pronged approach to hand analysis such as model development, reach analysis, and grasping prediction is also discussed.

All the chapters are written by more than 25 leading experts in the field with a hope to provide readers the gap between theory and practice and to trigger new research challenges in safety and risk in practices.

I am deeply indebted and wish to thank all of them for their contributions and cooperation. Thanks are also due to the Springer staff for their editorial work. I hope that the readers including engineers, teachers, scientists, postgraduates, researchers, managers, and practitioners will find this book as a state-of-the-references survey and as a valuable resource for understanding the latest developments in both qualitative and quantitative methods of safety and risk analysis and their applications in complex operating environments.

Piscataway, New Jersey, August 2010 Hoang Pham

Contents

Part III Applications

Contributors

Karim Abdel-Malek, Center for Computer Aided Design, University of Iowa, 111 Engineering Research Facility, Iowa City, Iowa 52242, USA

Terje Aven, University of Stavanger, Norway

Steve Beck, Center for Computer Aided Design, University of Iowa, 111 Engineering Research Facility, Iowa City, Iowa 52242, USA

Tamra Carpenter, DIMACS, Rutgers University, Piscataway, NJ 08854, USA

Subrata Chakraborty, Department of Civil Engineering, Bengal Engineering and Science University, Shibpur, Howrah, 711103, India

Jerry Cheng, DIMACS, Rutgers University, Piscataway, NJ 08854, USA

V. Gopika, Bhabha Atomic Research Centre, Mumbai, India

Faisal Goussous, Center for Computer Aided Design, University of Iowa, 111 Engineering Research Facility, Iowa City, Iowa 52242, USA

Ross Johnson, Center for Computer Aided Design, University of Iowa, 111 Engineering Research Facility, Iowa City, Iowa 52242, USA

Krzysztof Kołowrocki, Maritime University, ul. Morska 81-87, 81-225 Gdynia, Poland

James T. Luxhøj, Department of Industrial and Systems Engineering, Rutgers University, 08855-8018, Piscataway, USA

Tim Marler, Center for Computer Aided Design, University of Iowa, 111 Engineering Research Facility, Iowa City, Iowa 52242, USA

Laura A. McLay, Department of Statistical Sciences and Operations Research, Virginia Commonwealth University, 1015 Floyd Avenue, P.O. Box 843083, Richmond, Virginia 23284, USA. e-mail: lamclay@vcu.edu

Chris Murphy, Center for Computer Aided Design, University of Iowa, 111 Engineering Research Facility, Iowa City, Iowa 52242, USA

Yuichi Otsuka, Nagaoka University of Technology, Japan

Ahmet Oztekin, Department of Industrial and Systems Engineering, Rutgers University, Piscataway, USA

Minjae Park, College of Business Administration, Hongik University, Sangsu-dong, Mapogu, Seoul 121-791, Korea

Hoang Pham, Department of Industrial and Systems Engineering, Rutgers University, 96 Frelinghuysen Road, Piscataway, NJ 08855-8018, USA

K. Durga Rao, Paul Scherrer Institute, Villigen PSI, Switzerland

Fred Roberts, DIMACS, Rutgers University, Piscataway, NJ 08854, USA

Palash Chandra Sam, Department of Civil Engineering, Bengal Engineering and Science University, Shibpur, Howrah 711103, India

Norman Schneidewind, Naval Postgraduate School, 1411 Cunningham Road, Monterey, CA 93943-5219, USA

A. Srividya, Indian Institute of Technology Bombay, Mumbai, India

Charles S. Tapiero, Department of Finance and Risk Engineering, The Polytechnic Institute of New York University, Brooklyn, New York, USA

A. K. Verma, Indian Institute of Technology Bombay, Mumbai, India

Yaping Wang, Department of Industrial and Systems Engineering, Rutgers University, 96 Frelinghuysen Road, Piscataway, NJ 08855-8018, USA

Minge Xie, DIMACS, Rutgers University, Rutgers University, Piscataway, NJ 08854, USA

Enrico Zio, Politecnico di Milano, Italy

Part I
Safety and Reliability Modeling

Reliability Modelling of Complex Systems

Krzysztof Kołowrocki

1 Introduction

Many technical systems belong to the class of complex systems as a result of the large number of components they are built of and their complicated operating processes. As a rule these are series systems composed of large number of components. Sometimes the series systems have either components or subsystems reserved and then they become parallel–series or series–parallel reliability structures. We meet large series systems, for instance, in piping transportation of water, gas, oil and various chemical substances. Large systems of these kinds are also used in electrical energy distribution. A city bus transportation system composed of a number of communication lines each serviced by one bus may be a model series system, if we treat it as not failed, when all its lines are able to transport passengers. If the communication lines have at their disposal several buses we may consider it as either a parallel–series system or an "*m* out of *n*" system. The simplest example of a parallel system or an "*m* out of *n*" system may be an electrical cable composed of a number of wires, which are its basic components, whereas the transmitting electrical network may be either a parallel–series system or an "*m* out of *n*"-series system. Large systems of these types are also used in telecommunication, in rope transportation and in transport using belt conveyers and elevators. Rope transportation systems like port elevators and ship-rope elevators used in shipyards during ship docking are model examples of series–parallel and parallel–series systems.

In the case of large systems, the determination of the exact reliability functions of the systems leads us to complicated formulae that are often useless for reliability practitioners. One of the important techniques in this situation is the

K. Kołowrocki (✉)
Maritime University, ul. Morska 81-87, 81-225 Gdynia, Poland
e-mail: katmatkk@am.gdynia.pl

H. Pham (ed.), *Safety and Risk Modeling and Its Applications*,
Springer Series in Reliability Engineering, DOI: 10.1007/978-0-85729-470-8_1,

asymptotic approach to system reliability evaluation. In this approach, instead of the preliminary complex formula for the system reliability function, after assuming that the number of system components tends to infinity and finding the limit reliability of the system, we obtain its simplified form.

The mathematical methods used in the asymptotic approach to the system reliability analysis of large systems are based on limit theorems on order statistics distributions, considered in very wide literature, for instance in [4, 5, 8, 23]. These theorems have generated the investigation concerned with limit reliability functions of the systems composed of two-state components. The main and fundamental results on this subject that determine the three-element classes of limit reliability functions for homogeneous series systems and for homogeneous parallel systems have been established by Gniedenko in [6]. These results are also presented, sometimes with different proofs, for instance in subsequent works [1, 9]. The generalisations of these results for homogeneous "m out of n" systems have been formulated and proved by Smirnow in [19], where the seven-element class of possible limit reliability functions for these systems has been fixed. As it has been done for homogeneous series and parallel systems classes of limit reliability functions have been fixed by Chernoff and Teicher in [2] for homogeneous series–parallel and parallel–series systems. Their results were concerned with so-called "quadratic" systems only. They have fixed limit reliability functions for the homogeneous series–parallel systems with the number of series subsystems equal to the number of components in these subsystems, and for the homogeneous parallel–series systems with the number of parallel subsystems equal to the number of components in these subsystems. Kołowrocki has generalised their results for non-"quadratic" and non-homogeneous series–parallel and parallel–series systems in [9]. All these results may also be found, for instance, in [10].

The results concerned with the asymptotic approach to system reliability analysis have become the basis for the investigation concerned with domains of attraction [10, 14] for the limit reliability functions of the considered systems and the investigation concerned with the reliability of large hierarchical systems as well [3, 10]. Domains of attraction for limit reliability functions of two-state systems are introduced. They are understood as the conditions that the reliability functions of the particular components of the system have to satisfy in order that the system limit reliability function is one of the limit reliability functions from the previously fixed class for this system. Exemplary theorems concerned with domains of attraction for limit reliability functions of homogeneous series systems are presented here and the application of one of them is illustrated. Hierarchical series–parallel and parallel–series systems of any order are defined, their reliability functions are determined and limit theorems on their reliability functions are applied to reliability evaluation of exemplary hierarchical systems of order two.

All the results so far described have been obtained under the linear normalisation of the system lifetimes. The chapter contains the results described above and comments on their newest generalisations recently presented in [10].

The chapter is concerned with the application of limit reliability functions to the reliability evaluation of large systems. Two-state large non-repaired systems composed of independent components are considered. The asymptotic approach to the system reliability investigation and the system limit reliability function is defined. Two-state homogeneous series, parallel and series–parallel systems are defined and their exact reliability functions are determined. The classes of limit reliability functions of these systems are presented. The results of the investigation concerned with domains of attraction for the limit reliability functions of the considered systems and the investigation concerned with the reliability of large hierarchical systems as well are discussed in the paper. The chapter contains exemplary applications of the presented facts to the reliability evaluation of large technical systems. Moreover, series-"*m* out of *n*" systems and "*m* out of *n*"-series systems are defined, and exemplary theorems on their limit reliability functions are presented and applied to the reliability evaluation of a piping transportation system and a rope elevator. Applications of the asymptotic approach in large series systems reliability improvement are also presented. The paper is completed by showing the possibility of applying the asymptotic approach to the reliability analysis of large systems placed in their variable operation processes. In this scope, the asymptotic approach to reliability evaluation for a port grain transportation system related to its operation process is performed.

2 Reliability of Two-State Systems

We assume that

$$E_i, \quad i = 1, 2, \ldots, n, \quad n \in N,$$

are two-state components of the system having reliability functions

$$R_i(t) = P(T_i > t), \quad t \in (-\infty, \infty),$$

where

$$T_i, \quad i = 1, 2, \ldots, n,$$

are independent random variables representing the lifetimes of components E_i with distribution functions

$$F_i(t) = P(T_i \leq t), \quad t \in (-\infty, \infty),$$

The simplest two-state reliability structures are series and parallel systems. We define these systems first.

Definition 1 We call a two-state system series if its lifetime T is given by

$$T = \min_{1 \leq i \leq n} \{T_i\}.$$

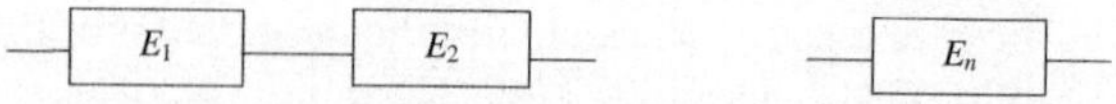

Fig. 1 The scheme of a series system

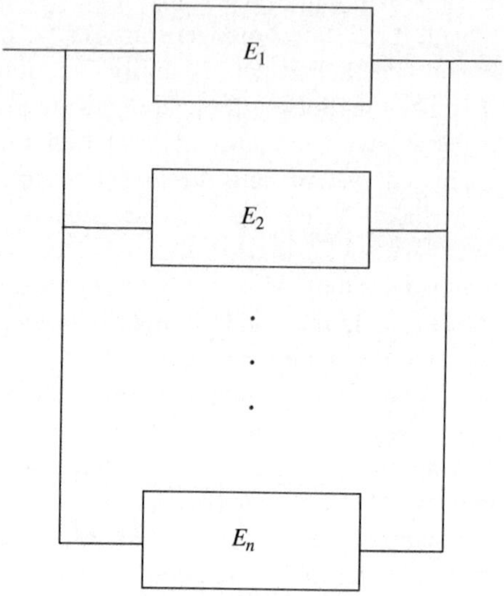

Fig. 2 The scheme of a parallel system

The scheme of a series system is given in Fig. 1.

Definition 1 means that the series system is not failed if and only if all its components are not failed and therefore its reliability function is given by

$$\overline{\boldsymbol{R}}_n(t) = \prod_{i=1}^{n} R_i(t), \quad t \in (-\infty, \infty). \tag{1}$$

Definition 2 We call a two-state system parallel if its lifetime T is given by

$$T = \max_{1 \le i \le n} \{T_i\}.$$

The scheme of a parallel system is given in Fig. 2.

Definition 2 means that the parallel system is failed if and only if all its components are failed and therefore its reliability function is given by

$$\boldsymbol{R}_n(t) = 1 - \prod_{i=1}^{n} F_i(t), \quad t \in (-\infty, \infty). \tag{2}$$

Another basic, a bit more complex, two-state reliability structure is a series–parallel system. To define it, we assume that

$$E_{ij}, \quad i = 1, 2, \ldots, k_n, \quad j = 1, 2, \ldots, l_i, k_n, \quad l_1, l_2, \ldots, l_{k_n} \in N,$$

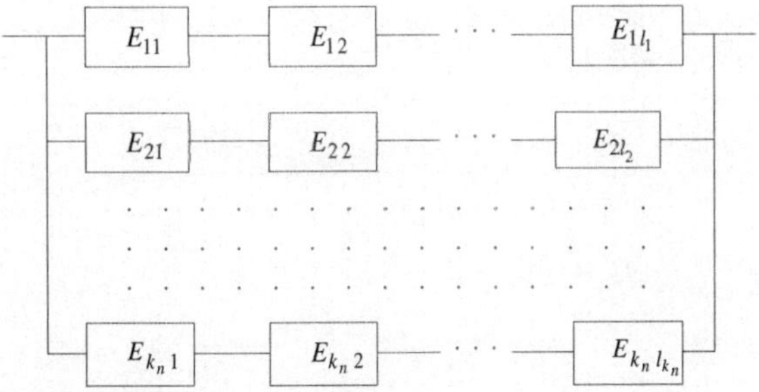

Fig. 3 The scheme of a series–parallel system

are two-state components of the system having reliability functions

$$R_{ij}(t) = P(T_{ij} > t), \quad t \in (-\infty, \infty),$$

where

$$T_{ij}, \quad i = 1, 2, \ldots, k_n, \quad j = 1, 2, \ldots, l_i,$$

are independent random variables representing the lifetimes of components E_{ij} with distribution functions

$$F_{ij}(t) = P(T_{ij} \leq t), \quad t \in (-\infty, \infty).$$

Definition 3 We call a two-state system series–parallel if its lifetime T is given by

$$T = \max_{1 \leq i \leq k_n} \{ \min_{1 \leq j \leq l_i} \{T_{ij}\}\}.$$

The scheme of a regular series–parallel system is given in Fig. 3.

By joining formulae (1) and (2) for the reliability functions of two-state series and parallel systems it is easy to conclude that the reliability function of the two-state series–parallel system is given by

$$\boldsymbol{R}_{k_n, l_1, l_2, \ldots, l_{k_n}}(t) = 1 - \prod_{i=1}^{k_n} \left[1 - \prod_{j=1}^{l_i} R_{ij}(t) \right], \quad t \in (-\infty, \infty), \tag{3}$$

where k_n is the number of series subsystems linked in parallel and l_i are the numbers of components in the series subsystems.

Definition 4 We call a two-state series–parallel system regular if

$$l_1 = l_2 = \cdots = l_{k_n} = l_n, \quad l_n \in N,$$

i.e., if the numbers of components in its series subsystems are equal. The scheme of a regular series–parallel system is given in Fig. 4.

Definition 5 We call a two-state system homogeneous if its component lifetimes have an identical distribution function $F(t)$, i.e., if its components have the same reliability function

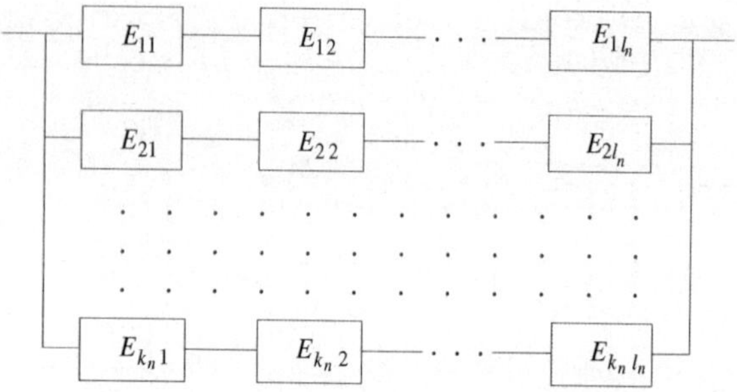

Fig. 4 The scheme of a regular series–parallel system

$$R(t) = 1 - F(t), \quad t \in (-\infty, \infty).$$

The above definition and equations (1–3) result in the simplified formulae for the reliability functions of the homogeneous systems stated in the following corollary.

Corollary 1 *The reliability function of the homogeneous two-state system is given by*

– *for a series system*

$$\overline{\boldsymbol{R}}_n(t) = [R(t)]^n, \quad t \in (-\infty, \infty), \tag{4}$$

– *for a parallel system*

$$\boldsymbol{R}_n(t) = 1 - [F(t)]^n, \quad t \in (-\infty, \infty), \tag{5}$$

– *for a regular series–parallel system*

$$\boldsymbol{R}_{k_n,l_n}(t) = 1 - \left[1 - [R(t)]^{l_n}\right]^{k_n}, \quad t \in (-\infty, \infty). \tag{6}$$

3 Asymptotic Approach to System Reliability

The asymptotic approach to the reliability of two-state systems depends on the investigation of limit distributions of a standardised random variable

$$(T - b_n)/a_n,$$

where T is the lifetime of a system and $a_n > 0$ and $b_n \in (-\infty, \infty)$ are suitably chosen numbers called normalising constants.

Since

$$P((T - b_n)/a_n > t) = P(T > a_n t + b_n) = \boldsymbol{R}_n(a_n t + b_n),$$

where $\boldsymbol{R}_n(t)$ is a reliability function of a system composed of n components, then the following definition becomes natural.

Definition 6 We call a reliability function $\Re(t)$ the limit reliability function of a system having a reliability function $\boldsymbol{R}_n(t)$ if there exist normalising constants $a_n > 0, b_n \in (-\infty, \infty)$ such that

$$\lim_{n\to\infty} \boldsymbol{R}_n(a_n t + b_n) = \Re(t) \quad \text{for } t \in C_{\Re},$$

where $C_{\Re}$ is the set of continuity points of $\Re(t)$.

Thus, if the asymptotic reliability function $\Re(t)$ of a system is known, then for sufficiently large n, the approximate formula

$$\boldsymbol{R}_n(t) \cong \Re((t - b_n)/a_n), \quad t \in (-\infty, \infty) \tag{7}$$

may be used instead of the system exact reliability function $\boldsymbol{R}_n(t)$.

3.1 Reliability of Large Two-State Series Systems

The investigations of limit reliability functions of homogeneous two-state series systems are based on the following auxiliary theorem.

Lemma 1 *If*

(i) $\overline{\Re}(t) = \exp\left[-\overline{V}(t)\right]$ *is a non-degenerate reliability function,*

(ii) $\overline{\boldsymbol{R}}_n(t)$ *is the reliability function of a homogeneous two-state series system defined by* (4),

(iii) $a_n > 0, b_n \in (-\infty, \infty)$,

then

$$\lim_{x\to\infty} \overline{\boldsymbol{R}}_n(a_n t + b_n) = \overline{\boldsymbol{R}}(t) \quad \textit{for } t \in C_{\overline{\Re}}$$

if and only if

$$\lim_{x\to\infty} nF(a_n t + b_n) = \overline{V}(t) \quad \textit{for } t \in C_{\overline{V}}$$

Proof The proof may be found in [1, 6, 9].

Lemma 1 is an essential tool in finding limit reliability functions of two-state series systems. It is also the basis for fixing the class of all possible limit reliability functions of these systems. This class is determined by the following theorem.

Theorem 1 *The only non-degenerate limit reliability functions of the homogeneous two-state series system are:*

$$\overline{\Re}_1(t) = \exp[-(-t)^{-\alpha}] \quad \textit{for } t<0, \qquad \overline{\Re}_1(t) = 0 \quad \textit{for } t \geq 0, \alpha > 0;$$

$$\bar{\mathfrak{R}}_2(t) = 1 \quad for \quad t<0, \qquad \bar{\mathfrak{R}}_2(t) = \exp[-t^{\alpha}] \quad for\ t \geq 0,\ \alpha > 0;$$

$$\bar{\mathfrak{R}}_3(t) = \exp[-\exp[t]] \quad for\ t \in (-\infty, \infty).$$

Proof The proof may be found in [1, 6, 9].

3.2 Reliability of Large Two-State Parallel Systems

The class of limit reliability functions for homogeneous two-state parallel systems may be determined on the basis of the following auxiliary theorem.

Lemma 2 *If*

(i) $\mathfrak{R}(t) = 1 - \exp[-V(t)]$ *is a non-degenerate reliability function,*
(ii) $\boldsymbol{R}_n(t)$ *is the reliability function of a homogeneous two-state parallel system defined by (5),*
(iii) $a_n > 0, b_n \in (-\infty, \infty)$,

then

$$\lim_{n\to\infty} R_n(a_n t + b_n) = \mathfrak{R}(t) \quad for\ t \in C_{\mathfrak{R}},$$

if and only if

$$\lim_{n\to\infty} nR(a_n t + b_n) = V(t) \quad for\ t \in C_V.$$

Proof The proof may be found in [1, 6, 8].

By applying *Lemma 2* it is possible to fix the class of limit reliability functions for homogeneous two-state parallel systems. However, it is easier to obtain this result using the duality property of parallel and series systems expressed in the relationship

$$R_n(t) = 1 - \bar{\boldsymbol{R}}_n(-t) \quad \text{for } t \in (-\infty, \infty),$$

that results in the following lemma [1, 6, 9, 10].

Lemma 3 *If* $\bar{\mathfrak{R}}(t)$ *is the limit reliability function of a homogeneous two-state series system with reliability functions of particular components* $\bar{\boldsymbol{R}}(t)$, *then*

$$\mathfrak{R}(t) = 1 - \bar{\mathfrak{R}}(-t) \quad for\ t \in C_{\bar{\mathfrak{R}}}$$

is the limit reliability function of a homogeneous two-state parallel system with reliability functions of particular components

$$R(t) = 1 - \bar{\boldsymbol{R}}(-t) \quad for\ t \in C_{\bar{\mathfrak{R}}}.$$

At the same time, if (a_n, b_n) *is a pair of normalising constants in the first case, then* $(a_n - b_n)$ *is such a pair in the second case.*

The application of Lemma 3 and Theorem 1 yields the following result.

Theorem 2 *The only non-degenerate limit reliability functions of the homogeneous parallel system are:*

$$\Re_1(t) = 1 \quad \text{for } t \le 0, \qquad \Re_1(t) = 1 - \exp[-t^{-\alpha}] \quad \text{for } t > 0, \alpha > 0;$$

$$\Re_2(t) = 1 - \exp[-(-t)^{\alpha}] \quad \text{for } t < 0, \qquad \Re_2(t) = 0 \quad \text{for } t \ge 0, \alpha > 0;$$

$$\Re_3(t) = 1 - \exp[-\exp[-t]] \quad \text{for } t \in (-\infty, \infty).$$

Proof The proof may be found in [1, 6, 9].

3.3 Reliability Evaluation of Large Two-State Series–Parallel Systems

The proofs of the theorems on limit reliability functions for homogeneous regular series–parallel systems and methods of finding such functions for individual systems are based on the following essential lemmas.

Lemma 4 *If*

(i) $k_n \to \infty$,
(ii) $\Re(t) = 1 - \exp[-V(t)]$ *is a non-degenerate reliability function,*
(iii) $R_{k_n,l_n}(t)$ *is the reliability function of a homogeneous regular two-state series–parallel system defined by* (6),
(iv) $a_n > 0, b_n \in (-\infty, \infty)$,

then

$$\lim_{n \to \infty} \boldsymbol{R}_{k_n,l_n}(a_n t + b_n) = \Re(t) \quad \text{for } t \in C_{\Re}$$

if and only if

$$\lim_{n \to \infty} k_n [R(a_n t + b_n)]^{l_n} = V(t) \quad \text{for } t \in C_V$$

Proof The proof may be found in [9].

Lemma 5 *If*

(i) $k_n \to k, k > 0, \ l_n \to \infty$,
(ii) $\Re(t)$ *is a non-degenerate reliability function,*
(iii) $\boldsymbol{R}_{k_n,l_n}(t)$ *is the reliability function of a homogeneous regular two-state series–parallel system defined by* (6),

(iv) $a_n > 0, b_n \in (-\infty, \infty)$,

then

$$\lim_{n\to\infty} \boldsymbol{R}_{k_n,l_n}(a_n t + b_n) = \Re(t) \quad \textit{for } t \in C_{\Re}$$

if and only if

$$\lim_{n\to\infty} [R(a_n t + b_n)]^{l_n} = \Re_0(t) \quad \textit{for } t \in C_{\Re_0},$$

where $\Re_0(t)$ *is a non-degenerate reliability function and moreover*

$$\Re(t) = 1 - [1 - \Re_0(t)]^k \quad \textit{for } t \in (-\infty, \infty).$$

Proof The proof may be found in [9].

The types of limit reliability functions of a series–parallel system depend on the system shape [8], i.e., on the relationships between the number k_n of its series subsystems linked in parallel and the number l_n of components in its series subsystems. The results based on *Lemma 4* and *Lemma 5* may be formulated in the form of the following theorem.

Theorem 3 *The only non-degenerate limit reliability functions of the homogeneous regular two-state series–parallel system are:*

Case 1

$$k_n = n, \quad |l_n - c\log n| \gg s, \quad s > 0, \quad c > 0.$$

$$\Re_1(t) = 1 \quad \textit{for } t \le 0, \qquad \Re_1(t) = 1 - \exp[-t^{-\alpha}] \quad \textit{for } t > 0, \alpha > 0;$$

$$\Re_2(t) = 1 - \exp[-(-t)^{\alpha}] \quad \textit{for } t < 0, \qquad \Re_2(t) = 0 \quad \textit{for } t \ge 0, \alpha > 0;$$

$$\Re_3(t) = 1 - \exp[-\exp[-t]] \quad \textit{for } t \in (-\infty, \infty);$$

Case 2

$$k_n = n, \quad l_n - c\log n \approx s, \quad s \in (-\infty, \infty), \quad c > 0.$$

$$\Re_4(t) = 1 \quad \textit{for } t < 0, \qquad \Re_4(t) = 1 - \exp[-\exp[-t^{\alpha} - s/c]] \quad \textit{for } t \ge 0, \alpha > 0;$$

$$\Re_5(t) = 1 - \exp[-\exp[(-t)^{\alpha} - s/c]] \quad \textit{for } t < 0, \qquad \Re_5(t) = 0 \quad \textit{for } t \ge 0, \alpha > 0;$$

$$\begin{aligned} \Re_6(t) &= 1 - \exp[-\exp[\beta(-t)^{\alpha} - s/c]] \quad \textit{for } t < 0, \\ \Re_6(t) &= 1 - \exp[-\exp[-t^{\alpha} - s/c]] \quad \textit{for } t \ge 0, \alpha > 0, \beta > 0; \end{aligned}$$

$$\begin{aligned} &\Re_7(t) = 1 \quad \textit{for } t < t_1, \quad \Re_7(t) = 1 - \exp[-\exp[-s/c]] \quad \textit{for } t_1 \le t < t_2, \\ &\Re_7(t) = 0 \quad \textit{for } t \ge t_2, t_1 < t_2; \end{aligned}$$

Case 3

$$k_n \to k, \quad k > 0, \quad l_n \to \infty.$$

$$\Re_8(t) = 1 - [1 - \exp[-(-t)^{-\alpha}]]^k \quad \textit{for } t<0, \qquad \Re_8(t) = 0 \quad \textit{for } t \geq 0, \alpha > 0;$$

$$\Re_9(t) = 1 \quad \textit{for } t<0, \qquad \Re_9(t) = 1 - [1 - \exp[-t^{\alpha}]]^k \quad \textit{for } t \geq 0, \alpha > 0;$$

$$\Re_{10}(t) = 1 - [1 - \exp[-\exp t]]^k \quad \textit{for } t \in (-\infty, \infty).$$

Proof The proof may be found in [9].

Using the duality property of parallel–series and series–parallel systems similar to this given in *Lemma 3* for parallel and series systems it is possible to prove that the only limit reliability functions of the homogeneous regular two-state parallel–series system are

$$\bar{\Re}_i(t) = 1 - \Re_i(-t) \quad \textit{for } t \in C_{\Re_i}, \quad i = 1, 2, \ldots, 10.$$

Applying *Lemma 2*, it is possible to prove the following fact [10].

Corollary 2 *If components of the homogeneous two-state parallel system have Weibull reliability functions*

$$R(t) = \exp[-\beta t^{\alpha}] \quad \textit{for } t \geq 0, \alpha > 0, \beta > 0$$

and

$$a_n = b_n/(\alpha \log n), \quad b_n = (\log n/\beta)^{1/\alpha},$$

then

$$\Re_3(t) = 1 - \exp[-\exp[-t]], \quad t \in (-\infty, \infty),$$

is its limit reliability function.

Example 1 (*a steel rope, durability*). Let us consider a steel rope composed of 36 strands used in ship-rope elevator and assume that it is not failed if at least one of its strands is not broken. Under this assumption we may consider the rope as a homogeneous parallel system composed of $n = 36$ basic components. The cross-section of this rope is shown in Fig. 5.

Further, assuming that the strands have Weibull reliability functions with parameters

$$\alpha = 2, \quad \beta = (7.07)^{-6}.$$

By (5), the rope's exact reliability function takes the form

$$\boldsymbol{R}_{36}(t) = 1 - [1 - \exp[-(7.07)^{-6} t^2]^{36} \quad \textit{for } t \geq 0.$$

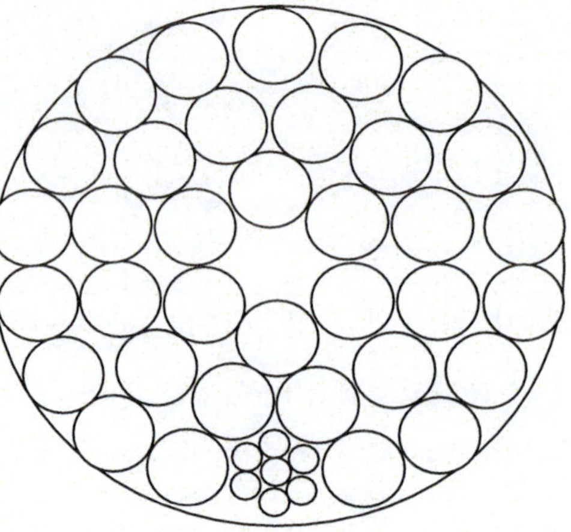

Fig. 5 The steel rope cross-section

Thus, according to *Corollary 2*, assuming

$$a_n = (7.07)^3/(2\sqrt{\log 36}), \quad b_n = (7.07)^3\sqrt{\log 36}$$

and applying (7), we arrive at the approximate formula for the rope reliability function of the form

$$\boldsymbol{R}_{36}(t) \cong \mathfrak{R}_3((t-b_n)/a_n) = 1 - \exp[-\exp[-0.01071t + 7.167]]$$
$$\textit{for } t \in (-\infty, \infty).$$

The mean value of the rope lifetime T and its standard deviation, in months, calculated on the basis of the above approximate result and according to the formulae

$$E[T] = Ca_n + b_n, \quad \sigma = \pi a_n/\sqrt{6},$$

where $C \cong 0.5772$ is Euler's constant, respectively, are:

$$E[T] \cong 723, \quad \sigma \cong 120.$$

The values of the exact and approximate reliability functions of the rope are presented in Table 1 and graphically in Fig. 6. The differences between them are not large, which means that the mistakes in replacing the exact rope reliability function by its approximate form are practically not significant.

4 Domains of Attraction for System Limit Reliability Functions

The problem of domains of attraction for the limit reliability functions of two-state systems solved completely in [14] we will illustrate partly for two-state series homogeneous systems only. From Theorem 1 it follows that the class of limit

Table 1 The values of the exact and approximate reliability functions of the steel rope

t	$R_{36}(t)$	$\Re_3\left(\frac{t-b_n}{a_n}\right)$	$\Delta = R_{36} - \Re_3$
0	1.000	1.000	0.000
400	1.000	1.000	0.000
500	0.995	0.988	−0.003
550	0.965	0.972	−0.007
600	0.874	0.877	−0.003
650	0.712	0.707	0.005
700	0.513	0.513	0.000
750	0.330	0.344	−0.014
800	0.193	0.218	−0.025
900	0.053	0.081	**−0.028**
1000	0.012	0.029	−0.017
1100	0.002	0.010	−0.008
1200	0.000	0.003	−0.003

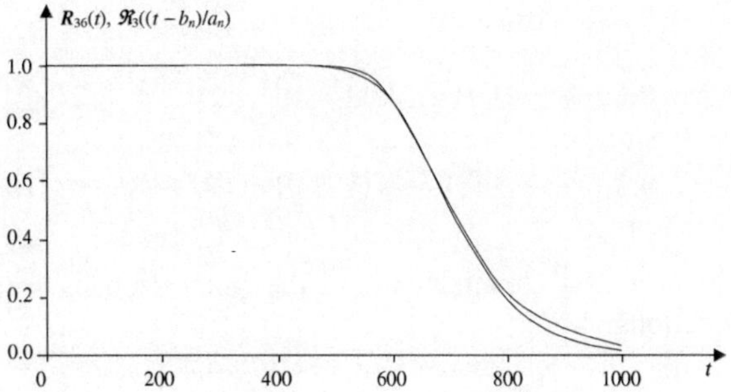

Fig. 6 The graphs of the exact and approximate reliability functions of the steel rope

reliability functions for a homogeneous series system is composed of three functions, $\bar{\Re}_i(t), i = 1, 2, 3$. Now we will determine domains of attraction $D_{\bar{\Re}_i}$ for these fixed functions, i.e., we will determine the conditions which the reliability functions $R(t)$ of the particular components of the homogeneous series system have to satisfy in order that the system limit reliability function is one of the reliability functions $\bar{\Re}_i(t), i = 1, 2, 3$.

Proposition 1 *If R(t) is a reliability function of the homogeneous series system components, then*

$$R(t) \in D_{\bar{\Re}_1}$$

if and only if

$$\lim_{r \to -\infty} \frac{1 - R(r)}{1 - R(rt)} = t^{\alpha} \quad for\ t > 0.$$

Proposition 2 *If R(t) is a reliability function of the homogeneous series system components, then*

$$R(t) \in D_{\overline{\Re_2}}$$

if and only if

(i) $\exists\, y \in (-\infty, \infty)\ R(y) = 1$ *and* $R(y + \varepsilon) < 1$ *for* $\varepsilon > 0$,
(ii) $\lim_{r \to 0^+} \frac{1-R(rt+y)}{1-R(r+y)} = t^{\alpha}$ *for* $t > 0$.

Proposition 3 *If R(t) is a reliability function of the homogeneous series system components, then*

$$R(t) \in D_{\overline{\Re_3}}$$

if and only if

$$\lim_{n \to \infty} n[1 - R(a_n t + b_n)] = e^t \quad \textit{for } t \in (-\infty, \infty)$$

with

$$b_n = \inf\{t : R(t + 0) \le 1 - \frac{1}{n} \le R(t - 0)\},$$

$$a_n = \inf\{t : R(t(1 + 0) + b_n) \le 1 - \frac{e}{n} \le R(t(1 - 0) + b_n)\}.$$

Example 2 If components of the homogeneous series system have reliability functions

$$R(t) = \begin{cases} 1, & t<0 \\ 1 - t, & 0 \le t < 1 \\ 0, & t \ge 1, \end{cases}$$

then

$$R(t) \in D_{\overline{\Re_2}}.$$

The results of the analysis on domains of attraction for limit reliability functions of two-state systems may automatically be transmitted to multi-state systems. To do this, it is sufficient to apply theorems about two-state systems such as the ones presented here to each vector co-ordinate of the multi-state reliability functions [10].

5 Reliability of Large Hierarchical Systems

Prior to defining the hierarchical systems of any order we once again consider a series–parallel system like a system presented in Fig. 3. This system here is called a series–parallel system of order 1. Its scheme is given in Fig. 7.

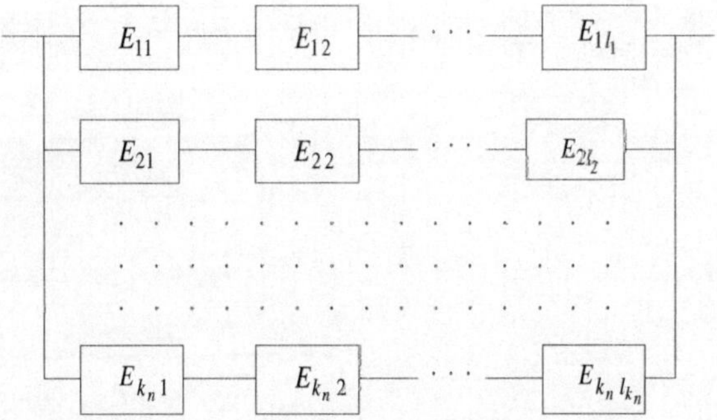

Fig. 7 The scheme of a series–parallel system of order 1

It is made up of components

$$E_{i_1j_1},\quad i_1 = 1,2,\ldots,k_n,\quad j_1 = 1,2,\ldots,l_{i_1},$$

with the lifetimes, respectively,

$$T_{i_1j_1},\quad i_1 = 1,2,\ldots,k_n,\quad j_1 = 1,2,\ldots,l_{i_1}.$$

Its lifetime is given by

$$T = \max_{1\le i_1\le k_n}\left\{\min_{1\le j_1\le l_{i_1}}\{T_{i_1j_1}\}\right\}. \tag{8}$$

Now we assume that each component

$$E_{i_1j_1},\quad i_1 = 1,2,\ldots,k_n,\quad j_1 = 1,2,\ldots,l_{i_1},$$

of the series–parallel system of order 1 is a subsystem composed of components

$$E_{i_1j_1i_2j_2},\quad i_2 = 1,2,\ldots,k_n^{(i_1j_1)},\quad j_2 = 1,2,\ldots,l_{i_2}^{(i_1j_1)},$$

and has a series–parallel structure. The interpretation of this assumption is illustrated in Fig. 8.

This means that each subsystem lifetime T_{i1j1} is given by

$$T_{i_1j_1} = \max_{1\le i_2\le k_n^{(i_1j_1)}}\left\{\min_{1\le j_2\le l_{i_2}^{(i_1j_1)}}\{T_{i_1j_1i_2j_2}\}\right\},\quad i_1 = 1,2,\ldots,k_n,\quad j_1 = 1,2,\ldots,l_{i_1}, \tag{9}$$

where

$$T_{i_1j_1i_2j_2},\quad i_2 = 1,2,\ldots,k_n^{(i_1j_1)},\quad j_2 = 1,2,\ldots,l_{i_2}^{(i_1j_1)},$$

are the lifetimes of the subsystem components $E_{i1j1i2j2}$.

The system defined this way is called a hierarchical series–parallel system of order 2. Its lifetime, from (8) and (9), is given by the formula

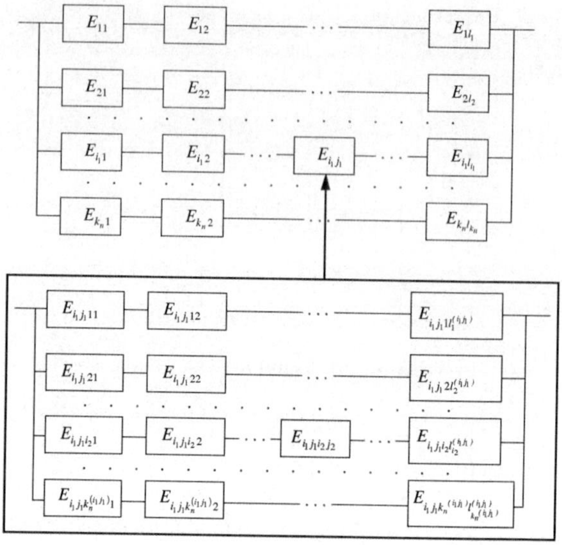

Fig. 8 The scheme of a series–parallel system of order 2

$$T = \max_{1 \le i_1 \le k_n} \left\{ \min_{1 \le j_1 \le l_{i_1}} \left[\max_{1 \le i_2 \le k_n^{(i_1j_1)}} \left(\min_{1 \le j_2 \le l_{i_2}^{(i_1j_1)}} T_{i_1j_1i_2j_2} \right) \right] \right\},$$

where k_n is the number of series systems linked in parallel and composed of series–parallel subsystems $E_{i_1j_1}$, l_{i_1} are the numbers of series–parallel subsystems $E_{i_1j_1}$ in these series systems, $k_n^{(i_1j_1)}$ are the numbers of series systems in the series–parallel subsystems $E_{i_1j_1}$ linked in parallel, and $l_{i_2}^{(i_1j_1)}$ are the numbers of components in these series systems of the series–parallel subsystems $E_{i_1j_1}$.

In an analogous way it is possible to define two-state parallel–series systems of order 2.

Generally, in order to define hierarchical series–parallel and parallel–series systems of any order r, $r \ge 1$, we assume that

$$E_{i_1j_1\ldots i_rj_r},$$

where

$$i_1 = 1,2,\ldots,k_n, \quad j_1 = 1,2,\ldots,l_{i_1}, \quad i_2 = 1,2,\ldots,k_n^{(i_1j_1)}, \quad j^2 = 1,2\ldots,l_{i_2}^{(i_1j_1)},\ldots,$$
$$i_r = 1,2,\ldots,k_n^{(i_1j_1\ldots i_{r-1}j_{r-1})}, \quad j_r = 1,2,\ldots,l_{i_r}^{(i_1j_1\ldots i_{r-1}j_{r-1})}$$

and

$$k_n, l_{i_1}, k_n^{(i_1j_1)}, l_{i_2}^{(i_1j_1)},\ldots,k_n^{(i_1j_1\ldots i_{r-1}j_{r-1})}, l_{i_r}^{(i_1j_1\ldots i_{r-1}j_{r-1})} \in N,$$

are two-state components having reliability functions

$$R_{i_1 j_1 \ldots i_r j_r}(t) = P(T_{i_1 j_1 \ldots i_r j_r} > t), \quad t \in (-\infty, \infty),$$

and random variables

$$T_{i_1 j_1 \ldots i_r j_r},$$

where

$$i_1 = 1,2,\ldots,k_n, \;\; j_1 = 1,2,\ldots,l_{i_1}, \;\; i_2 = 1,2,\ldots,k_n^{(i_1 j_1)}, \;\; j_2 = 1,2,\ldots,l_{i_2}^{(i_1 j_1)},\ldots,$$
$$i_r = 1,2,\ldots,k_n^{(i_1 j_1 \ldots i_{r-1} j_{r-1})}, \;\; j_r = 1,2,\ldots,l_{i_r}^{(i_1 j_1 \ldots i_{r-1} j_{r-1})},$$

are independent random variables with distribution functions

$$F_{i_1 j_1 \ldots i_r j_r}(t) = P(T_{i_1 j_1 \ldots i_r j_r} \le t), \quad t \in (-\infty, \infty),$$

representing the lifetimes of the components $E_{i_1 j_1 \ldots i_r j_r}$.

Definition 7 A two-state system is called a series–parallel system of order r if its lifetime T is given by

$$T = \max_{1 \le i_1 \le k_n} \left\{ \min_{1 \le j_1 \le l_{i_1}} \left\{ \max_{1 \le i_1 \le k_n^{(i_1 j_1)}} \left\{ \min_{1 \le j_2 \le l_{i_2}^{(i_1 j_1)}} \ldots \right.\right.\right.$$
$$\left.\left.\left. \left[\max_{1 \le i_r \le k_n^{(i_1 j_1 \ldots i_{r-1} j_{r-1})}} \left(\min_{1 \le j_r \le l_{i_r}^{(i_1 j_1 \ldots i_{r-1} j_{r-1})}} T_{i_1 j_1 \ldots i_r j_r} \right) \right] \ldots \right\}\right\}\right\},$$

where $k_n, k_n^{(i_1 j_1)}, \ldots, k_n^{(i_1 j_1 i_2 j_2 \ldots i_{r-1} j_{r-1})}$ are the numbers of suitable series systems of the system composed of series–parallel subsystems and linked in parallel, $l_{i_1}, l_{i_2}^{(i_1 j_1)}, \ldots, l_{i_{r-1}}^{(i_1 j_1 i_2 j_2 \ldots i_{r-1} j_{r-1})}$ are the numbers of suitable series–parallel subsystems in these series systems, and $l_{i_r}^{(i_1 j_1 i_2 j_2 \mathrm{K} i_{r-1} j_{r-1})}$ are the numbers of components in the series systems of the series–parallel subsystems.

Definition 8 A two-state series–parallel system of order r is called homogeneous if its component lifetimes $T_{i_1 j_1 \ldots i_r j_r}$ have an identical distribution function

$$F(t) = P(T_{i_1 j_1 \ldots i_r j_r} \le t), \quad t \in (-\infty, \infty),$$

where

$$i_1 = 1,2,\ldots,k_n, \;\; j_1 = 1,2,\ldots,l_{i_1}, \;\; i_2 = 1,2,\ldots,k_n^{(i_1 j_1)}, \;\; j_2 = 1,2,\ldots,l_{i_2}^{(i_1 j_1)},\ldots,$$
$$i_r = 1,2,\ldots,k_n^{(i_1 j_1 \ldots i_{r-1} j_{r-1})}, \;\; j_r = 1,2,\ldots,l_{i_r}^{(i_1 j_1 \ldots i_{r-1} j_{r-1})},$$

i.e., if its components $E_{i_1 j_1 \ldots i_r j_r}$ have the same reliability function

$$R(t) = 1 - F(t), \quad t \in (-\infty, \infty)$$

Definition 9 A two-state series–parallel system of order r is called regular if

$$l_{i_1} = l_{i_2}^{(i_1 j_1)} = \cdots = l_{i_r}^{(i_1 j_1 \ldots i_{r-1} j_{r-1})} = l_n$$

and

$$k_n^{(i_1 j_1)} = \cdots = k_n^{(i_1 j_1 \ldots i_{r-1} j_{r-1})} = k_n,$$

where k_n is the number of series systems in the series–parallel subsystems and l_n are the numbers of series–parallel subsystems or, respectively, the numbers of components in these series systems.

Using mathematical induction it is possible to prove that the reliability function of the homogeneous and regular two-state hierarchical series–parallel system of order r is given by [3]

$$\boldsymbol{R}_{k,k_n,l_n}(t) = 1 - [1 - [\boldsymbol{R}_{k-1,k_n,l_n}(t)]^{l_n}]^{k_n} \quad \text{for } k = 2, 3, \ldots, r$$

and

$$R_{1,k_n,l_n}(t) = 1 - [1 - [R(t)]^{l_n}]^{k_n}, \quad t \in (-\infty, \infty),$$

where k_n and l_n are defined in *Definition 9.*

The following results are also proved in [3].

Corollary 3 *If components of the homogeneous and regular two-state hierarchical series–parallel system of order r have an exponential reliability function*

$$R(t) = \exp[-\lambda t] \quad \textit{for } t \geq 0, \lambda > 0,$$

then its reliability function is given by

$$R_{k,k_n,l_n}(t) = 1 - [1 - \exp[R_{k-1,k_n,l_n}(t)]^{l_n}]^{k_n} \quad \textit{for } t \geq 0$$

for $k = 2, 3, \ldots, r$ *and*

$$R_{1,k_n,l_n}(t) = 1 - [1 - \exp[-\lambda l_n t]]^{k_n} \quad \textit{for } t \geq 0.$$

Theorem 4 *If*

(i) $\Re(t) = 1 - \exp[-V(t)]$, $t \in (-\infty, \infty)$, *is a non-degenerate reliability function,*

(ii) $\lim_{n \to \infty} l_n^{r-1} k_n^{-\frac{1}{l_n}} = 0 \quad \textit{for } r \geq 1,$

(iii) $\lim_{n \to \infty} k_n^{l_n^{r-1} + \cdots + 1} [R(a_n t + b_n)]^{l_n^r} = V(t) \quad \textit{for } t \in C_V, r \geq 1, t \in (-\infty, \infty),$

then

$$\lim_{n\to\infty} \boldsymbol{R}_{r,k_n,l_n}(a_n t + b_n) = \Re(t) \quad \textit{for } t \in C_{\Re}, r \geq 1, t \in (-\infty, \infty).$$

Proposition 4 *If components of the homogeneous and regular two-state hierarchical series–parallel system of order r have an exponential reliability function*

$$R(t) = \exp[-\lambda t] \quad \textit{for } t \geq 0, \lambda > 0,$$

$$\lim_{n\to\infty} l_n^{r-1} k_n^{-\frac{1}{l_n}} = 0 \quad \textit{for } r \geq 1.$$

And

$$a_n = \frac{1}{\lambda l_n^r}, \quad b_n = \frac{1}{\lambda}\left(\frac{1}{l_n} + \frac{1}{l_n^2} + \cdots + \frac{1}{l_n^r}\right) \log k_n,$$

then

$$\Re_3(t) = 1 - \exp[-\exp[-t]] \quad \textit{for } t \in (-\infty, \infty), \tag{10}$$

is its limit reliability function.

Example 3 A hierarchical regular series–parallel homogeneous system of order $r = 2$ is such that $k_n = 200$, $l_n = 3$. The system components have identical exponential reliability functions with the failure rate $\lambda = 0.01$.

Under these assumptions its exact reliability function, according to *Corollary* 3, is given by

$$R_{2,200,3}(t) = 1 - [1 - [1 - [1 - \exp[-0.01 \cdot 3t]]^{200}]^3]^{200} \quad \text{for } t \geq 0.$$

Next applying *Proposition 4* with normalising constants

$$a_n = \frac{1}{0.01.9} = 11.1,\ b_n = \frac{1}{0.01}\left(\frac{1}{3} + \frac{1}{9}\right) \log 200 = 235.5,$$

we conclude that the system limit reliability function is given by

$$\Re_3(t) = 1 - \exp[-\exp[-t]] \quad \text{for } t \in (-\infty, \infty),$$

and from (7), the following approximate formula is valid

$$\boldsymbol{R}_{2,200,3}(t) \cong \Re_3(0.09t - 21.2) = 1 - \exp[-\exp[-0.09t + 21.2]]$$
$$\text{for } t \in (-\infty, \infty).$$

The accuracy of this approximation is illustrated in Table 2 and Fig. 9.

Definition 10 A two-state system is called a parallel–series system of order *r* if its lifetime *T* is given by

Table 2 Values of exact and approximate reliability functions of a hierarchical regular series–parallel homogeneous system of order 2

t	$\boldsymbol{R}$	$\Re_3$	$\Delta = \boldsymbol{R} - \Re_3$
200	0.999996	1.000000	–0.000004
210	0.997281	0.999953	–0.002672
220	0.934547	0.982668	–0.048121
230	0.705338	0.807704	–0.102366
240	0.414313	0.488455	**–0.074142**
250	0.205450	0.238551	–0.033101
260	0.092891	0.104885	–0.011994
270	0.040069	0.044050	–0.003981
280	0.016873	0.018149	–0.001276
290	0.007014	0.007419	–0.000405
300	0.002895	0.003023	–0.000128
310	0.001189	0.001230	–0.000041
320	0.000487	0.000500	–0.000013
330	0.000199	0.000203	–0.000116
340	0.000081	0.000083	–0.000002

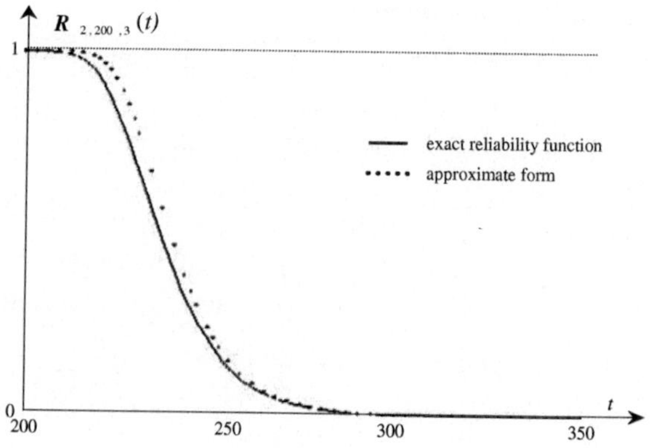

Fig. 9 Graphs of exact and approximate reliability functions of a hierarchical regular series–parallel homogeneous system of order 2

$$T = \min_{1 \le i_1 \le k_n} \left\{ \max_{1 \le j_1 \le l_{i_1}} \left\{ \min_{1 \le j_2 \le k_{i_2}^{(i_1 j_1)}} \left\{ \max_{1 \le j_2 \le l_{i_2}^{(i_1 j_1)}} \cdots \left[\min_{1 \le i_r \le k_n^{(i_1 j_1 \ldots i_{r-1} j_{r-1})}} \left(\max_{1 \le j_r \le l_{i_r}^{(i_1 j_1 \ldots i_{r-1} j_{r-1})}} T_{i_1 j_1 \ldots i_r j_r} \right) \right] \cdots \right\} \right\} \right\},$$

where $k_n, k_n^{(i_1 j_1)}, \ldots, k_n^{(i_1 j_1 i_2 j_2 \ldots i_{r-1} j_{r-1})}$ are the numbers of suitable parallel systems of the system composed of parallel–series subsystems and linked in series, $l_{i_1}, l_{i_2}^{(i_1 j_1)}, \ldots, l_{i_{r-1}}^{(i_1 j_1 i_2 j_2 \ldots i_{r-2} j_{r-2})}$ are the numbers of suitable parallel–series subsystems in these parallel systems, and $l_{i_r}^{(i_1 j_1 i_2 j_2 \ldots i_{r-1} j_{r-1})}$ are the numbers of components in the parallel systems of the parallel–series subsystems.

Definition 11 A two-state parallel–series system of order r is called homogeneous if its component lifetimes $T_{i_1 j_1 \ldots i_r j_r}$ have an identical distribution function

$$F(t) = P(T_{i_1 j_1 \ldots i_r j_r} \leq t),$$

where

$$i_1 = 1, 2, \ldots, k_n, \quad j_1 = 1, 2, \ldots, l_{i_1}, \quad i_2 = 1, 2, \ldots, k_n^{(i_1 j_1)}, \quad j_2 = 1, 2, \ldots, l_{i_2}^{(i_1 j_1)}, \ldots,$$
$$i_r = 1, 2, \ldots, k_n^{(i_1 j_1 \ldots i_{r-1} j_{r-1})}, \quad j_r = 1, 2, \ldots, l_{i_r}^{(i_1 j_1 \ldots i_{r-1} j_{r-1})},$$

i.e., if its components $E_{i_1 j_1 \ldots i_r j_r}$ have the same reliability function

$$R(t) = 1 - F(t), \; t \in (-\infty, \infty).$$

Definition 12 A two-state parallel–series system of order r is called regular if

$$l_{i_1} = l_{i_2}^{(i_1 j_1)} = \cdots = l_{i_r}^{(i_1 j_1 \ldots i_{r-1} j_{r-1})} = l_n$$

and

$$k_n^{(i_1 j_1)} = \cdots = k_n^{(i_1 j_1 \ldots i_{r-1} j_{r-1})} = k_n$$

where k_n is the number of parallel systems in the parallel–series subsystems and l_n are the numbers of parallel–series subsystems or, respectively, the numbers of components in these parallel systems.

Applying mathematical induction it is possible to prove that the reliability function of the homogeneous and regular two-state hierarchical parallel–series system of order r is given by [3]

$$\bar{\boldsymbol{R}}_{k,k_n,l_n}(t) = [1 - [1 - \bar{\boldsymbol{R}}_{k-1,k_n,l_n}(t)]^{l_n}]^{k_n} \quad \text{for } k = 2, 3, \ldots r$$

and

$$\bar{\boldsymbol{R}}_{1,k_n,l_n}(t) = \left[1 - [F(t)]^{l_n}\right]^{k_n}, t \in (-\infty, \infty),$$

where k_n and l_n are defined in Definition 12.

The following results are also proved in [3].

Corollary 4 *If components of the homogeneous and regular two-state hierarchical parallel–series system of order r have an exponential reliability function*

$$R(t) = \exp[-\lambda t] \quad \textit{for } t \geq 0, \lambda > 0,$$

then its reliability function is given by

$$\bar{\boldsymbol{R}}_{k,k_n,l_n}(t) = \left[1 - \left[1 - \bar{\boldsymbol{R}}_{k-1,k_n,l_n}(t)\right]^{l_n}\right]^{k_n} \quad \textit{for } k = 2, 3, \ldots, r$$

and

$$\bar{\boldsymbol{R}}_{1,k_n,l_n}(t) = [1 - [1 - \exp[-\lambda t]]^{l_n}]^{k_n} \quad \textit{for } t \geq 0.$$

Theorem 5 *If*

(i) $\overline{\boldsymbol{R}}(t) = \exp[-\bar{V}(t)]$, $t \in (-\infty, \infty)$, *is a non-degenerate reliability function,*

(ii) $\lim\limits_{n\to\infty} l_n^{r-1} k_n^{-\frac{1}{l_n}} = 0 \quad \textit{for } r \geq 1,$

(iii) $\lim\limits_{n\to\infty} k_n^{l_n^{r-1}+\cdots+1}[F(a_n t + b_n)^{l_n^r} = \bar{V}(t) \quad \textit{for } t \in C_v, r \geq 1 t \in (-\infty, \infty),$

then

$$\lim_{n\to\infty} \bar{\boldsymbol{R}}_{r,k_n,l_n}(a_n t + b_n) = \bar{\Re}(t) \quad \textit{for } t \in C_{\bar{\Re}}, r \geq 1,\ t \in (-\infty, \infty).$$

Proposition 5 *If components of the homogeneous and regular two-state hierarchical parallel–series system of order r have an exponential reliability function*

$$R(t) = \exp[-\lambda t] \quad \textit{for } t \geq 0, \lambda > 0,$$

$$\lim_{n\to\infty} l_n^{r-1} k_n^{-\frac{1}{l_n}} = 0 \quad \textit{for } r \geq 1,\ \lim_{n\to\infty} l_n = l, l \in N,$$

and

$$a_n = \frac{1}{\lambda} \frac{1}{k_n^{\frac{1}{l_n}+\cdots+\frac{1}{l_n^r}}}, \quad b_n = 0,$$

then

$$\bar{\Re}_2(t) = \exp[-t^{l^r}] \quad \textit{for } t \geq 0 \tag{11}$$

is its limit reliability function.

Example 3 We consider a hierarchical regular parallel–series homogeneous system of order $r = 2$ such that $k_n = 200$, $l_n = 3$, whose components have identical exponential reliability functions with the failure rate $\lambda = 0.01$.

Its exact reliability function, according to *Corollary* 4, is given by

$$\overline{\boldsymbol{R}}_{2,200,3}(t) = 1 - [1 - [1 - [1 - \exp[-0.01 \cdot 3t]]^{200}]^3]^{200} \quad \text{for } t \geq 0.$$

Next applying *Proposition 5* with normalising constants

$$a_n = \frac{1}{0.01} \cdot \frac{1}{200^{1/3+1/9}} = 9.4912, \quad b_n = 0,$$

we conclude that

Table 3 Values of exact and approximate reliability functions of a hierarchical regular parallel–series homogeneous system of order 2

t	$\bar{R}$	$\bar{\Re}_2$	$\Delta = \bar{R} - \bar{\Re}_2$
0	1.000000	1.000000	0.000000
2	0.999999	0.999999	0.000000
4	0.999656	0.999579	0.000077
6	0.988445	0.983952	0.004493
8	0.876948	0.806167	0.070781
10	0.451036	0.200822	**0.250214**
12	0.040806	0.000253	0.040553
14	0.000070	0.000000	0.000070
16	0.000000	0.000000	0.000000

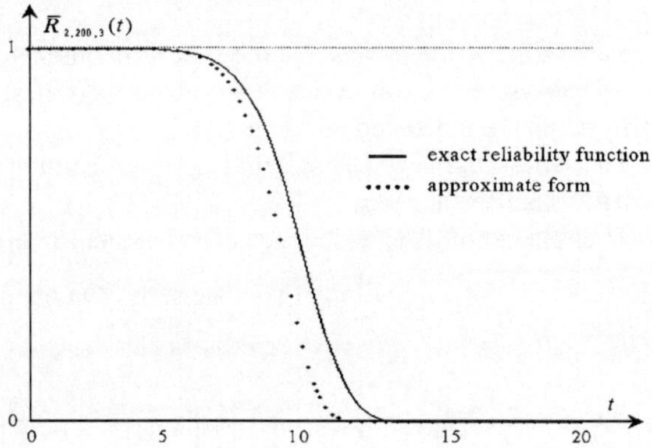

Fig. 10 Graphs of exact and approximate reliability functions of a hierarchical regular parallel–series homogeneous system of order 2

$$\bar{\Re}_2(t) = \exp[-t^9] \quad \text{for } t \geq 0$$

is the system limit reliability function, and from (7), the following approximate formula is valid

$$\bar{\boldsymbol{R}}_{2,200,3}(t) \cong \bar{\Re}_2(0.1054t) = \exp[-(0.1054t)^9] \quad \textit{for } t \geq 0.$$

The accuracy of this approximation is illustrated in Table 3 and Fig. 10.

6 Reliability of Large Series-"*m* out of *n*" Systems

Definition 13 A two-state system is called a series-"m out of k_n" system if its lifetime T is given by

$$T = T_{(k_n - m+1)}, \quad m = 1, 2, \ldots, k_n,$$

where $T^{(k_n - m+1)}$ is the mth maximal order statistic in the set of random variables

Fig. 11 The scheme of a series-"m out of k_n" system

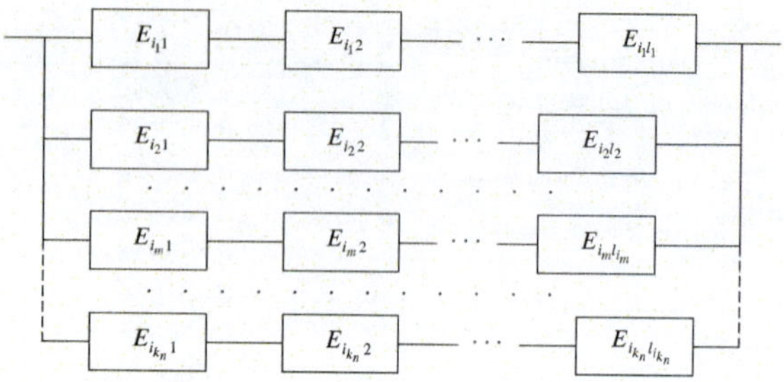

$$T_i = \min_{1 \le j \le l_i} \{T_{ij}\},\ i = 1, 2, \ldots, k_n.$$

The above definition means that the series-"m out of k_n" system is composed of k_n series subsystems and it is not failed if and only if at least m out of its k_n series subsystems are not failed.

The series-"m out of k_n" system is a series–parallel for $m = 1$ and it becomes a series system for $m = k_n$.

The scheme of a series-"m out of k_n" system is given in Fig. 11.

The reliability function of the two-state series-"m out of k_n" system is given either by [18]

$$\boldsymbol{R}^{(m)}_{k_n,l_1,l_2,\ldots,l_{k_n}}(t) = 1 - \sum_{\substack{r_1,r_2,\ldots,r_{k_n}=0 \\ r_1+r_2+\ldots+r_{k_n} \le m-1}}^{1} \prod_{i=1}^{k_n} \left[\prod_{j=1}^{l_i} R_{ij}(t)\right]^{r_i} \left[1 - \prod_{j=1}^{l_i} R_{ij}(t)\right]^{1-r_i} \quad \text{for } t \in (-\infty, \infty)$$

or by

$$\bar{\boldsymbol{R}}^{(\bar{m})}_{k_n,l_1,l_2,\ldots,l_{k_n}}(t) = \sum_{\substack{r_1,r_2,\ldots,r_{k_n}=0 \\ r_1+r_2+\ldots+r_{k_n} \le \bar{m}}}^{1} \prod_{i=1}^{k_n} \left[1 - \prod_{j=1}^{l_i} R_{ij}(t)\right]^{r_i} \left[1 - \prod_{j=1}^{l_i} R_{ij}(t)\right]^{1-r_i} \quad \text{for } t \in (-\infty, \infty),\ \bar{m} = k_n - m.$$

Definition 14 The series-"m out of k_n" system is called regular if

$$l_1 = l_2 = \cdots = l_{k_n} = l_n, \quad l_n \in N.$$

The scheme of a regular series-"m out of k_n" system is given in Fig. 12.

Definition 15 The series-"m out of k_n" system is called homogeneous if its component lifetimes T_{ij} have an identical distribution function

$$F(t) = P(T_{ij} \le t), \quad t \in (-\infty, \infty), \quad i = 1, 2, \ldots, k_n, \quad j = 1, 2, \ldots, l_i,$$

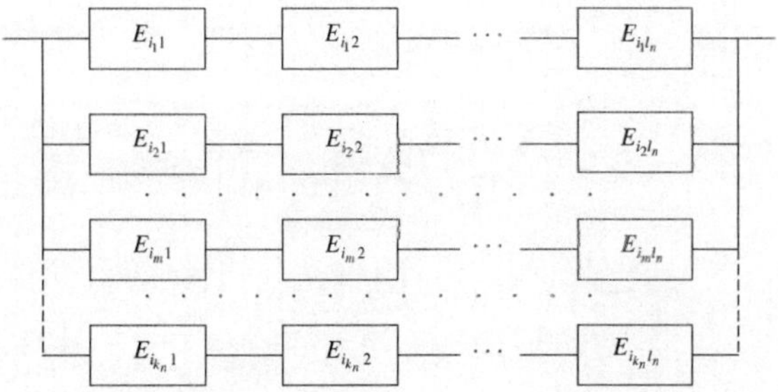

Fig. 12 The scheme of a regular series-"*m* out of k_n" system

i.e., if its components E_{ij} have the same reliability function

$$R(t) = 1 - F(t), \quad t \in (-\infty, \infty).$$

From the above definitions it follows that the reliability function of the homogeneous and regular series-"*m* out of k_n" system is given either by

$$\boldsymbol{R}_{k_n,l_n}^{(m)}(t) = 1 - \sum_{i=0}^{m-1} \binom{k_n}{i} [R^{l_n}(t)]^i [1 - R^{l_n}(t)]^{k_n - i} \quad \mathit{for}\, t \in (-\infty, \infty)$$

or by

$$\bar{\boldsymbol{R}}_{k_n,l_n}^{(\bar{m})}(t) = \sum_{i=0}^{\bar{m}} \binom{k_n}{i} [1 - R^{l_n}(t)^i [R^{l_n}(t)]^{k_n - i} \quad \text{for}\, t \in (-\infty, \infty), \quad \bar{m} = k_n - m,$$

where k_n is the number of series subsystems in the "*m* out of k_n" system and l_n is the number of components of the series subsystems.

The following results are proved in [18].

Corollary 5 *If components of the homogeneous and regular two-state series-"m out of k_n" system have Weibull reliability function*

$$R(t) = \exp[-\beta t^{\alpha}] \quad \mathit{for}\, t \geq 0, \alpha > 0, \beta > 0,$$

then its reliability function is given either by

$$\boldsymbol{R}_{k_n,l_n}^{(m)}(t) = 1 - \sum_{i=0}^{m-1} \binom{k_n}{i} [\exp[-i l_n \beta t^{\alpha}]][1 - \exp[-l_n \beta t^{a}]]^{k_n - i} \quad \mathit{for}\, t \geq 0 \tag{12}$$

or by

$$\overline{\boldsymbol{R}}_{k_n,l_n}^{(\overline{m})}(t) = \sum_{i=o}^{\overline{m}} \binom{k_n}{i} [1 - \exp[-l_n \beta t^{\alpha}]]^i [\exp[-(k_n - i) l_n \beta t^{a}]] \quad \mathit{for}\, t \geq 0, \overline{m} = k_n - m. \tag{13}$$

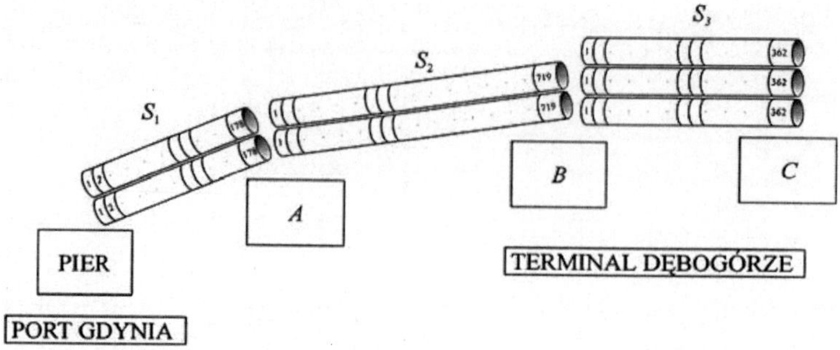

Fig. 13 The scheme of the oil transportation system

Proposition 6 *If components of the two-state homogeneous and regular series-"m out of k_n" system have Weibull reliability function*

$$R(t) = \exp[-\beta t^{\alpha}] \quad for\ t \geq 0, \alpha > 0,\ \beta > 0,$$

and

$$\lim_{n\to\infty} k_n = k, k > 0,\ 0 < m \leq k, \quad \lim_{n\to\infty} l_n = \infty,$$
$$a_n = (\beta l_n)^{-\frac{1}{\alpha}},\ b_n = 0,$$

then

$$\Re_9^{(2)}(t) = 1 - \sum_{i=0}^{m-1} \binom{k}{i} \exp[-it^{\alpha}][1 - \exp[-t^{\alpha}]]^{k-i} \quad for\ t \geq 0$$

is its limit reliability function, i.e., for $t \geq 0$, we have

$$\boldsymbol{R}_{k_n,l_n}^{(m)}(t) \cong \Re_9^{(2)}\left(\frac{t - b_n}{a_n}\right) = 1 - \sum_{i=0}^{m-1} \binom{k}{i} \exp[-i\beta l_n t^{\alpha}][1 - \exp[-\beta l_n t^{\alpha}]]^{k-i}. \quad (14)$$

Example 4 The piping transportation system is set up to receive from ships, store and send by carriages or cars oil products such as petrol, driving oil and fuel oil. The scheme of the oil transportation system is shown in Fig. 13.

Three terminal parts A, B and C fulfil these purposes. They are linked by the piping transportation systems. The unloading of tankers is performed at the pier. The pier is connected to terminal part *A* through the transportation subsystem S_1 built of two piping lines. In part *A* there is a supporting station fortifying tankers' pumps and making possible further transport of oil by means of subsystem S_2 to terminal part *B*. Subsystem S_2 is built of two piping lines. Terminal part *B* is connected to terminal part *C* by subsystem S_3. Subsystem S_3 is built of three piping lines. Terminal part *C* is set up for loading the rail cisterns with oil products and for the wagon carrying these to the railway station.

We will analyse the reliability of the subsystem S_3 only. This subsystem consists of $k_n = 3$ identical piping lines, each composed of $l_n = 360$ steel pipe segments. In each of these lines there are pipe segments with Weibull reliability function

$$R(t) = \exp[-0.0000000008t^4] \quad \text{for } t \geq 0.$$

We suppose that the system is good if at least two of its piping lines are not failed. Thus, according to Definitions 14–15, it may be considered as a homogeneous and regular series-"2 out of 3" system, and according to *Proposition 6*, assuming

$$a_n = \frac{1}{(\beta l_n)^{1/\alpha}} = \frac{1}{(0.000000288)^{1/4}}, \quad b_n = 0,$$

and using (3), its reliability function is given by

$$\boldsymbol{R}^{(2)}_{3,360}(t) \cong \Re^{(2)}_9\left(\frac{t}{a_n}\right) = \sum_{i=0}^{1}\binom{3}{i}\exp\left[-i \cdot 0.000000288t^4\right]$$
$$\times \left[1 - \exp[-0.000000288t^4]\right]^{3-i} \quad \text{for } t \geq 0.$$

7 Reliability of Large "*m* out of *n*"-Series Systems

Definition 16 A two-state system is called an "m_i out of l_i"-series system if its lifetime T is given by

$$T = \min_{1 \leq i \leq k_n}\left\{T_{(l_i - m_i + 1)}\right\}, \quad m_i = 1, 2, \ldots, l_i,$$

where $T_{(l_i - m_i + 1)}$ is the m_ith maximal order statistic in the set of random variables

$$T_{i1}, T_{i2}, \ldots, T_{il_i}, \quad i = 1, 2, \ldots, k_n.$$

The above definition means that the "m_i out of l_i"-series system is composed of k_n subsystems that are "m_i out of l_i" systems and it is not failed if all its "m_i out of l_i" subsystems are not failed.

The "m_i out of l_i"-series system is a parallel–series system if $m_1 = m_2 = \cdots = m_{k_n} = 1$ and it becomes a series system if $m_i = l_i$ for all $i = 1, 2, \ldots, k_n$.

The scheme of an "m_i out of l_i"-series system is given in Fig. 14.

The reliability function of the two-state "m_i out of l_i"-series system is given either by [18]

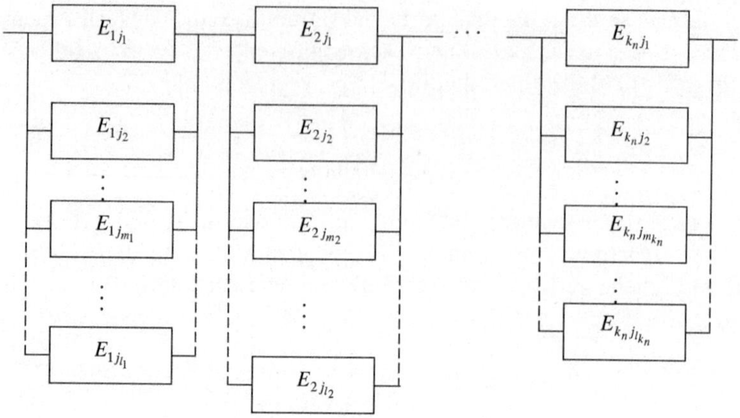

Fig. 14 The scheme of an "m_i out of l_i"-series system

Fig. 15 The scheme of a regular "m out of l_n"-series system

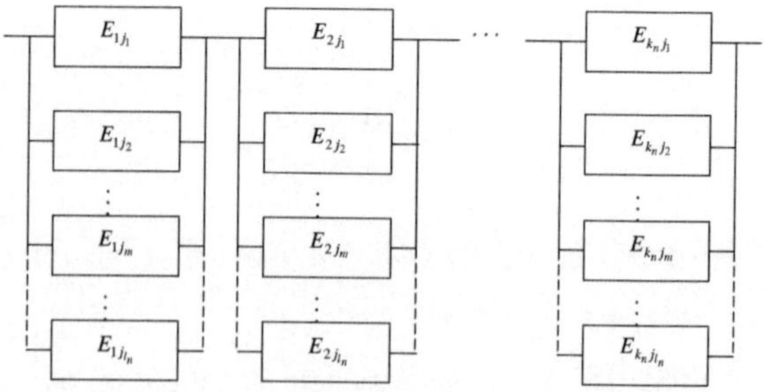

$$\overline{\boldsymbol{R}_{k_n,l_1,l_2,\ldots,l_{k_n}}^{(m1,m2,\ldots,m_{k_n})}}(t) = \prod_{i=1}^{k_n}\left[1 - \sum_{\substack{r1,r2,\ldots rl_i=0\\ r_1+r2+\ldots+rl_i \le m_i-1}}^{1} \left[\prod_{j=1}^{l_i} R_{ij}(t)\right]^{r_i}\left[1-\prod_{j=1}^{l_i} R_{ij}(t)\right]^{1-r_i}\right]$$

$$\text{for } t \in (-\infty, \infty)$$

or by

$$\overline{\boldsymbol{R}_{k_n,l_1,l_2,\ldots,l_{k_n}}^{(\overline{m}_1,\overline{m}_2,\ldots,\overline{m}_{k_n})}}(t) = \prod_{i=1}^{k_n}\left[\sum_{\substack{r1,r2,\ldots,rl_i=0\\ r1+r2+\ldots+rl_i \le \overline{m}_i}}^{1}\left[1-\prod_{j=1}^{l_i} R_{ij}(t)\right]^{r_i}\left[\prod_{j=1}^{l_i} R_{ij}(t)\right]^{1-r_i}\right] \text{ for } t \in (-\infty,\infty),$$

where $\bar{m}_i = l_i - m_i, i = 1, 2, \ldots, k_n$.

The scheme of a regular "m out of l_n"-series system is given in Fig. 15.

Definition 17 The two-state "m_i out of l_i"-series system is called homogeneous if its component lifetimes T_{ij} have an identical distribution function

$$F(t) = P(T_{ij} \leq t), \quad t \in (-\infty, \infty), \quad i = 1, 2, \ldots, k_n, \quad j = 1, 2, \ldots, l_i,$$

i.e., if its components E_{ij} have the same reliability function

$$R(t) = 1 - F(t), \quad t \in (-\infty, \infty).$$

Definition 18 The "m_i out of l_i"-series system is called regular if

$$l_1 = l_2 = \cdots = l_{k_n} = l_n$$

and

$$m_1 = m_2 = \cdots = m_{k_n} = m, \quad \text{where } l_n, m \in N, \ m \leq l_n.$$

The reliability function of the two-state homogeneous and regular "m out of l_n"-series system is given either by

$$\overline{\boldsymbol{R}^{(m)}_{k_n,l_n}}(t) = \left[1 - \sum_{i=0}^{m-1} \binom{l_n}{i} [R(t)]^i [1 - R(t)]^{l_n - i}\right]^{k_n} \quad \text{for } t \in (-\infty, \infty)$$

or by

$$\overline{\bar{\boldsymbol{R}}^{(\bar{m})}_{k_n,l_n}}(t) = \left[\sum_{i=0}^{\bar{m}} \binom{l_n}{i} [1 - R(t)]^i [R(t)]^{l_n - i}\right]^{k_n} \quad \text{for } t \in (-\infty, \infty), \bar{m} = l_n - m$$

where k_n is the number of "m out of l_n" subsystems linked in series and l_n is the number of components in the "m out of l_n" subsystems.

The following results are proved in [18].

Corollary 6 *If the components of the two-state homogeneous and regular "m out of l_n"-series system have Weibull reliability function*

$$R(t) = \exp[-\beta t^{\alpha}] \quad \textit{for } t \geq 0, \alpha > 0, \ \beta > 0,$$

then its reliability function is given either by

$$\overline{\boldsymbol{R}^{(m)}_{k_n,l_n}}(t) = \left[1 - \sum_{i=0}^{m-1} \binom{l_n}{i} \exp[-i\beta t^{\alpha}][1 - \exp[-\beta t^{\alpha}]]^{l_n - i}\right]^{k_n} \quad \textit{for } t \geq 0 \tag{15}$$

or by

$$\overline{\bar{\boldsymbol{R}}^{(\bar{m})}_{k_n,l_n}}(t) = \left[\sum_{i=0}^{\bar{m}} \binom{l_n}{i} [1 - \exp[-\beta t^{\alpha}]]^i \exp[-(l_n - i)\beta t^{\alpha}]\right]^{k_n} \quad \textit{for } t \geq 0, \bar{m} = l_n - m.$$

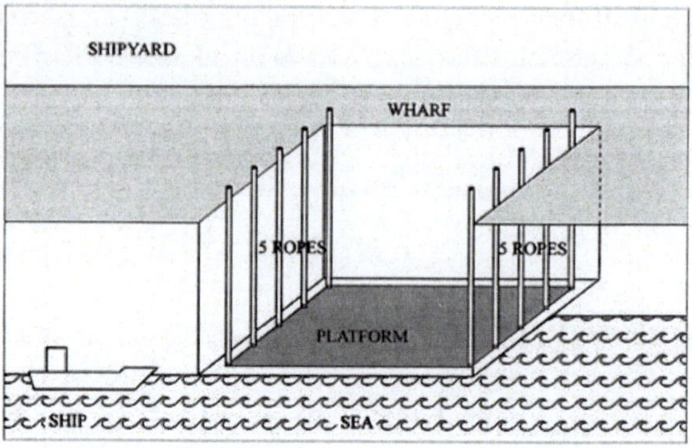

Fig. 16 The scheme of the ship-rope transportation system

Proposition 7 *If components of the two-state homogeneous and regular "m out of l_n"-series system have Weibull reliability function*

$$R(t) = \exp[-\beta t^{\alpha}] \quad \textit{for } t \geq 0, \alpha > 0,\ \beta > 0,$$

and

$$\lim_{n\to\infty} k_n = k, \quad k > 0, \quad 0 < m \leq k, \quad \lim_{n\to\infty} l_n = \infty,$$
$$a_n = \frac{b_n}{\alpha \log n}, \quad b_n \left[\frac{\log n}{\beta}\right]^{\frac{1}{\alpha}}, \tag{16}$$

then

$$[\bar{\mathfrak{R}}_3^{(0)}(t)]^k = \left[1 - \exp[-\exp[-t]] \sum_{i=0}^{m-1} \frac{\exp[-it]}{i!}\right]^k$$

for $t \in (-\infty, \infty)$, is its limit reliability function, i.e.,

$$\begin{aligned}\bar{\boldsymbol{R}}_{k_n,l_n}^{(m)}(t) &\cong \left[\bar{\mathfrak{R}}_3^{(0)}\left(\frac{t - b_n}{a_n}\right)\right]^k \\ &= \left[1 - \exp\left[-\exp\left[-\frac{t - b_n}{a_n}\right]\right] \sum_{i=0}^{m-1} \frac{\exp\left[-i\frac{t-b_n}{a_n}\right]}{i!}\right]^k \textit{for } t \in (-\infty, \infty),\end{aligned} \tag{17}$$

where a_n and b_n are defined by (16).

Example 2 Let us consider the ship-rope transportation system (elevator). The elevator is used to dock and undock ships coming into shipyards for repairs. The scheme of the ship-rope transportation system is shown in Fig. 16.

The elevator is composed of a steel platform carriage placed in its syncline (hutch). The platform is moved vertically with 10 rope hoisting winches fed by

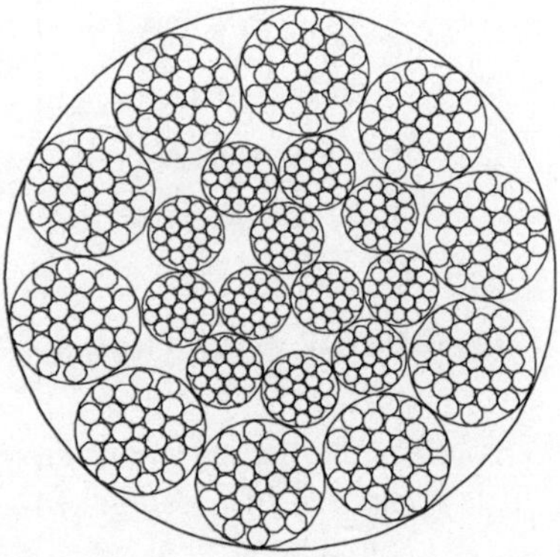

Fig. 17 The cross-section of the rope

separate electric motors. During ship docking the platform, with the ship settled in special supporting carriages on the platform, is raised to the wharf level (upper position). During undocking, the operation is reversed. While the ship is moving into or out of the syncline and while stopped in the upper position the platform is held on hooks and the loads in the ropes are relieved.

In our further analysis we will discuss the reliability of the rope system only. The system under consideration is in order if all its ropes do not fail. Thus, we may assume that it is a series system composed of 10 components (ropes). Each of the ropes is composed of 22 strands. The cross-section of the rope is shown in Fig. 17.

Thus, considering the strands as basic components of the system and assuming that each of the ropes is not failed if at least $m = 5$ out of its strands are not failed, according to Definitions 17–18, we conclude that the rope elevator is the two-state homogeneous and regular "5 out of 22"-series system. It is composed of $k_n = 10$ series-linked "5 out of 22" subsystems (ropes) with $l_n = 22$ components (strands). Assuming additionally that strands have Weibull reliability functions with parameters $\alpha = 2$, $\beta = 0.05$, i.e.,

$$R(t) = \exp[-0.05t^2] \quad \text{for } E_{i1j1},$$

from (15), we conclude that the elevator reliability function is given by

$$\overline{\boldsymbol{R}}^{(5)}_{10,22}(t) = \left[1 - \sum_{i=0}^{4} \binom{22}{i} \exp[-i0.05t^2][1 - \exp[-0.05t^2]]^{22-i}\right]^{10} \quad \text{for } t \geq 0.$$

Next, applying *Proposition 7* with

$$a_n = \frac{7.8626}{2\log 22} \cong 1.2718, \quad b_n = \left[\frac{\log 22}{0.05}\right]^{\frac{1}{2}} \cong 7.8626,$$

and (17) we get the following approximate formula for the elevator reliability function

$$\bar{\boldsymbol{R}}_{k_n,l_n}^{(m)}(t) \cong \left[\bar{\Re}_3^{(0)}\left(\frac{t-b_n}{a_n}\right)\right]^k = \left[1 - \exp\left[-\exp\left[-\frac{t-b_n}{a_n}\right]\right]\sum_{i=0}^{m-1}\frac{\exp\left[-i\frac{t-b_n}{a_n}\right]}{i!}\right]^k$$
$$for\, t \in (-\infty, \infty),$$

8 Asymptotic Approach to Systems Reliability Improvement

We consider the homogeneous series system illustrated in Fig. 18.

It is composed of n components E_{i1}, $i = 1, 2, \ldots, n$, having lifetimes T_{i1}, $i = 1, 2, \ldots, n$, and exponential reliability functions

$$R(t) = \exp[-\lambda t] \quad for\, t \geq 0, \;\; \lambda > 0.$$

Its lifetime and its reliability function, respectively, are given by

$$T^{(0)} = \min_{1 \leq i \leq n} \{T_{i1}\},$$

$$\boldsymbol{R}_n(t) = [R(t)]^n = \exp[-\lambda n t], \; t \geq 0.$$

In order to improve the reliability of this series system the following exemplary methods can be used:

- replacing the system components by the improved components with reduced failure rates by a factor ρ, $0 < \rho < 1$,
- a warm duplication (a single reservation) of system components,
- a cold duplication of system components,
- a mixed duplication of system components,
- a hot system duplication,
- a cold system duplication.

It is supposed here that the reserve components are identical to the basic ones.

The results of these methods of system reliability improvement are briefly presented below, giving the system schemes, lifetimes and reliability functions [18–17].

Case 1 Replacing the system components by the improved components E'_{i1}, $i = 1, 2, \ldots, n$, with reduced failure rates by a factor ρ, $0 < \rho < 1$, having lifetimes $T'_{i1}, i = 1, 2, \ldots, n$, and exponential reliability functions (Fig. 19)

$$R(\rho t) = \exp[-\rho\lambda\, t] \quad \text{for } t \geq 0, \;\; \lambda > 0.$$

$$T^{(1)} = \min_{1 \leq i \leq n} \{T'_{i1}\},$$

$$\boldsymbol{R}_n^{(1)}(t) = [R(\rho t)]^n = \exp[-\rho\lambda n t], \; t \geq 0.$$

Case 2 A hot reservation of the system components (Fig. 20)

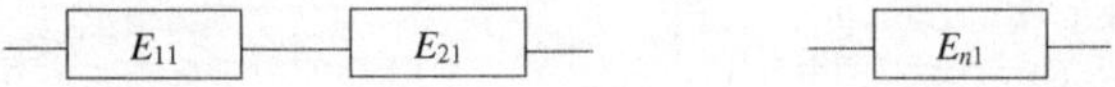

Fig. 18 The scheme of a series system

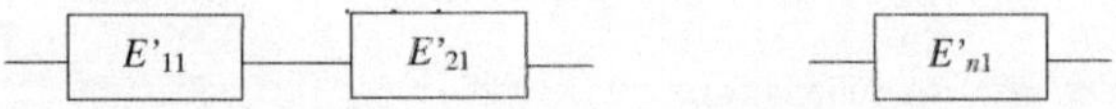

Fig. 19 The scheme of a series system with improved components

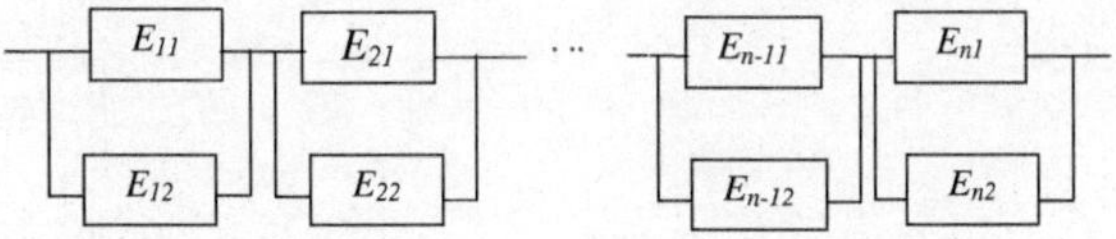

Fig. 20 The scheme of a series system with components having hot reservation

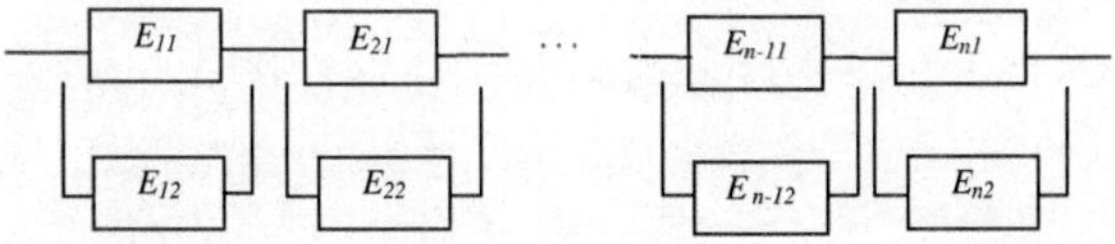

Fig. 21 The scheme of a series system with components having cold reservation

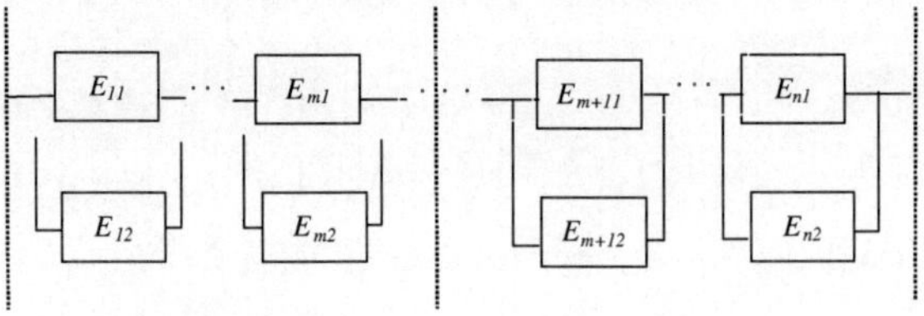

Fig. 22 The scheme of a series system with components having mixed reservation

$$T^{(2)} = \min_{1 \le i \le n} \{ \max_{1 \le j \le 2} \{T_{ij}\}\},$$

$$\boldsymbol{R}_n^{(2)}(T) = [1 - [F(t)]^2]^n = [1 - [1 - \exp[-\lambda t]]^2]^n, \quad t \ge 0.$$

Case 3 A cold reservation of the system components (Fig. 21)

$$T^{(3)} = \min_{1 \le i \le n} \left\{ \sum_{j=1}^{2} T_{ij} \right\},$$

$$\boldsymbol{R}_n^{(3)}(t) = [1 - [F(t)] * [F(t)]]^n = [1 + \lambda t]^n \exp[-n\lambda t], \quad t \ge 0.$$

Case 4 A mixed reservation of the system components (Fig. 22)

$$T^{(4)} = \min\left\{ \min_{1 \le i \le m} \left\{ \sum_{j=1}^{2} T_{ij} \right\}, \quad \min_{m+1 \le i \le n} \left\{ \max_{1 \le j \le 2} \{T_{ij}\} \right\} \right\},$$

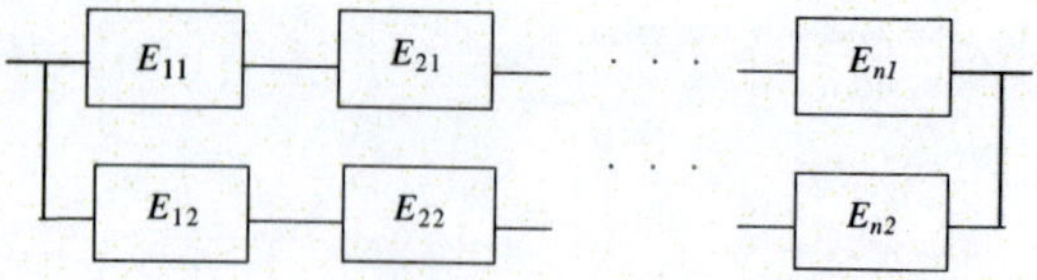

Fig. 23 The scheme of a series system with hot reservation

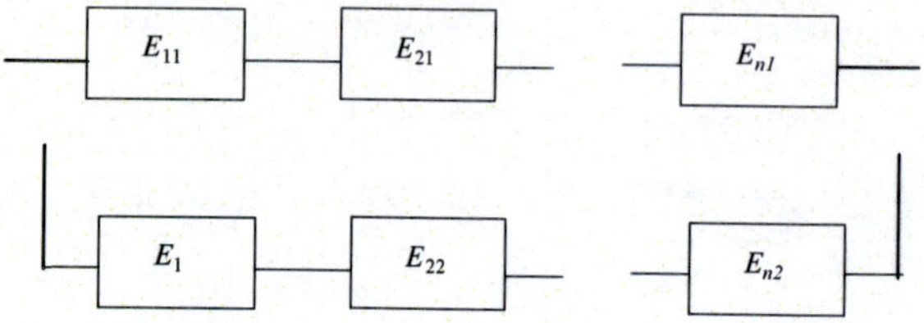

Fig. 24 The scheme of a series system with cold reservation

$$R_n^{(4)}(t) = [1 - [F(t)] * [F(t)]]^m [1 - R^2 t]^{n-m}$$
$$= [1 + \lambda t]^m \exp[-\lambda nt][2 - \exp[-\lambda t]]^{n-m}, \quad t \geq 0.$$

Case 5 A hot system reservation (Fig. 23)

$$T^{(5)} = \max_{1 \leq j \leq 2} \left\{ \min_{1 \leq j \leq n} \{T_{ij}\} \right\},$$

$$\boldsymbol{R}_n^{(5)}(t) = 1 - [1 - [R(t)]^n]^2 = 1 - [1 - \exp[-n\lambda t]]^2, \quad t \geq 0.$$

Case 6 A cold system reservation (Fig. 24)

$$T^{(6)} = \sum_{j=1}^{2} \min_{1 \leq i \leq n} \{T_{ij}\},$$

$$\boldsymbol{R}_n^{(6)}(t) = 1 - [1 - [R(t)]^n] * [1 - [R(t)]^n] = [1 + n\lambda t] \exp[-n\lambda t], \; t \geq 0.$$

The difficulty arises when selecting the right method of improvement of reliability for a large system. This problem may be simplified and approximately solved by the application of the asymptotic approach. Comparisons of the limit reliability functions of the systems with different types of reserve and such systems with improved components allow us to find the value of the components' decreasing failure rate factor ρ, which gives rise to an equivalent effect on the system reliability improvement. Similar results are obtained under comparison of the system lifetime mean values. As an example we will present the asymptotic approach to the above methods of improving reliability for homogeneous two-state series systems. The following results are proved in [15–17].

Proposition 8 *Case 1 If*

$$a_n = 1/\lambda\rho n, \quad b_n = 0,$$

then

$$\Re^{(1)}(t) = \exp[-t] \quad \text{for } t \geq 0,$$

is the limit reliability function of the homogeneous exponential series system with reduced failure rates of its components, i.e.,

$$\boldsymbol{R}_n^{(1)}(t) = \Re^{(1)}(\lambda\rho nt) = \exp[-\lambda\rho nt] \quad \text{for } t \geq 0$$

and

$$\boldsymbol{T}^{(1)} = E[T^{(1)}] = \frac{1}{\lambda\rho n}.$$

Case 2 If

$$a_n = 1/\lambda\sqrt{n}, \quad b_n = 0,$$

then

$$\Re^{(2)}(t) = \exp[-t^2] \quad \text{for } t \geq 0,$$

is the limit reliability function of the homogeneous exponential series system with hot reservation of its components, i.e.,

$$\boldsymbol{R}_n^{(2)}(t) \cong \Re^{(2)}(\lambda\sqrt{n}t) = \exp[-\lambda^2 nt^2] \quad \text{for } t \geq 0$$

and

$$\boldsymbol{T}^{(2)} = E[T^{(2)}] \cong \Gamma\left(\frac{3}{2}\right)\frac{1}{\lambda\sqrt{n}}.$$

Case 3 If

$$a_n = \sqrt{2}/\lambda\sqrt{n}, \quad b_n = 0,$$

then

$$\Re^{(3)}(t) = \exp[-t^2] \quad \text{for } t \geq 0,$$

is the limit reliability function of the homogeneous exponential series system with cold reservation of its components, i.e.,

$$\boldsymbol{R}_n^{(3)}(t) \cong \Re^{(3)}\left(\lambda\sqrt{\frac{nt}{2}}\right) = \exp\left[-\frac{1}{2}\lambda^2 nt^2\right] \quad \text{for } t \geq 0$$

and

$$T^{(3)} = E[T^{(3)}] \cong \Gamma\left(\frac{3}{2}\right)\frac{1}{\lambda}\sqrt{\frac{2}{n}}.$$

Case 4 If

$$a_n = \frac{1}{\lambda}\sqrt{\frac{2}{2n-m}}, \quad b_n = 0,$$

then

$$\Re^{(4)}(t) = \exp[-t^2] \; for \; t \geq 0,$$

is the limit reliability function of the homogeneous exponential series system with mixed reservation of its components, i.e.,

$$\boldsymbol{R}_n^{(4)}(t) \cong \Re^{(4)}\left(\lambda\sqrt{\frac{2n-m}{2}}\right) = \exp\left[-\frac{2n-m}{2}\lambda^2 t^2\right] \quad for \; t \geq 0$$

and

$$T^{(4)} = E[T^{(4)}] \cong \Gamma\left(\frac{3}{2}\right)\frac{1}{\lambda}\sqrt{\frac{2}{2n-m}}.$$

Case 5 If

$$a_n = \frac{1}{\lambda n}, \quad b_n = 0,$$

then

$$\Re^{(5)}(t) = 1 - [1 - \exp[-t]]^2 \quad for \; t \geq 0,$$

is the limit reliability function of the homogeneous exponential series system with hot reservation, i.e.,

$$\boldsymbol{R}_n^{(5)}(t) = \Re^{(5)}(\lambda n t) = 1 - [1 - \exp[-\lambda n t]]^2 \quad for \; t \geq 0$$

and

$$T^{(5)} = E[T^{(5)}] = \frac{3}{2\lambda n}.$$

Case 6 If

$$a_n = \frac{1}{\lambda n}, \quad b_n = 0,$$

then

$$\Re^{(6)}(t) = [1+t]\exp[-t] \quad \textit{for } t \geq 0,$$

is the limit reliability function of the homogeneous exponential series system with cold reservation, i.e.,

$$\boldsymbol{R}_n^{(6)}(t) = \Re^{(6)}(\lambda nt) = [1+\lambda nt]\exp[-\lambda nt] \quad \textit{for } t \geq 0$$

and

$$\boldsymbol{T}^{(6)} = E[T^{(6)}] = \frac{2}{\lambda n}.$$

Corollary 7 *Comparison of the system reliability functions*

$$\Re^{(i)}(t) = \Re^{(1)}(t), \quad i = 2, 3, \ldots, 6,$$

results, respectively, in the following values of the factor ρ*:*

$$\rho = \rho(t) = \lambda t \quad \textit{for } i = 2,$$

$$\rho = \rho(t) = \frac{1}{2}\lambda t \quad \textit{for } i = 3,$$

$$\rho = \rho(t) = \frac{2n-m}{m} \quad \textit{for } i = 4,$$

$$\rho = \rho(t) = 1 - \log[2 - \exp[-\lambda nt]] \quad \textit{for } i = 5,$$

$$\rho = \rho(t) = 1 - \frac{1}{\lambda nt}\log[1+\lambda nt] \quad \textit{for } i = 6,$$

while comparison of the system lifetimes

$$\boldsymbol{T}^{(i)}(t) = \boldsymbol{T}^{(1)}(t), \quad i = 2, 3, \ldots, 6$$

results, respectively, in the following values of the factor ρ*:*

$$\rho = \frac{1}{\Gamma(\frac{3}{2})\sqrt{n}} \quad \textit{for } i = 2,$$

$$\rho = \frac{1}{\Gamma(\frac{3}{2})\sqrt{2n}} \quad \textit{for } i = 3,$$

$$\rho = \frac{1}{\Gamma(\frac{3}{2})n\sqrt{\frac{2}{2n-m}}} \quad \textit{for } i = 4,$$

$$\rho = \frac{2}{3} \quad for\, i = 5,$$

$$\rho = \frac{1}{2} \quad for\, i = 6.$$

Example 5 We consider a simplified bus service company composed of 81 communication lines. We suppose that there is one bus operating on each communication line and that all buses are of the same type with the exponential reliability function

$$R(t) = \exp[-\lambda t] \quad \text{for}\, t \geq 0, \lambda > 0.$$

Additionally we assume that this communication system is working when all its buses are not failed, i.e. it is failed when any of the buses is failed. The failure rate of the buses evaluated on statistical data coming from the operational process of bus service company transportation system is assumed to be equal to 0.0049 h^{-1}.

Under these assumptions the considered transportation system is a homogeneous series system made up of components with a reliability function

$$R(t) = \exp[-0.0049t] \quad \text{for}\, t \geq 0.$$

Here we will use four sensible methods from those considered for system reliability improvement. Namely, we apply the four previously considered cases.

Case 1 Replacing the system components by the improved components with reduced failure rates by a factor ρ.

Applying *Proposition 8* with normalising constants

$$a_{81} = \frac{1}{0.0049 \cdot 81\rho} = \frac{1}{0.397\rho}, \quad b_{81} = 0,$$

we conclude that

$$\Re^{(1)}(t) = \exp[-t] \quad \text{for}\, t \geq 0,$$

is the limit reliability function of the system, i.e.

$$\boldsymbol{R}_n^{(1)}(t) \cong \Re^{(1)}(0.397\rho t) = \exp[-0.397\rho t] \quad \text{for}\, t \geq 0$$

and

$$\boldsymbol{T}^{(1)} = E[T^{(1)}] \cong \frac{1}{0.397\rho}\text{h}.$$

Case 2 Improving the reliability of the system by a single hot reservation of its components. This means that each of 81 communication lines has at its disposal two identical buses it can use and its task is performed if at least one of the buses is not failed.

Applying *Proposition 8* with normalising constants

$$a_{81} = \frac{1}{0.0049 \cdot \sqrt{81}} = \frac{1}{0.0441}, \quad b_{81} = 0.$$

We conclude that

$$\mathfrak{R}^{(2)}(t) = \exp[-t^2], \quad t \geq 0.$$

is the limit reliability function of the system, i.e.,

$$\boldsymbol{R}_{81}^{(2)}(t) \cong \mathfrak{R}^{(2)}(0.0441t) \cong \exp[-0.0019t^2], \quad t \geq 0,$$

and

$$T^{(2)} = E[T^{(2)}] \cong \Gamma\left(\frac{3}{2}\right)\frac{1}{0.0049\sqrt{81}} \cong 20.10\,\text{h}.$$

Case 4 Improving the reliability of the system by a single mixed reservation of its components.

This means that each of 81 communication lines has at its disposal two identical buses. There are $m = 50$ communication lines with small traffic which are using one bus permanently and after its failure it is replaced by the second bus (a cold reservation) and $n - m = 81 - 50 = 31$ communication lines with large traffic which are using two buses permanently (a hot reservation).

Applying *Proposition 8* with normalising constants

$$a_n = \frac{1}{0.0049}\sqrt{\frac{2}{112}} = \frac{1}{0.0367}, \quad b_n = 0,$$

we conclude that

$$\mathfrak{R}^{(4)}(t) = \exp[-t^2] \quad \text{for } t \geq 0,$$

is the limit reliability function of the system, i.e.,

$$\boldsymbol{R}_n^4(t) \cong \mathfrak{R}^{(4)}(0.0367t) = \exp[-0.00135t^2] \quad \text{for } t \geq 0$$

and

$$T^{(4)} = E[T^{(4)}] \cong \Gamma\left(\frac{3}{2}\right)\frac{1}{0.0049}\sqrt{\frac{2}{112}} \cong 24.15\,\text{h}.$$

Case 5 Improving the reliability of the system by a single hot reservation.

This means that the transportation system is composed of two independent companies, each of them operating on the same 81 communication lines and having at their disposal one identical bus for use on each line.

Applying *Proposition 8* with normalising constants

$$a_{81} = \frac{1}{0.0049 \cdot 81} = \frac{1}{0.397}, \quad b_{81} = 0.$$

We conclude that

$$\mathfrak{R}^{(5)}(t) = 1 - [1 - \exp[-t]]^2 \quad \text{for } t \geq 0,$$

is the limit reliability function of the system, i.e.,

$$\boldsymbol{R}_n^5(t) \cong \mathfrak{R}^{(5)}(0.397t) = 1 - [1 - \exp[-0.397t]]^2 \quad \text{for } t \geq 0$$

and

$$\boldsymbol{T}^{(5)} = E[T^{(5)}] \cong \frac{3}{2 \cdot 0.0049 \cdot 81} \cong 3.78\,\text{h}.$$

Comparing the system reliability functions for considered cases of improvement, from *Corollary 7*, results in the following values of the factor ρ:

$$\rho = \rho(t) = 0.0049t \quad \text{for } i = 2,$$

$$\rho = 0.0340t \quad \text{for } i = 4,$$

$$\rho = \rho(t) = 1 - \log[2 - \exp[-0.397t]] \quad \text{for } i = 5,$$

while comparison of the system lifetimes results, respectively, in:

$$\rho = 0.1254 \quad \text{for } i = 2,$$

$$\rho = 0.1043 \quad \text{for } i = 4,$$

$$\rho = 0.0667 \quad \text{for } i = 5.$$

Methods of system reliability improvement presented here supply practitioners with simple mathematical tools, which can be used in everyday practice. The methods may be useful not only in the operation processes of real technical objects but also in designing new operation processes and especially in optimising these processes. Only the case of series systems made up of components having exponential reliability functions with single reservations of their components and subsystems is considered. It seems to be possible to extend these results to systems that have more complicated reliability structures, and made up of components with different from the exponential reliability functions.

9 Reliability of Large Systems in Their Operation Processes

This section proposes an approach to the solution of the practically very important problem of linking systems' reliability and their operation processes. To connect the interactions between the systems' operation processes and their reliability structures that are changing in time a semi-Markov model [7] of the system operation processes is applied. This approach gives a tool that is practically important and not difficult for everyday use for evaluating reliability of systems with changing reliability structures during their operation processes [13, 11,20–22]. Application of the proposed methods is illustrated here in the reliability evaluation of the port grain transportation system.

We assume that the system during its operation process is taking different operation states. We denote by $Z(t), t \in \langle 0, \infty \rangle$, the system operation process that may assume v different operation states from the set

$$Z = \{z_1, z_2, \ldots, z_v\}.$$

In practice a convenient assumption is that $Z(t)$ is a semi-Markov process [7] with its conditional sojourn times θ_{bl} at the operation state z_b when its next operation state is z_l, $b, l = 1, 2, \ldots, v$, $b \neq l$. In this case this process may be described by:

- the vector of probabilities of the initial operation states $[p_b(0)]_{1xv}$,
- the matrix of the probabilities of its transitions between the states $[p_{bl}]_{vxv}$,
- the matrix of the conditional distribution functions $[H_{bl}(t)]_{vxv}$ of the sojourn times θ_{bl}, $b \neq l$, where

$$H_{bl}(t) = P(\theta_{bl} < t) \quad \text{for } b, l = 1, 2, \ldots, v, b \neq l,$$

and

$$H_{bb}(t) = 0 \quad \text{for } b = 1, 2, \ldots, v.$$

Under these assumptions, the lifetime θ_{bl} mean values are given by

$$M_{bl} = E[\theta_{bl}] = \int_0^\infty t dH_{bl}(t), \quad b, l = 1, 2, \ldots, v, b \neq l. \tag{18}$$

The unconditional distribution functions of the sojourn times θ_b of the process $Z(t)$ at the states $z_b, b = 1, 2, \ldots, v$, are given by

$$H_b(t) = \sum_{l=1}^{v} p_{bl} H_{bl}(t), \; b = 1, 2, \ldots, v.$$

The mean values $E[\theta_b]$ of the unconditional sojourn times θ_b are given by

$$M_b = E[\theta_b] = \sum_{l=1}^{v} p_{bl} M_{bl}, \quad b = 1, 2, \ldots, v, \tag{19}$$

where M_{bl} are defined by (18).

Limit values of the transient probabilities at the states

$$p_b(t) = P(Z(t) = z_b), \quad t \in <0, \infty), \quad b = 1, 2, \ldots, v,$$

are given by [7]

$$p_b = \lim_{t \to \infty} p_b(t) = \frac{\pi_b M_b}{\sum_{l=1}^{v} \pi_l M_l}, \quad b = 1, 2, \ldots, v, \tag{20}$$

where the probabilities π_b of the vector $[\pi_b]_{1xv}$ satisfy the system of equations

$$\begin{cases} [\pi_b] = [\pi_b][p_{bl}] \\ \sum_{l=1}^{v} \pi_l = 1. \end{cases}$$

We consider a series–parallel system and we assume that the changes of its operation process $Z(t)$ states have an influence on the system components E_{ij} reliability and on the system reliability structure as well. Thus, we denote [22] the conditional reliability function of the system component E_{ij} while the system is at the operational state z_b, $b = 1, 2, \ldots, v$, by

$$[R^{(i,j)}(t)]^{(b)} = P(T_{ij}^{(b)} \geq t / Z(t) = z_b),$$

for $t \in <0, \infty)$, $b = 1, 2, \ldots, v$, and the conditional reliability function of the non-homogeneous regular series–parallel system while the system is at the operational state z_b, $b = 1, 2, \ldots, v$, by

$$\begin{aligned} [\boldsymbol{R}_{k_n, l_n}(t)]^{(b)} &= P(T^{(b)} \geq t / Z(t) = z_b) \\ &= 1 - \prod_{i=1}^{a} [1 - [[R^{(i)}(t)]^{(b)}]^{l_n}]^{q_i k_n} \quad \text{for } t \in <0, \infty) \end{aligned} \tag{21}$$

and

$$[R^{(i)}(t)]^{(b)} = \prod_{j=1}^{e_i} [[R^{(i,j)}(t)]^{(b)}]^{p_{ij}}, \quad i = 1, 2, \ldots, \text{a}. \tag{22}$$

The reliability function $[R^{(i,j)}(t)]^{(b)}$ is the conditional probability that the component E_{ij} lifetime $T_{ij}^{(b)}$ is not less than t, while the process $Z(t)$ is at the operation state z_b. Similarly, the reliability function $[\boldsymbol{R}_{k_n, l_n}(t)]^{(b)}$ is the conditional probability that the series–parallel system lifetime $T^{(b)}$ is not less than t, while the process $Z(t)$ is at the operation state z_b. In the case when the system operation time

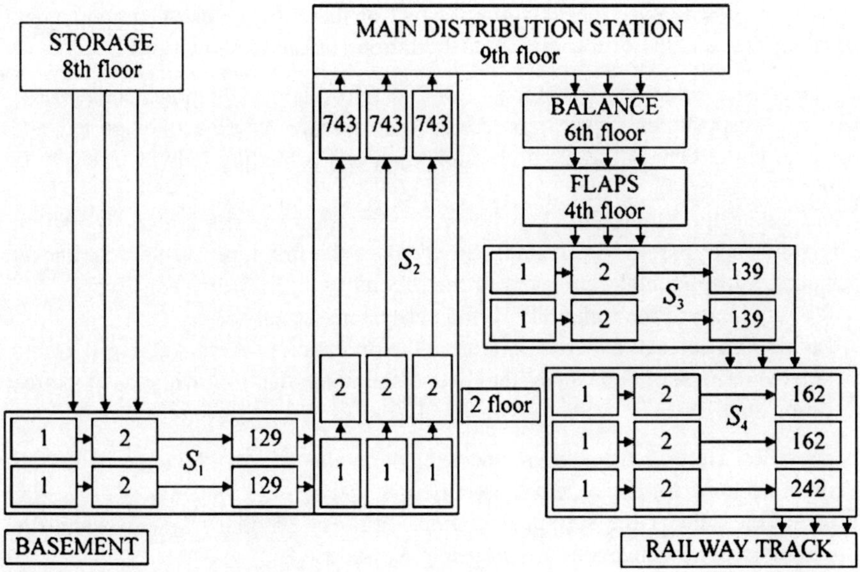

Fig. 25 The scheme of the grain transportation system

is large enough, the unconditional reliability function of the series–parallel system is given by [22]

$$\boldsymbol{R}_{k_n,l_n}(t) = P(T > t) \cong \sum_{b=1}^{v} p_b[\boldsymbol{R}_{k_n,l_n}(t)]^{(b)} \quad \text{for } t \geq 0 \tag{23}$$

and T is the unconditional lifetime of the series–parallel system.

The mean values and variances of the series–parallel system lifetimes are

$$M \cong \sum_{b=1}^{v} p_b M_b, \tag{24}$$

where

$$M_b = \int_0^\infty [\boldsymbol{R}_{k_n,l_n}(t)]^{(b)} dt, \tag{25}$$

and

$$D[T^{(b)}] = 2\int_0^\infty t[\boldsymbol{R}_{k_n l_n}(t)]^{(b)} dt - [M_b]^2 \quad \text{for } b = 1, 2, \ldots, v. \tag{26}$$

Example 6 We analyse the reliability of one of the subsystems of the port grain elevator. The scheme of the grain transportation system is shown in Fig. 25.

The considered system is composed of four two-state non-homogeneous series–parallel transportation subsystems assigned to handle clearing of exported and imported grain. One of the basic elevator functions is loading railway trucks with grain.

In loading the railway trucks with grain the following elevator transportation subsystems take part: S_1–horizontal conveyors of the first type, S_2–vertical bucket elevators, S_3–horizontal conveyors of the second type, S_4–worm conveyors.

We will analyse the reliability of the subsystem S_4 only.

Taking into account experts' opinion in the operation process $Z(t)$, $t \geq 0$, of the considered transportation subsystem we distinguish the following as its three operation states:

an operation state z_1–the system operation with the largest efficiency when all components of the subsystem S_4 are used,
an operation state z_2–the system operation with less efficiency system when the first and second conveyors of subsystem S_4 are used,
an operation state z_3–the system operation with least efficiency when the first conveyor of subsystem S_4 is used.

On the basis of data coming from experts, the probabilities of transitions between the subsystem S_4 operation states are given by

$$[p_{bl}] = \begin{bmatrix} 0 & 0.357 & 0.643 \\ 0.8 & 0 & 0.2 \\ 0.385 & 0.615 & 0 \end{bmatrix},$$

and their mean values, from (19), are

$$M_1 = E[\theta_1] = 0.357 \cdot 0.36 + 0.643 \cdot 0.2 \cong 0.257,$$

$$M_2 = E[\theta_2] = 0.8 \cdot 0.05 + 0.2 \cdot 0.2 \cong 0.08,$$

$$M_3 = E[\theta_3] = 0.385 \cdot 0.08 + 0.615 \cdot 0.05 \cong 0.062.$$

Since from the system of equations

$$\begin{cases} [\pi_1, \pi_2, \pi_3] = [\pi_1, \pi_2, \pi_3] \begin{bmatrix} 0 & 0.357 & 0.643 \\ 0.8 & 0 & 0.2 \\ 0.385 & 0.615 & 0 \end{bmatrix}. \\ \pi_1 + \pi_2 + \pi_3 = 1 \end{cases}$$

We get

$$\pi_1 = 0.374, \quad \pi_2 = 0.321, \quad \pi_3 = 0.305,$$

then the limit values of the transient probabilities $p_b(t)$ at the operation states z_b, according to (20), are given by

$$p_1 = 0.684, \quad p_2 = 0.183, \quad p_3 = 0.133. \tag{27}$$

The subsystem S_4 consists of three chain conveyors. Two of these are composed of 162 components and the remaining one is composed of 242 components. Thus it is a non-regular series–parallel system. In order to make it a regular system we conventionally complete two first conveyors having 162 components with 80 components that do not fail. After this supplement subsystem S_4 consists of $k_n = 3$ conveyors, each composed of $l_n = 242$ components. In two of them there are:

– two driving wheels with reliability functions

$$R^{(1,1)}(t) = \exp[-0.0798t],$$

– 160 links with reliability functions

$$R^{(1,2)}(t) = \exp[-0.124t],$$

– 80 components with "reliability functions"

$$R^{(1,3)}(t) = \exp[-\lambda_1(1)t], \quad \text{where } \lambda_1(1) = 0.$$

The third conveyer is composed of:

– two driving wheels with reliability functions

$$R^{(2,1)}(t) = \exp[-0.167t]$$

– 240 links with reliability functions

$$R^{(2,2)}(t) = \exp[-0.208t].$$

At the operation state z_1, the subsystem S_4 becomes a non-homogeneous regular series–parallel system with parameters

$k_n = 3$, $l_n = 242$, $a = 2$, $q_1 = 2/3$, $q_2 = 1/3$,
$e_1 = 3$, $e_2 = 2$,
$p_{11} = 2/242$, $p_{12} = 160/242$, $p_{13} = 80/242$,
$p_{21} = 2/242$, $p_{22} = 240/242$,

and from (21) to (22) the reliability function of this system is given by

$$\begin{aligned}[\boldsymbol{R}_{3,242}(t)]^{(1)} &= 1 - [1 - \exp[-19.9892t]]^2[1 - \exp[-50.2628t]] \\ &= 2\exp[-19.9892t] - 2\exp[-70.252t] + \exp[-50.2628t] \\ &\quad + \exp[-90.2412t] - \exp[-39.9784t] \quad \text{for } t \ge 0. \end{aligned} \tag{28}$$

According to (25–26), the subsystem lifetime mean value and the standard deviation are

$$M_1 \cong 0.078, \quad \sigma_1 \cong 0.054. \tag{29}$$

At the operation state z_2, the subsystem S_4 becomes a non-homogeneous regular series–parallel system with parameters

$k_n = 2,\ l_n = 162,\ a = 1,\ q_1 = 1,\ e_1 = 2,$
$p_{11} = 2/162,\ p_{12} = 160/162,$

and from (21) to (22) the reliability function of this system is given by

$$\begin{aligned}\left[\boldsymbol{R}_{2,162}(t)\right]^{(2)} &= 1 - [1 - \exp[-20.007t]]^2 \\ &= 2\exp[-20.007t] - \exp[-40.014t] \quad \textit{for } t \geq 0.\end{aligned} \tag{30}$$

According to (25–26), the subsystem lifetime mean value and the standard deviation are

$$M_2 \cong 0.075, \quad \sigma_2 \cong 0.056. \tag{31}$$

At the operation state z_3, the subsystem S_4 becomes a non-homogeneous regular series–parallel (series) system with parameters

$k_n = 1,\ l_n = 162,\ q_1 = 1,\ e_1 = 3,$
$p_{11} = 2/162,\ p_{12} = 160/162,$

and from (21) to (22) the reliability function of this system is given by

$$[\boldsymbol{R}_{1,162}(t)]^{(3)} = \exp[-19.999t] \quad \text{for } t \geq 0. \tag{32}$$

According to (25–26), the system lifetime mean value and the standard deviation are

$$M_{(3)} \cong 0.050, \quad \sigma_3 \cong 0.050. \tag{33}$$

Finally, considering (23), the subsystem S_4 unconditional reliability is given by

$$\boldsymbol{R}(t) \cong 0.684 \cdot [\boldsymbol{R}_{3,242}(t)]^{(1)} + 0.183 \cdot [\boldsymbol{R}_{2,162}(t)]^{(2)} + 0.133 \cdot [\boldsymbol{R}_{1,162}(t)]^{(3)}, \tag{34}$$

where $[\boldsymbol{R}_{3,242}(t)]^{(1)}$, $[\boldsymbol{R}_{2,162}(t)]^{(2)}$, $[\bar{\boldsymbol{R}}_{1,162}(t)]^{(3)}$ are given by (28), (30), (32).

Hence, applying (27) and (29), (31), (33) and (34), we get the mean values and standard deviations of the subsystem unconditional lifetimes given by

$$M \cong 0.684 \cdot 0.078 + 0.183 \cdot 0.075 + 0.133 \cdot 0.050 \cong 0.074, \ \sigma(1) \cong 0.054.$$

10 Conclusions

Generalisations of the results on limit reliability functions of two-state homogeneous systems [12] for these and other systems in case they are non-homogeneous, are mostly given in Kołowrocki [9] and [10]. These results allow us to evaluate reliability characteristics of homogeneous and non-homogeneous series–parallel and parallel–series systems with regular reliability structures, i.e., systems composed of subsystems having the same numbers of components. However, this fact does not restrict the completeness of the performed analysis, since by conventional joining of a suitable number of components which do not fail, in series subsystems of the non-regular series–parallel systems, leads us to the regular non-homogeneous series–parallel systems. Similarly, conventional joining of a suitable number of failed components in parallel subsystems of the non-regular parallel–series systems we get the regular non-homogeneous parallel–series systems. Thus the problem has been analysed exhaustively.

The results concerned with the asymptotic approach to system reliability analysis, in a natural way, have led to investigation of the speed of convergence of the system reliability function sequences to their limit reliability functions [10]. These results have also initiated the investigations of limit reliability functions of "m out of n"-series, series-"m out of n" systems, the investigations on the problems of the system reliability improvement and on the reliability of systems with varying in time their structures and their components reliability described briefly in [10].

More general and practically important complex systems composed of multi-state and degrading in time components are considered in wide literature, for instance in [24]. An especially important role they play in the evaluation of technical systems reliability and safety and their operating process effectiveness is described in [10] for large multi-state systems with degrading components. The most important results regarding generalisations of the results on limit reliability functions of two-state systems dependent on transferring them to series, parallel, "m out of n", series–parallel and parallel–series multi-state systems with degrading components are given in [10]. Some practical applications of the asymptotic approach to the reliability evaluation of various technical systems are contained in [10] as well.

The proposed method offers enough simplified formulae to allow significant simplifying of large systems' reliability evaluating and optimising calculations.

11 Appendix: Notations

E_i	Components of series and parallel systems
E_{ij}	Components of series–parallel and parallel–series systems

T_i	Component lifetimes of two-state series and parallel systems
T_{ij}	Component lifetimes of two-state series–parallel and parallel–series systems
T	A two-state system lifetime
$R(t)$	A component reliability function of a two-state homogeneous system
$F(t)$	A component lifetime distribution function of a two-state homogeneous system
$\bar{\boldsymbol{R}}_n(t)$	A reliability function of a two-state homogeneous series system
$\boldsymbol{R}_n(t)$	A reliability function of a two-state homogeneous parallel system
$\bar{\boldsymbol{R}}_{k_n l_n}(t)$	A reliability function of a two-state homogeneous parallel–series system
$\boldsymbol{R}_{k_n l_n}(t)$	A reliability function of a two-state homogeneous series–parallel system
$\bar{\Re}(t)$	A limit reliability function of two-state homogeneous series and parallel–series systems
$\Re(t)$	A limit reliability function of two-state homogeneous parallel and series–parallel systems
$E(T)$	A mean lifetime of a two-state system
$\sigma(T)$	A lifetime standard deviation of a two-state system
$D_{\bar{\Re}i}$	Domains of attraction of limit reliability functions $\bar{\Re}_i(t)$ of two-state homogeneous series system
$\boldsymbol{R}_{r,k_n,l_n}(t)$	A reliability function of a two-state series–parallel system of order r
$\Re_i(t)$	A limit reliability function of a two-state series–parallel system of order r
$\bar{\boldsymbol{R}}_{r,k_n,l_n}(t)$	A reliability function of a two-state parallel–series system of order r
$\bar{\Re}_i(t)$	A limit reliability function of a two-state parallel–series system of order r
$\boldsymbol{R}^{(m)}_{k_n,l_1,l_2,\ldots,l_{k_n}}(t)$	A reliability function of a homogeneous two-state series-"m out of k_n" system
$\bar{\boldsymbol{R}}^{(\bar{m})}_{k_n,l_1,l_2,\ldots,l_{k_n}}(t)$	A reliability function of a homogeneous two-state series-"m out of k_n" system
$\boldsymbol{R}^{(m)}_{k_n,l_n}(t)$	A reliability function of a homogeneous and regular two-state series-"m out of k_n" system
$\overline{\boldsymbol{R}}^{(\overline{m})}_{k_n,l_n}(t)$	A reliability function of a homogeneous and regular two-state series-"m out of k_n" system
$\overline{\boldsymbol{R}^{(m_1,m_2,\ldots,m_{k_n})}_{k_n,l_1,l_2,\ldots,l_{k_n}}}(t)$	A reliability function of a two-state "m_i out of l_i"-series system
$\overline{\overline{\boldsymbol{R}}^{(\overline{m}_1,\overline{m}_2,\ldots,\overline{m}_{k_n})}_{k_n,l_1,l_2,\ldots,l_{k_n}}}(t)$	A reliability function of a two-state "m_i out of l_i"-series system

$\overline{\boldsymbol{R}_{k_n,l_n}^{(m)}}(t)$	A reliability function of a homogeneous and regular two-state "m out of k_n"-series system
$\overline{\boldsymbol{R}}_{k_n,l_n}^{(\overline{m})}(t)$	A reliability function of a homogeneous and regular two-state "m out of k_n"-series system
$\Re_i^{(m)}(t)$	A limit reliability function of a homogeneous and regular two-state series-"m out of k_n" system
$\Re_i^{(\overline{m})}(t)$	A limit reliability function of a homogeneous and regular two-state series-"m out of k_n" system
$\overline{\Re_i^{(m)}}(t)$	A limit reliability function of a homogeneous and regular two-state "m out of k_n"-series system
$\overline{\Re}_i^{(\overline{m})}(t)$	A limit reliability function of a homogeneous and regular two-state "m out of k_n"-series system
ρ	A factor reducing a component failure rate
$\boldsymbol{R}_n^{(1)}(t)$	A reliability function of a two-state series system with components improved by reducing their failure rates by a factor ρ
$\boldsymbol{R}_n^{(2)}(t)$	A reliability function of a two-state series system with a single hot reservation of its components
$\boldsymbol{R}_n^{(3)}(t)$	A reliability function of a two-state series system with a single cold reservation of its components
$\boldsymbol{R}_n^{(4)}(t)$	A reliability function of a two-state series system with a single mixed reservation of its components
$\boldsymbol{R}_n^{(5)}(t)$	A reliability function of a two-state series system with its single hot reservation
$\boldsymbol{R}_n^{(6)}(t)$	A reliability function of a two-state series system with its single cold reservation
$\Re^{(1)}(t)$	A limit reliability function of a two-state series system with components improved by reducing their failure rates by a factor ρ
$\Re^{(2)}(t)$	A limit reliability function of a two-state series system with a single hot reservation of its components
$\Re^{(3)}(t)$	A limit reliability function of a two-state series system with a single cold reservation of its components
$\Re^{(4)}(t)$	A limit reliability function of a two-state series system with a single mixed reservation of its components
$\Re^{(5)}(t)$	A limit reliability function of a two-state series system with its single hot reservation
$\Re^{(6)}(t)$	A limit reliability function of a two-state series system with its single cold reservation
$\boldsymbol{T}^{(1)}$	A lifetime mean value of a two-state series system with components improved by reducing their failure rates by a factor ρ
$\boldsymbol{T}^{(2)}$	A lifetime mean value of a two-state series system with a single hot reservation of its components

$\boldsymbol{T}^{(3)}$	A lifetime mean value of a two-state series system with a single cold reservation of its components
$\boldsymbol{T}^{(4)}$	A lifetime mean value of a two-state series system with a single mixed reservation of its components
$\boldsymbol{T}^{(5)}$	A lifetime mean value of a two-state series system with its single hot reservation
$\boldsymbol{T}^{(6)}$	A lifetime mean value of a two-state series system with its single cold reservation
z_k	A system operational state
$Z(t)$	A process of changing system operational states
θ_{kl}	Conditional sojourn times of a process $Z(t)$at operational states
$[H(t)]_{vxv}$	A matrix of conditional distribution functions of sojourn times θ_{kl}
$[p_k(0)]_{1xv}$	A vector of probabilities of process $Z(t)$ initial states
$E[\theta_{kl}]$	Mean values of sojourn times θ_{kl}
θ_k	Unconditional sojourn times of process $Z(t)$ at states z_k
$H_k(t)$	Unconditional distribution functions of sojourn times θ_k
$E[\theta_k]$	Mean values of unconditional sojourn times θ_k
M^k	Mean values of unconditional sojourn times θ_k
$p_k(t)$	Transient probabilities of process $Z(t)$ at states z_k
p_k	Limit values of transient probabilities $p_k(t)$
$\boldsymbol{R}^{(k)}(t)$	Conditional reliability functions of a two-state system at operational states z_k
$\boldsymbol{R}(\boldsymbol{t})$	An unconditional reliability function of a two-state system
$T_{ij}^{(k)}$	Conditional lifetimes of system components E_{ij} of a non-homogeneous two-state series–parallel system at operational states z_k
$[\boldsymbol{R}^{(i,j)}(t)]^{(k)}$	Conditional reliability functions of system components E_{ij} of a non-homogeneous two-state series–parallel system at operational states z_k
$T^{(k)}$	Conditional lifetimes of a non-homogeneous two-state series–parallel system at operational states z_k
$\boldsymbol{R}_{k_n,l_n}^{(k)}(t)$	Conditional reliability functions of a non-homogeneous two-state series–parallel system at operational states z_k
T	An unconditional lifetime of a non-homogeneous two-state series–parallel system
$\boldsymbol{R}(t)$	Unconditional reliability functions of a non-homogeneous two-state series–parallel system
m	An unconditional mean value of a non-homogeneous two-state series–parallel system lifetime
σ^2	An unconditional variance of a non-homogeneous two-state series–parallel system lifetime

References

1. Barlow RE, Proschan F (1975) Statistical theory of reliability and life testing. Probability models. Holt Rinehart and Winston, Inc., New York
2. Chernoff H, Teicher H (1965) Limit distributions of the minimax of independent identically distributed random variables. Proc Am Math Soc 116:474–491
3. Cichocki A (2003) Wyznaczanie granicznych funkcji niezawodności systemów hierarchicznych przy standaryzacji potęgowej. Ph.D. Thesis. Maritime University, Gdynia–Systems Research Institute, Polish Academy of Sciences, Warzawa
4. Fisher RA, Tippett LHC (1928) Limiting forms of the frequency distribution of the largest and smallest member of a sample. Proc Cambr Phil Soc 24:180–190
5. Frechet M (1928) Sur la loi de probabilite de l'ecart maximum. Ann de la Soc Polonaise de Math 6:93–116
6. Gniedenko BW (1943) Sur la distribution limite du terme maximum d'une serie aleatoire. Ann Math 44:432–453
7. Grabski F (2002) Semi-Markov models of systems reliability and operations. Systems Research Institute, Polish Academy of Sciences, Warsaw
8. Gumbel EJ (1962) Statistics of extremes. New York
9. Kołowrocki K (1993) On a class of limit reliability functions for series–parallel and parallel–series systems. Monograph Maritime University Press, Gdynia
10. Kołowrocki K (2004) Reliability of large systems. Elsevier, Amsterdam
11. Kołowrocki K (2006) Reliability and risk evaluation of complex systems in their operation processes. Int J Mater Struct Reliab 4(2):129–147
12. Kołowrocki K (2008) Reliability of large systems. In: Everrit R, Melnick E (eds) Encyclopedia of quantitative risk assessment, vol 4. Wiley, New York, pp 1466–1471
13. Kołowrocki K, Soszyńska J (2005) Reliability and availability of complex systems. Qual Reliab Eng Int 22(1):79–99 (Wiley)
14. Kurowicka D (2001) Techniques in representing high dimensional distributions. Ph.D. Thesis. Maritime University, Gdynia–Delft University
15. Kwiatuszewska-Sarnecka B (2003) Analiza niezawodnościowa efektywności rezerwowania w systemach szeregowych. Ph.D. Thesis. Maritime Academy, Gdynia–Systems Research Institute, Polish Academy of Sciences, Warsaw
16. Kwiatuszewska-Sarnecka B (2006) Reliability improvement of large multi-state parallel–series systems. Int J Autom Comput 2:157–164
17. Kwiatuszewska-Sarnecka B (2006) Reliability improvement of large parallel–series systems. Int J Mater Struct Reliab 4(2):149–159
18. Milczek B (2004) Niezawodność dużych systemów szeregowo progowych k z n. Ph.D. Thesis. Maritime Academy, Gdynia–Systems Research Institute, Polish Academy of Sciences, Warsaw
19. Smirnow NW (1949) Predielnyje Zakony Raspredielenija dla Czlienow Wariacjonnogo Riada. Trudy Matem Inst im WA Stieklowa
20. Soszyńska J (2006) Reliability of large series–parallel system in variable operation conditions. Int J Autom Comput 2(3):199–206
21. Soszyńska J (2006) Reliability evaluation of a port oil transportation system in variable operation conditions. Int J Press Vessels Pip 83(4):304–310
22. Soszyńska J (2007) Analiza niezawodnościowa systemów w zmiennych warunkach eksploatacyjnych. Ph.D. Thesis. Maritime Academy, Gdynia–Systems Research Institute, Polish Academy of Sciences, Warsaw
23. Von Mises R (1936) La distribution de la plus grande de n valeurs. Revue Mathematique de l'Union Interbalkanique 1:141–160
24. Xue J, Yang K (1995) Dynamic reliability analysis of coherent multi-state systems. IEEE Trans Reliab 44(4):683–688

The Price of Safety and Economic Reliability

Charles S. Tapiero

1 Introduction

Risk results from the direct and indirect adverse consequences of outcomes and events that were not accounted for or that we were ill prepared for, and concerns their effects on individuals, firms or the society at large. It results from many reasons, internally induced or occurring externally. In the former case, consequences are the result of failures, misjudgment or intentional (or non-intentional) acts perpetrated by some parties. In the latter case, consequences are the result of uncontrollable events or actions we were not able to apprehend [37, 38, 40]. Risk involves in general four factors:

1. Probabilities and their distributions associated to well- or ill-defined states (whether they are predictable, rare, black swans or "normal" states).
2. The consequences associated to predictable or unpredictable events, which may be known or defined by random magnitudes (cost, damages, illness, etc.).
3. Risk attitudes or a preference function, expressing risk sensitivities, usually associated to a propensity to manage risks and set both either ex-ante risk preventive procedures and ex-post (recovery).
4. Risk sharing and the pricing of risks (when risks can be exchanged), based on risk preferences and financial markets that allow an exchange of risks by both investors and speculators.

C. S. Tapiero (✉)
Department of Finance and Risk Engineering,
The Polytechnic Institute of New York University,
Brooklyn, New York
e-mail: ctapiero@poly.edu

H. Pham (ed.), *Safety and Risk Modeling and Its Applications*,
Springer Series in Reliability Engineering, DOI: 10.1007/978-0-85729-470-8_2,

These are relevant to a broad number of fields, each providing a different approach to the measurement, the valuation and the management of risk which *is motivated by psychological needs and the need to deal with problems that result from uncertainty and the adverse consequences they may induce* (Tapiero 2003).

Safety, by contrast, is both a consequence and a potential source of risk which is either objective, measured in terms of the probabilities and the consequences that define risk or perceptive—reflecting a state of mind, objective or conditioned. As a result, safety assumes many and compounded forms, such as being protected from consequential events or from being exposed to something that causes a loss. Practically, the word safety is used in many contexts. It may refer to *home safety* and allude to protective measures taken against external and harmful events (like weather, home invasion, etc.), computer safety in the sense of cyber security or to specific elements in use (stairs, cars, food, etc.). For example, see [2, 5, 15]. By the same token, safety may be inherent to a car's reliability and to designs constructed to prevent accidental losses (or "safe at any speed" and thereby lead to some drivers to drive recklessy).

Managing risk and safety consists then in defining, measuring, estimating, analyzing, valuing-pricing and integrating all facets of risk and their safety-consequential effects (real or not, external, internally induced or use dependent) into a whole system which can contribute to their design, economy, controls and management. The purpose of this paper is to provide an economic approach to reliability design and safety based on financial and economic considerations regarding a system design, its safety consequences which depend on the system reliability, the proficiency of the user and his risk behaviors who may be at fault in operating unsafely the system (whether he assumes or not its consequences). The paper provides some specific examples that can be treated analytically and that highlight approaches for financial reliability and safety design. These include essentially the utility-based approach which is subjective and based on a model of risk attitudes and the market price approach which is based on an equilibrium and rational economic model which is forward looking (implied by predictable future states or future losses). Extensions to information and power asymmetries between the system designer (say the selling firm) and the user (say the buyer), as well as conflicting objectives between these parties, are topics to be considered in a subsequent paper (see also [4]).

Reliability and systems safety designs are defined mostly from an engineering quantitative-statistical perspective based on specific assumptions regarding components reliabilities and safety and integrating them in a "compound systems architecture" that meets engineering and economic specifications for risk and safety [7, 11, 20, 23, 35, 36, 40]. In such an approach, safety is well defined in terms of default probabilities and their direct and indirect consequences on the system at hand and to persons directly or indirectly associated to these systems (risk externalities). For example, Truccoa et al. [46] present an approach that integrates Human and organizational Factors (HOF) into risk analysis using BBN (Bayesian Belief Networks [21]) and applied their approach to Maritime Reliability and Safety. Additional references including the International Maritime

Organization ([24]) and other institutions and authors have attracted attention to these issues [16–18]. By the same token, Netjasov and Janic [29] provide an up-to- date review of risk and safety modeling in civil aviation (see also [19, 22]). An extensive regulatory structure has been established to this effect to supplement the individual efforts of airlines, air navigation systems and construct incentive to limit the risks of flying. Netjasov and Janic [29] focus on models for aircraft and air traffic control management operations, collision risks, human factors and third-party risks. In their review, they emphasize, as we do in this paper, safety is a risk consequence and safety is commensurate to the risks sustained (although ignoring the feedback effects of safety by protected parties). The literature in both Air and Maritime safety is of course extensive and with numerous approaches and techniques applied, due to the commercial importance of safety (for example, [1, 2, 6, 8–11, 13, 25–26, 30, 32, 44, 45]). Expectedly different approaches are taken in the food industry [3, 4, 14] where regulatory standards for safety take three forms: process or design (reliability—my addition), performance or product and combined standards. Performance standards are controls that regulate maximum tolerance level of risk in food (such as zero tolerance of certain pathogens) while process standards specify the risk controls to be instituted by firms. Modeling the economic risk-reliability-safety paradigm has resisted any clear development.

Model imbedded in RAM (Reliability, Availability, Maintainability) as well as quality control, risk management, insurance, industrial psychology, chemical analyses, root cause analysis, statistical destructive and non-destructive controls testing, stress testing, training and education combined with industry-wide and government regulation, etc. provide ample ground to handle the many facets and issues that one is concerned with risk and safety—albeit, each model providing an approach that faces a specific aspect of risk-reliability-safety (see also Rouvroye and Van den Blieck [33]).

In our approach, safety is a consequence imbedded in both a system's reliability and architecture and in the behavior, the state of mind of users (whether assessed or imparted by information, misinformation or by purposeful deceit). In this context, risk as safety is a risk consequence which depends on the system, on the behavior of the person-user (who may turn out to be a mis-user) and on elements that may or may not be motivated to instill an unsafe state (and therefore potentially a risk source due to an excessive feeling of safety). These aspects of risk and safety have been generally neglected however, and their modeling remains a challenging task (for a related methodological approach See [41]). They result however in a "reliability in fact and use" which is random expressing the conditions, the behavior and the vicissitudes users are subjected to.

When risks are neutral (in the sense that they are not purposefully created), safety depends on two factors: the reliability of product or a process which is designed to a set of specifications including the product architecture, its fail-safe mechanisms, etc., and the user behavior and risk attitude whose safety depends on his own actions–the manners in which he operates the product-system.

For example, car safety depends on both the car and the driver driving a car. Safety in such cases would preempt unintentional actions by the driver who would face adverse consequences. For these reasons, engineers and industrial managers design processes to be robust. They do so to reduce both the failure probabilities and reduce processes' sensitivity to external users, behavior and uncontrollable events (which includes meeting technical specifications under a broader set of parametric assumptions). Human errors and purposeful human actions may nonetheless create harm to both themselves and others, even if systems can be robust and may have a built-in self-recovery capability [7, 31, 34].

Managing risk and safety consists in defining, measuring, estimating, analyzing, valuing-pricing and integrating all facets of risk and their safety-consequential effects (real or not, external, internally induced or use dependent) into a whole system which can contribute to their design, economy, controls and management. The purpose of this paper is to provide an economic approach to reliability design and safety based on economic considerations regarding a system design, its safety consequences which depend on both system reliability and the proficiency of the user who may be at fault in operating unsafely the system. The paper provides some specific examples that highlight the approach proposed for reliability and safety design and contrasts a number of approaches—risk and finance based, which we consider. Extensions to asymmetric information between the system design and the user, conflicting objectives and controls for safety (which we call counterparty risks), are natural extensions of this paper, and a rich avenue of further research.

2 Reliability-Safety and Economic Modeling

To contrast the risk-based approach and the traditional approach we consider first a traditional model for reliability design which is formulated as follows. We let $R(.)$ be a system reliability, defined in terms of m components types, some of which have a built-in redundancy $R(.) = R(n_1, n_2, n_3, \ldots, n_m)$ with a marginal production cost given by: $C(R(.), D)$, where D is the optimal quantity produced for a reliability level $R(.)$ given by:

$$\frac{\partial \Phi(R(.), D)}{\partial D} = C(R(.), D) \tag{1}$$

where $\Phi(R(.), D)$ is the total production cost for a quantity D and reliability design $R(.)$. Note that $R(.)$ is usually a function of time and a function of a product architecture (usually complex). However, for simplicity, we shall ignore this time dependency. If the firm (for a given reliability) is a price taker, and if aggregate profits are $\pi D - \Phi(R(.), D)$, then the optimal quantity produced is defined by equating the product price and its marginal cost, or, $\pi = C(R(.); D)$. In typical reliability models, two problems are assumed (which we amend to meet the economic marginal pricing rule considered here) that seek to determine the product configuration. A first problem consists in minimizing units marginal costs (or price) subject to a reliability specification constraint, or maximizing the reliability subject to a profitability constraint. Explicitly, we have the following:

$$\begin{aligned}&\underset{R(n_1,\ldots n_m)}{\text{Min}}\ C(R(.),D)\ \text{Subject to: } R(.)\geq R_c\\&\underset{n_1,\ldots n_m}{\text{Max}}\ R(.)\ \text{Subject to:}\quad \pi - C(R(.);D)\geq 0\ \text{and}\ R(.)\geq R_c\end{aligned} \tag{2}$$

The former constraint, R_c, is a "contracted or regulated reliability" while the latter is a profitability constraint. The solutions of these problems are technical, providing some managerial insights by an interpretation of the problems' Lagrange multipliers. These problems express the firm's viewpoint, ignoring the safety-consequential effects of the product reliability and users' behaviour and responsibility when using the product. If we consider for example, the profitability constraint and assume an optimal quantity produced, then at the (quantity) margin, we have $-\pi + C(R,D^*) = 0$ and as a result (by implicit differentiation):

$$\frac{d\pi}{dR} = \frac{\partial C(R,D^*)}{\partial R} > 0,\ \frac{d\pi}{dD^*} = \frac{\partial C(R,D^*)}{\partial D^*} < 0, \tag{3}$$

In this case, a firm reliability design, motivated by an aggregate profit, will be to solve:

$$\underset{R(.)}{\text{Max}} \prod = \pi D^* - \Phi(R,D^*)\ \text{Subject to: } \pi - C(R,D^*)\geq 0 \tag{4}$$

And as a result (assuming the Lagrange optimization problem with price equal the production marginal cost), we have (with $\Lambda = \pi D^* - \Phi(R,D^*) + \lambda(\pi - C(R,D^*))$):

$$-\Phi_R(R,D^*) + \pi_R D^* + \lambda(\pi_R - C_R(R,D^*)) = 0\ \text{and}\ \pi = C(R,D^*) \tag{5}$$

where π_R and C_R are partial differentials. Since, at the marginal price $\pi_R = C_R$, the reliability design is set to the average marginal cost, and the quantity-reliability problem is solved by solving:

$$\pi_R = \frac{\Phi_R(R,D^*)}{D^*};\ \pi = \Phi_D(R,D^*) \tag{6}$$

and therefore,

$$\frac{\pi_R}{\pi} = \frac{\Phi_R(R,D^*)}{D^*\Phi_D(R,D^*)}\ \text{or}\ \frac{\pi}{\pi_c} = \exp\left\{\int_{R_c}^{R} \frac{\Phi_R(r,D^*)}{D^*\Phi_D(r,D^*)}dr\right\} \tag{7}$$

where π_c is the marginal price, equated to the marginal cost at the regulated reliability R_c. For example, say that aggregate production costs are given by a Cobb–Douglas production function,

$$\Phi(R, D^*) = aR^{\alpha}D^{\beta} \text{ then } \pi_R = \frac{\alpha\Phi(.)}{R} \text{ and } \pi = \frac{\beta\Phi(.)}{D} \tag{8}$$

As a result,

$$\frac{\pi}{\pi_c} = \left(\frac{R}{R_c}\right)^{\frac{\alpha}{\beta}} \tag{9}$$

Note that if production costs are an increasing function of the reliability, then prices increase with reliability growth. Further,

$$\begin{aligned} d\left(\frac{\pi}{\pi_c}\right)/d\left(\frac{R}{R_c}\right) &= \frac{\alpha}{\beta}\left(\frac{R}{R_c}\right)^{\frac{\alpha}{\beta}-1} > 0 \\ d^2\left(\frac{\pi}{\pi_c}\right)/d\left(\frac{R}{R_c}\right)^2 &= -\left(1-\frac{\alpha}{\beta}\right)\frac{\alpha}{\beta}\left(\frac{R}{R_c}\right)^{\frac{\alpha}{\beta}-2}, \quad \text{if } \frac{\alpha}{\beta} > 1 \end{aligned} \tag{10}$$

which is the case for reasonable parameters. Note that while the first parameter reflects the costs effects of reliability improvement, the second parameter reflects the economies of scale effects. For example, if we set the parameters (1.1, 0.98), we have then:

$$\frac{1.1}{0.98} > 1 \text{ and } \frac{d^2(\pi/\pi_c)}{d(R/R_c)^2} = \left(\frac{1.1}{0.98} - 1\right)\frac{1.1}{0.98}\left(\frac{R}{R_c}\right)^{-\frac{0.86}{0.98}} > 0$$

In other words, the market price of a product in a competitive market determines as well its reliability. A reliability design is thus defined, implicitly, by its market price. Explicitly, in our example, for an n-components redundant system, we have a reliability $R(n_1, n_2, \ldots, n_m) = \prod_{i=1}^{m} [1 - (1 - r_i)^{n_i}]$. If a regulated reliability is as stated above R_c at a regulated price π_c then a feasible reliability design is defined by the market price of reliability which is explicitly given by Eq. 10, or:

$$0 \le R_c\left(\frac{\pi}{\pi_c}\right)^{\frac{\beta}{\alpha}} = \prod_{i=1}^{m} [1 - (1 - r_i)^{n_i}] < 1, \ \frac{\pi}{\pi_c} > 1$$

with a price given by the marginal pricing condition $\pi = \Phi_D(R, D^*) = a\beta R^{\alpha}D^{\beta-1}$. In this case, the total demand determines the price. In competitive markets, the demand faced by a firm is independent of the price and therefore, both the reliability and the quantity to be met by the firm are a function of the industry marginal cost. If the firm's marginal cost is smaller than that of the market (either due to its economies of scale or to its technology in place compared to other firms), the optimal quantity to be produced will be larger and defined by a solution of:

$$0 \le R_c\left(\frac{\Phi_D\left(\prod_{i=1}^{m}[1-(1-r_i)^{n_i}],D^*\right)}{\pi_c}\right)^{\frac{\beta}{\alpha}} = \prod_{i=1}^{m}[1-(1-r_i)^{n_i}] < 1.$$

3 Safety and Reliability and Consequential Costs

The approach outlined above neglects however two essential aspects: the consequential effects of reliability risks and their safety considerations as well as the "behavior" and the user reliability. Namely, say that an individual user has a probability p of using safely the system acquired at price π and an uncertain consequential cost if it fails. The effective system reliability will then be a series system with reliability $Q = pR(.)$. This implies that the producer and the user have both a responsibility and a stake in reliability and in safety design. The producer's responsibility is to provide a reliable and robust product and information and education for the system and its proper use while the user's responsibility is to learn how to use the product and use it safely. Further, the transaction leading to an exchange between a buyer and a seller is necessarily an exchange of mutual responsibility, with both parties potentially liable wholly or partly for the consequential costs. For example, a car well designed with a number of safe gadgets can be very unsafe if these safe gadgets are misunderstood and poorly used. Further, while the producer may only design the product-system reliability, it is the user whose behavior defines the product safety by his own actions (and therefore reliability). The reliability-safety design problem thus requires that it considers as well the risk-safety consequences of both the product design and the consumer-user behavior when using this product. To account for such a situation, we shall define the probability of an adverse event (due as well to the use of the product with probability $1 - \tilde{p}R(.)$, where $\tilde{p}$ is the user reliability—at present a random variable assumed by the system-firm and the consequence of such an event defined by a random cost $\tilde{Z}$. The firm effective problem consists then in the selection of a reliability that meets now both the reliability and profitability constraints. The firm profit is a random variable and the design is a risk-based financial problem determined by the price of exchange between the firm and the user (and its market price) and the respective behavior by the user and the firm. In this paper, we shall consider only the problem from a firm's viewpoint, given by a random profit defined by Eq. 11 which it will seek to manage financially. A number of approaches are then possible consisting of a "simple" expected utility optimization, a VaR precautionary risk constraint optimization, a Mean–Variance Risk constraint approach and finally, a financial market pricing approach. Each of these approaches is outlined in brief terms in this paper. Let the firm profit be a random variable given as stated above by:

$$\tilde{\prod} = \pi D - \Phi(R(.), D) - \sum_{j=1}^{\tilde{M}} (1 - \theta)\tilde{Z}_j \tag{11}$$

where the user's reliability is not known, defined by a random variable $\tilde{p}$ hypothesized by the firm, with known mean and variance, say $\hat{p} = E(\tilde{p}), \sigma_p^2 = \text{var}(\tilde{p})$ and with a number of independent consequential and costly events $\tilde{M} \sim \binom{D}{j}(1 - \tilde{p}R(.))^j (\tilde{p}R(.))^{D-j}$ with random magnitudes $\tilde{Z}_j$. Note that $\tilde{M}$ is a binomial mixture (Lexian) distribution, and therefore,

$$E(\tilde{M}) = D(1 - \hat{p}R(.)),\ \text{var}(\tilde{M}) = D(1 - \hat{p}R(.))\hat{p}R(.) + D(D-1)R^2(.)\sigma_p^2, \tag{12}$$

The expected profit and its variance providing two objectives and financial indicators are thus:

$$E\left(\tilde{\prod}\right) = \pi D - \Phi(R(.), D) - (1 - \theta)E\left(\tilde{M}\right)E\left(\tilde{Z}_j\right) \text{ and} \tag{13}$$

$$\text{var}\left(\tilde{\prod}\right) = (1-\theta)^2 \text{var}\left(\sum_{j=1}^{\tilde{M}} \tilde{Z}_j\right) = (1-\theta)^2\left\{E(\tilde{M})\text{var}\left(Z_j\right) + \text{var}(\tilde{M})\left[E\left(Z_j\right)\right]^2\right\} \tag{14}$$

The analysis of these random streams underlies a risk-based approach to economic reliability and safety. Note that the effects of reliability on both the expected profit and its variance can be either positive or negative, since:

$$\frac{\partial E\left(\tilde{\Pi}\right)}{\partial R(.)} = -\frac{\partial \Phi(R(.), D)}{\partial R(.)} + (1-\theta)D\hat{p}E\left(\tilde{Z}_j\right) \tag{15}$$

$$\frac{\partial \text{var}\left(\tilde{\Pi}\right)}{\partial R(.)} = (1-\theta)^2 D\left\{-\hat{p}\text{var}\left(Z_j\right) + \left(\left(\hat{p} - 2\hat{p}^2 R(.)\right) + 2(D-1)R(.)\sigma_p^2\right)\left[E\left(Z_j\right)\right]^2\right\} \tag{16}$$

while the variance effects on the safety parameter's variance are always positive and quadratic in the size of the demand (in other words, "more is not always better"):

$$\begin{aligned} \frac{\partial \text{var}\left(\tilde{\Pi}\right)}{\partial \sigma_p^2} &= (1-\theta)^2 D(D-1)R^2(.)\left[E\left(Z_j\right)\right]^2 > 0 \\ \frac{\partial \text{var}\left(\tilde{\Pi}\right)}{\partial \hat{p}} &= \frac{(1-\theta)^2 DR(.)}{\text{var}\left(Z_j\right)}\left\{-1 + \frac{(1 - 2\hat{p}R(.))\left[E\left(Z_j\right)\right]^2}{\text{var}\left(Z_j\right)}\right\} \end{aligned} \tag{17}$$

We shall use these results to study their effects on the price of reliability and safety using the expected utility approach. However, we see from Eqs. 13–16 the effects of reliability on the profit function as well as the effects of the user's heterogeneity (expressed by the variance σ_p^2). For example, a more reliable product can increase or decrease expected profits [see (15)]. Profits increase in expectation if the external costs are larger than the marginal cost of production with respect to reliability. Explicit results can be obtained by using a quadratic utility function of profits which seeks greater expected profits and less profits variance as will be seen below.

This case is considered below:

3.1 The Expected Utility Approach and the Certain Equivalent

Consider a firm's utility for money given by $u(.)$ with a certain equivalent for a return a period discounted at the risk free rate by:

$$Eu\left(\tilde{\prod}\right) = u(CE) \tag{18}$$

where the profits a period hence are random and their certain equivalent is given by definition in Eq. 18. As a result, the present value of such a certain equivalent (in the next period) can be priced today at its discounted value using a risk free rate, or:

$$CE_0 = \frac{CE}{1 + R_f} \tag{19}$$

where R_f is a risk free rate (applied to discount a sum that has no randomness and therefore no risk—the certain equivalent). For simplicity, assume a quadratic utility function given by:

$$u\left(\tilde{\prod}\right) = \tilde{\prod} - \rho\left(\tilde{\prod} - E\left(\tilde{\prod}\right)\right)^2, \quad \rho > 0 \tag{20}$$

where $\rho > 0$ defines an index of risk aversion for profit variability. As a result, using (18), we have

$$CE - \rho\left(CE - E\tilde{\prod}\right)^2 = E\tilde{\prod} - \rho\,\mathrm{var}\left(\tilde{\prod}\right) \tag{21}$$

Since, $\chi = E\prod - CE$ is a risk premium, we have:

$$0 = \rho\chi^2 + \chi - \rho\,\mathrm{var}\left(\tilde{\prod}\right) \tag{22}$$

The risk premium is therefore:

$$0 < \chi = -\frac{1}{2\rho} + \sqrt{\left[\frac{1}{2\rho}\right]^2 + \mathrm{var}\left(\tilde{\Pi}\right)} \tag{23}$$

Thus, if a design is planned initially and realized at a future time T, maximizing the discounted certain equivalent is given by:

$$\mathrm{Max}\, CE_0 = \frac{E\Pi - \chi}{1 + R_f} = \frac{E\tilde{\Pi} - \left(\sqrt{\left[\frac{1}{2\rho}\right]^2 + \mathrm{var}\left(\tilde{\Pi}\right)} - \frac{1}{2\rho}\right)}{1 + R_f} \tag{24}$$

which we optimize subject to any present-day constraints (such as budget design investment and cost expenditures borne initially). In this case, if an initial budget constraint B is imposed (say, a fixed cost) we require necessarily that $CE_0 > B$ in selecting both the quantity planned and the reliability of the underlying product. Such a constraint will be neglected here however. In addition, note that if the index of risk aversion is very large, then $1/2\rho$ is very small and therefore, the current certain equivalent is:

$$CE_0 = \frac{E(\tilde{\Pi}) - \sigma\left(\tilde{\Pi}\right)}{1 + R_f} \text{ where } \rho \to +\infty \text{ and } \sigma\left(\tilde{\Pi}\right) = \sqrt{\mathrm{var}\left(\tilde{\Pi}\right)} \tag{25}$$

which "prices returns and their volatility" equally! However, if the index of risk aversion is extremely small, then $1/2\rho$ is extremely large and therefore:

$$CE_0 = \frac{E(\tilde{\Pi})}{1 + R_f} \text{ where } \rho \to 0 \tag{26}$$

and the price of volatility is (as expected) null. In this sense, the true price CE_0 of the future profits is bounded by (25) and (26) or:

$$\frac{E\left(\tilde{\Pi}\right) - \sigma\left(\tilde{\Pi}\right)}{1 + R_f} \leq CE_0 \leq \frac{E\left(\tilde{\Pi}\right)}{1 + R_f} \tag{27}$$

which we write conveniently by:

$$CE_0 = \frac{E\left(\tilde{\Pi}\right) - \lambda\sigma\left(\tilde{\Pi}\right)}{1 + R_f},\ \lambda \in [0, 1] \tag{28}$$

The design of reliability can thus be conceived as the maximization of the present certain worth of the future profits defined by (28). In other words, if the firm selects optimally its reliability level, then:

$$0 = \frac{\partial CE_0}{\partial R} = \frac{1}{1+R_f}\left(\frac{\partial E\left(\tilde{\Pi}\right)}{\partial R} - \lambda \frac{\partial \sigma\left(\tilde{\Pi}\right)}{\partial R}\right), \lambda \in [0,1] \tag{29}$$

and therefore,

$$\lambda = \frac{\partial E(\tilde{\Pi})}{\partial R} \Big/ \frac{\partial \sigma(\tilde{\Pi})}{\partial R} \tag{30}$$

Thus, the price of future profits implied by a firm's reliability choice is thus given explicitly by:

$$CE_0 = \frac{E(\tilde{\Pi})}{1+R_f} - \frac{1}{1+R_f}\left(\frac{\partial E(\tilde{\Pi})}{\partial R} \Big/ \frac{\partial \sigma\left(\tilde{\Pi}\right)}{\partial R}\right)\sigma\left(\tilde{\Pi}\right) \tag{31}$$

Similarly, an optimal quantity is defined by optimization of (28) and therefore,

$$(1+R_f)\frac{\partial CE_0}{\partial D} = 0 \to \frac{\partial E\left(\tilde{\Pi}\right)}{\partial D} = \lambda \frac{\partial \sigma\left(\tilde{\Pi}\right)}{\partial D} \tag{32}$$

or,

$$\frac{\dfrac{\partial E\left(\tilde{\Pi}\right)}{\partial D}}{\dfrac{\partial \sigma\left(\tilde{\Pi}\right)}{\partial D}} = \frac{\dfrac{\partial E(\tilde{\Pi})}{\partial R}}{\dfrac{\partial \sigma\left(\tilde{\Pi}\right)}{\partial R}} \tag{33}$$

In this sense, the index of risk aversion of the firm designing a reliable product is implied in its choices. In the example treated above, we have instead of Eq. 33:

$$\frac{\dfrac{\partial E\left(\tilde{\Pi}\right)}{\partial D}}{\dfrac{\partial \sigma\left(\tilde{\Pi}\right)}{\partial D}} = \frac{1}{2}\sigma\left(\tilde{\Pi}\right) \times \left[\frac{-\dfrac{\partial \Phi(R(.),D)}{\partial R(.)} + (1-\theta)D\hat{p}E(\tilde{Z}_j)}{(1-\theta)^2 D\hat{p}\left\{-\text{var}(Z_j) + \left(1 - 2\hat{p}R(.) + \dfrac{2(D-1)R(.)\sigma_p^2}{\hat{p}}\right)[E(Z_j)]^2\right\}}\right] \tag{34}$$

where:

$$\frac{\partial E\left(\tilde{\Pi}\right)}{\partial D} = \pi - \frac{\partial \Phi(R(.),D)}{\partial D} - (1-\theta)(1-\hat{p}R(.))E(\tilde{Z}_j) \text{ and} \tag{35}$$

$$\frac{\partial \mathrm{var}\left(\tilde{\Pi}\right)}{\partial D} = (1-\theta)^2(1-\hat{p}R(.))\left\{\mathrm{var}\left(Z_j\right) + \hat{p}R(.) + (2D-1)\sigma_p^2 R^2\left[E\left(Z_j\right)\right]^2\right\} \quad (36)$$

and therefore (34) is reduced to:

$$\frac{\pi - \dfrac{\partial\Phi(R(.),D)}{\partial D} - (1-\theta)(1-\hat{p}R(.))E\left(\tilde{Z}_j\right)}{(1-\theta)^2(1-\hat{p}R(.))\left\{\mathrm{var}\left(Z_j\right) + \hat{p}R(.) + (2D-1)\sigma_p^2 R^2\left[E\left(Z_j\right)\right]^2\right\}} = \frac{-\dfrac{\partial\Phi(R(.),D)}{\partial R(.)} + (1-\theta)D\hat{p}E\left(\tilde{Z}_j\right)}{(1-\theta)^2 D\hat{p}\left\{-\mathrm{var}\left(Z_j\right) + \left(1 - 2\hat{p}R(.) + \dfrac{2(D-1)R(.)\sigma_p^2}{\hat{p}}\right)\left[E\left(Z_j\right)\right]^2\right\}} \quad (37)$$

which is one equation in the quantity and in the reliability, function of the consequential costs, the sharing agreement and the behavior of the user imbedded in the relevant parameters. Equation (37) together with Eq. 31 thus defines the current price of a reliability policy as well as the price of safety. Namely, set

$$\sigma_p^2 = 0,\ \hat{p} = 1 \quad (38)$$

Then,

$$\frac{\pi - \dfrac{\partial\Phi(R(.),D)}{\partial D} - (1-\theta)(1-R(.))E\left(\tilde{Z}_j\right)}{(1-\theta)^2(1-R(.))\left\{\mathrm{var}\left(Z_j\right) + R(.)\right\}} = \left[\frac{-\dfrac{\partial\Phi(R(.),D)}{\partial R(.)} + (1-\theta)DE\left(\tilde{Z}_j\right)}{(1-\theta)^2 D\left\{-\mathrm{var}\left(Z_j\right) + (1-2R(.))\left[E\left(Z_j\right)\right]^2\right\}}\right] \quad (39)$$

$$\frac{\dfrac{\partial E\left(\tilde{\Pi}\right)}{\partial D}}{\dfrac{\partial \sigma\left(\tilde{\Pi}\right)}{\partial D}} = 2\sigma\left(\tilde{\Pi}\right)\left[\frac{-\dfrac{\partial\Phi(R(.),D)}{\partial R(.)} + (1-\theta)DE\left(\tilde{Z}_j\right)}{(1-\theta)^2 D\left\{-\mathrm{var}\left(Z_j\right) + (1-2R(.))\left[E\left(Z_j\right)\right]^2\right\}}\right] \quad (40)$$

while (31) is reduced to:

$$CE_0(\sigma_p^2 = 0, \hat{p} = 1) = \frac{E\left(\tilde{\Pi}\right)}{1+R_f} - \frac{2\sigma^2\left(\tilde{\Pi}\right)}{1+R_f}\left[\frac{-\dfrac{\partial\Phi(R(.),D)}{\partial R(.)} + (1-\theta)DE\left(\tilde{Z}_j\right)}{(1-\theta)^2 D\left\{-\mathrm{var}\left(Z_j\right) + (1-2R(.))\left[E\left(Z_j\right)\right]^2\right\}}\right] \quad (41)$$

As a result, the price of safety is the increment we pay due to

$$\text{Price of safety} = CE_0(\sigma_p^2 = 0, \hat{p} = 1) - CE_0(\sigma_p^2, \hat{p}) \tag{42}$$

while the price sensitivity to the parameters defining the user behavior can be determined by the partial derivatives of this price with respect to these parameters.

3.2 The Quantile and Safety Risk Constraint

Alternatively, a firm may define its optimization problem in terms of a quantile risk reliability constraint which recognizes the user's uncertain reliability, or $\Pr\{\tilde{p}R(.) \geq R_c\}$. In this case, the firm assumes that it expects to meet an "effective and in the field" reliability constraint which includes the user's safe use of the underlying equipment. This problem is then reduced to a risk-constrained design which can be stated as a certain equivalent constraint, such as:

$$\begin{aligned} &\text{Max } CE_0 = E(\tilde{\Pi}) - \lambda\sigma(\tilde{\Pi}) \\ &\text{Subject to:} \quad \Pr\{\tilde{p}R(.) \geq R_c\} \geq 1 - \xi \end{aligned} \tag{43}$$

When the index of risk aversion is not known (which is mostly the case), this optimization problem can be reduced to two complementary problems as stated earlier with a quantile risk $\Pr\left\{\frac{R_c}{R(.)} \leq \tilde{p}\right\} \geq 1 - \xi$. Thus, if the probability distribution of the user's reliability is $f(\tilde{p})$ and letting the inequality constraint be for simplicity and equality, we have:

$$F\left(\frac{R_c}{R(.)}\right) = 1 - \xi \text{ and } R(.) = \frac{R_c}{F^{-1}(1-\xi)} \tag{44}$$

As a result,

$$CE_0 = \frac{E\left(\tilde{\Pi}\left(R = \frac{R_c}{F^{-1}(1-\xi)}\right)\right) - \lambda\sqrt{\operatorname{var}\left(\tilde{\Pi}\left(R = \frac{R_c}{F^{-1}(1-\xi)}\right)\right)}}{1 + R_f} \tag{45}$$

which we optimize to determine the optimal quantity as indicated above and obtain the parameter λ which is inserted in Eq. 35.

3.3 The Quantile Value at Risk Approach

The quantile risk approach has assumed a renewed interest in the framework of a VaR (Value at Risk) approach which sets aside a certain quantity of money to meet a risk exposure with a specified probability (see also [42, 39]). This approach replaces the utility approach (defined in terms of a risk aversion index) by a quantile risk constraint defined as follows:

$$P\left(\tilde{\prod} \leq -\,\mathrm{VaR}\right) \leq \xi \tag{46}$$

where ξ is the risk exposure that a firm sustains when its prospective future profits fall below an amount set, the Value at Risk, which it is willing risk in its design development. Assuming again an equality in the VaR constraint and letting the firm profit be defined by an elliptic distribution function $g(.)$, then the quantity set aside (the VaR) can be written as a linear function of the expected profits and their variance or:

$$\mathrm{VaR} \leq E\left(\tilde{\prod}\right) - q_{\xi}\sigma\left(\tilde{\prod}\right) \tag{47}$$

where $q_{1-\xi}$ expresses "a price" for the cost variability and is calculated by the following equations:

$$G(q_{1-\xi}) = \xi \quad \text{where } G(s) = \frac{\pi^{\frac{n-1}{2}}}{\Gamma\left(\frac{n-1}{2}\right)} \int_{s}^{-\infty} \int_{w^2}^{+\infty} \left(u - w^2\right)^{\frac{n-3}{2}} g(u)\,du\,dw \tag{48}$$

When the distribution is normal we have as a special case (replacing the inequality by an equality):

$$\mathrm{VaR} = E\left(\tilde{\prod}\right) - Q_{\xi}\sigma\left(\tilde{\prod}\right) \tag{49}$$

where $Q_{1-\xi}$ is the quantile (tail) probability of a standard normal distribution. While the VaR is the amount of money set aside to meet contingent losses. In this particular case, the amount of money at risk is proportional to the profits standard deviation and its price is:

$$CE_0 = CE_0^{(1)} + \mathrm{VaR} = \frac{1}{1+R_f}\left(CE_1^{(1)} + \mathrm{VaR}(1+R_f)\right)$$

where $CE_0^{(1)}$ is the certain equivalent of the amount of money at risk and therefore equal to:

$$CE_0^{(1)} = \frac{1}{1+R_f}\left(E\left(\tilde{\prod}_1^{(1)}\right) - \lambda_1\sigma\left(\tilde{\prod}_1^{(1)}\right)\right)$$

with a VaR given by:

$$\mathrm{VaR} = E\left(\tilde{\prod}_1^{(1)}\right) - q_{\xi}\sigma\left(\tilde{\prod}_1^{(1)}\right)$$

As a result, the amount at risk is:

$$CE_0 - \mathrm{VaR} = CE_0^{(1)} = \frac{1}{1+R_f}\left(CE_1^{(1)}\right)$$

or

$$\frac{1}{1+R_f}(E(\tilde{\Pi}) - \lambda\sigma(\tilde{\Pi})) - E\left(\tilde{\prod}_1^{(1)}\right) + q_\xi\sigma\left(\tilde{\prod}_1^{(1)}\right)$$
$$= \frac{1}{1+R_f}\left(E\left(\tilde{\prod}_1^{(1)}\right) - \lambda_1\sigma\left(\tilde{\prod}_1^{(1)}\right)\right)$$

and therefore:

$$\frac{1}{1+R_f}\left(E(\tilde{\Pi}) - E\left(\tilde{\prod}_1^{(1)}\right) - \lambda\sigma(\tilde{\Pi}) + \lambda_1\sigma\left(\tilde{\prod}_1^{(1)}\right)\right) = E\left(\tilde{\prod}_1^{(1)}\right) - q_\xi\sigma\left(\tilde{\prod}_1^{(1)}\right)$$

providing a relationship between the VaR profits (and their implied reliability) and the approach which does not set moneys aside (the VaR) to meet contingent liabilities and claims arising from unreliability.

This approach may be expanded in numerous directions to include as well the cost of holding money to meet risk contingencies (the VaR) as well as in estimating better the probability distributions of the consequential costs associated to unsafe events.

3.4 The Mean–Variance (Markowitz) Approach

Assume that the firm's objective is structured as a mean–variance objective. Then, our problems are to maximize the returns subject to a risk (variance) constraint, and minimize the risk (variance) subject to an expected profit constraints. In our reliability-safety problems, this results in a demand-reliability schedule which is a function of these two constraints, or to a schedule of pairs which generalizes Eq. 23 and given by: $\{D^*, R^* | v, \mu\}$ where (v, μ) are the variance and expected profit constraints associated to the two objectives problems. For an efficient schedule, meeting the reliability risk constraints, we have also a schedule: $\{D^*, R^* | v, \mu, R^* \geq R_c\}$

A quantitative formulation of these problems is thus:

$$\underset{D,R(.)}{\text{Max}}\, E\left(\tilde{\prod}\right) \text{ Subject to: } \quad \text{var}\left(\tilde{\prod}\right) \leq v \tag{50}$$

$$\underset{D,R(.)}{\text{Min}}\, \text{var}\left(\tilde{\prod}\right) \text{Subject to: } \quad E\left(\tilde{\prod}\right) \geq \mu \tag{51}$$

And their solutions provide a set of demand and reliability schedules constrained to their efficient solutions:

$$\{D(\mu), R(\mu) | R(\mu) \geq R_c\} \text{ and } \{D(v), R(v) | R(v) \geq R_c\} \tag{52}$$

The set of results (52) indicates for example, all demand and reliability levels (with a reliability larger than the regulated or design reliability sought) consistent with the expected profit constraint. A solution of this problem for a family of such constraint provides a variance efficiency curve. The calculation of these efficiency curves is given by the following: The conditions for optimality for the first and the second problem, defining the efficient reliability set are:

$$\text{Problem 1: } \frac{\dfrac{\partial E(\tilde{\Pi})}{\partial D}}{\dfrac{\partial \text{var}(\tilde{\Pi})}{\partial D}} = \frac{\dfrac{\partial E(\tilde{\Pi})}{\partial R}}{\dfrac{\partial \text{var}(\tilde{\Pi})}{\partial R}} \text{ and } \text{var}\left(\tilde{\prod}\right) = v \tag{53}$$

and

$$\text{Problem 2: } \frac{\dfrac{\partial E(\tilde{\Pi})}{\partial D}}{\dfrac{\partial \text{var}(\tilde{\Pi})}{\partial D}} = \frac{\dfrac{\partial E(\tilde{\Pi})}{\partial R}}{\dfrac{\partial \text{var}(\tilde{\Pi})}{\partial R}} \text{ and } \mu - E\left(\tilde{\prod}\right) = 0 \tag{54}$$

with equations defining the means and the variances as well as their differentials as seen above. This is a straightforward optimization of the Lagrange optimization problem above. Note that in these problems, we have:

$$2\sigma\left(\tilde{\prod}\right)\frac{\partial \sigma(\tilde{\Pi})}{\partial D} = \frac{\partial \text{vaR}(\tilde{\Pi})}{\partial D},\ 2\sigma\left(\tilde{\prod}\right)\frac{\partial \sigma(\tilde{\Pi})}{\partial R} = \frac{\partial \text{vaR}(\tilde{\Pi})}{\partial R} \tag{55}$$

and therefore,

$$\frac{\dfrac{\partial E(\tilde{\Pi})}{\partial D}}{\dfrac{\partial \sigma(\tilde{\Pi})}{\partial D}} = \frac{\dfrac{\partial E(\tilde{\Pi})}{\partial R}}{\dfrac{\partial \sigma(\tilde{\Pi})}{\partial R}} = \lambda \text{ and } \text{var}\left(\tilde{\prod}\right) = v;\text{ and } \mu - E\left(\tilde{\prod}\right) = 0 \tag{56}$$

which reduces to the problems considered earlier when using the certain equivalent with λ denoting the price (in an expected profit sense) of a unit standard deviation (a price smaller than one however for a risk averse firm).

For example, assume that $\Phi(R(.), D) = aR^{\alpha}D^{\beta}$ and $E(Z_j) = \text{var}(Z_j)$ then in Eq. 37, we will have:

$$\frac{\dfrac{\pi - a\beta R^{\alpha}D^{\beta-1}}{(1-\hat{p}R(.))E(\tilde{Z}_j)} - (1-\theta)}{(1-\theta)^2\left\{1 + \hat{p}R(.) + (2D-1)\sigma_p^2 R^2 E(Z_j)\right\}}$$
$$= \frac{1 - \theta - \dfrac{a\alpha R^{\alpha-1}D^{\beta}}{D\hat{p}E(\tilde{Z}_j)} +}{(1-\theta)^2\left\{-1 + \left(1 - 2\hat{p}R(.) + \dfrac{2(D-1)R(.)\sigma_p^2}{\hat{p}}\right)E(Z_j)\right\}} \tag{57}$$

which we can solve numerically for the reliability as a function of the quantity produced, and determine λ.

3.5 Safety, Reliability and Market Pricing

Financial assets are traded in financial markets and priced by these markets. All assets converted into financial instruments that are broadly traded can thus be priced. If assets cannot be transformed into such instruments and have one or all of their characteristics replicated by such instruments, then we cannot clearly and uniquely set their price. In this context, the Certain Equivalent profit considered in this paper is equivalent to a risk free bond and was thus, priced by a risk free rate. Thus, the "Certain Equivalent" profits defined, have allowed our implying and equating their price to that of a bond which is widely traded. By the same token, if it were possible to define optional instruments based on the cash flows of the firm profits then, "an implied market price" can be defined based on the "equivalent" option price (if it is traded).

For example, let the firm sell an obligation to investors who will pay a certain sum, say S in exchange for a share of profits, say above a strike and given by $\mathrm{Max}\left(\upsilon\left(\tilde{\prod}-K\right),0\right)$. If we consider for simplicity a one period problem and assume that markets are complete (for example, see [43]), and assume that future returns have a binomial value, say $\prod^{+}>K$ and $\prod^{-}<K$ then by risk neutral pricing (for example, see [43]), the current price is the risk free expectation of the future gain, namely:

$$S=\frac{1}{1+R_f}E^{*}\,\mathrm{Max}\left(\upsilon\left(\tilde{\prod}-K\right),0\right)=\frac{1}{1+R_f}p^{*}\upsilon\left(\prod^{+}-K\right) \quad (58)$$

and therefore, the implied probability of the return $\prod^{+}$ is p^{*}:

$$p^{*}=\frac{S(1+R_f)}{\upsilon(\prod^{+}-K)} \quad (59)$$

Thus, by the same principle of risk neutral pricing, and using the implied probability, we have:

$$\prod\nolimits_0=\frac{1}{1+R_f}\left(p^{*}\prod^{+}+(1-p^{*})\prod^{-}\right) \quad (60)$$

and therefore, the current (implied) profit of the firm priced by the market is:

$$\prod\nolimits_0=\frac{1}{1+R_f}\left(\frac{S(1+R_f)}{\upsilon(\prod^{+}-K)}\prod^{+}+\left(1-\frac{S(1+R_f)}{\upsilon(\prod^{+}-K)}\right)\prod^{-}\right) \quad (61)$$

Note that if there are two similar contracts (rather than one (*S*, *K*)), with say, *M*, *L*, we have then an additional equation:

$$\prod_0 = \frac{1}{1+R_f}\left(\frac{M(1+R_f)}{\upsilon(\prod^+ - L)}\prod^+ + \left(1 - \frac{M(1+R_f)}{\upsilon(\prod^+ - L)}\right)\prod^-\right) \tag{62}$$

which can be used to calculate the implied returns $(\prod^+, \prod^-)$. Further, if $p^* = p_{\text{user}}R$, then the implied probability of the user is:

$$p_{\text{user}} = \frac{S(1+R_f)}{\upsilon R(\prod^+ - K)} \tag{63}$$

In other words, financial observation of contracted prices provides an important source of information regarding the implied physical characteristics that underlie behavior of user and their reliability. These observations are not however "statistical" resulting from prior observations or past data but resulting from a current observation of prices of assets and their derivatives that in a complete market, are defined by their future consequences and an exchange between buyers and sellers on these returns that result in a unique equilibrium price—the current price. Since all derivatives products have outcomes which are a function of the future (and predictable) outcomes, derivative prices that are traded based on these returns can be used to infer these future returns. Since these returns are based on an exchange of parties that need to have the same risk attitudes, the resulting future states are virtual estimates defined by what market prices indicate what the beliefs of these states by the trading parties say they are (and not what they may be statistically or subjectively defined to be).

If one accepts current prices as reflection of future states imbedded in a market equilibrium model in which prices defined uniquely for set levels of risk (and other conditions that define such markets, we call complete), then a design of reliability and safety-consequential events will then be based on the maximization of all values actualized in expectation based on some implied probabilities defined by an appropriate and corresponding economic model (including of course both regulation and other external constraints used to construct the model of future "states"). Of course, if we assume that investors and speculators have an implied utility with an implied risk aversion, with an insurance price to remove all risks defined, then equivalence between the current certain equivalent and the current implied market price can be presumed. For example, in the case treated earlier with a quadratic utility, then if the full insurance price a party may be willing to pay to reduce the returns risk is $\lambda\sigma(\prod)$ and therefore the discounted certain equivalent (if equated to the current market price ought to be):

$$\prod_0 = CE_0 = \frac{E\Pi - \lambda\sigma(\prod)}{(1+R_f)} \tag{64}$$

and therefore, the implied "market" risk aversion in parameter λ can be expressed by a solution of the equation. This relationship thus provides a direct relationship between the risk neutral probability used in pricing future profits (assuming that markets are complete, i.e., with no arbitrage, a unique price, no transaction cost, etc.). This equivalence will be misleading however if the utility is not of the quadratic type and if it expresses specific parties' preferences rather than an "abstract" market preference risk.

This synthetic approach to market pricing requires therefore information regarding the future profits (the predictability of future states) and an exchange between many parties resulting in the definition of a price—unique, agreed on, with no party having a personal information or power it can exercise for its personal advantage. In such circumstances, markets are said to be complete while in all other cases, they are said to be incomplete. In practice, the future states may be unpredictable, information is not commonly available to all and equally, and parties to any exchange may have a power or a knowledge which is not shared. In such cases, a counterparty risk arises with some of the parties potentially profiting from the advantage they can derive. In an environment where the reliability is inherent in a product manufacture (say the selling party) and the consumer behavior (say the buying of the product), there are counter party risks at play that render the predictability of future states more difficult and thereby their pricing (and the price of safety to mitigate certain undesirable states) more difficult to determine. This approach provides nonetheless an important avenue of research and applications to pricing safety and risk in general [43].

4 Conclusion and Discussion

Safety and reliability are intimately related. They depend on both the system architecture and reliability and on the user assumed to transform the system reliability into a random reliability system in series. The implications that users (buyers) are unreliable, potentially (and mostly) at fault for the unsafe use of products and their consequences may be misleading. Rather, both buyers and sellers and buyers of risk-prone and unsafe products may be at fault (as indicated here). The uncertainty regarding user's reliability and user's heterogeneity with broadly varying reliabilities and care they assume in the use of an equipment or product has led to extremely large costs (when the expected costs of unsafe use are indeed very large), beyond the marginal costs of production. While reliability is a design parameter expressing a performance in a controlled setting, the combined reliability of the system and the user is "a reliability in fact", which is at best a compound random event, expressing our lack of knowledge regarding user's use, user's heterogeneity as well as the counterparty risks latent in such exchanges. This heterogeneity has an important effect on both firm's policies and firm's sustainability while counterparty risks have an important effect on the inequities that exist in markets for risk and safety.

Safety was defined in this paper as a consequential risk which has both direct and indirect effects. In the former case, safety is usually well defined and therefore accounted for, while in the latter case, it may be neglected with consequences that depend on whether the perpetrator of risk assumes or not his responsibilities. When risks are shared and in particular exchanged between parties (of broadly different information and risk attitudes), a market for risk might be defined in which case, the risk implied in a designed safety-reliability system can be defined in terms of risk prices.

When the parties are well defined and have different and conflicting objectives, this defines a counterparty—or strategic risk [41]. When the parties are not defined and exchange and risk sharing occurs through a financial market, then this defines a market for risk, independent of any person but resulting of "many persons, investors and speculators", pursuing their own self-betterment. This results in a price at which risk is exchanged which we called the market price of risk. This price is both model and rationality based—which can be violated in practice from many reasons. This paper has sought to merge the particular concerns of firms seeking to design economically safe and reliable systems. To do so a number of economic approaches have been contrasted and applied to deal with the traditional reliability design problems. In particular, I emphasized the risk-based approach and devised a number of solutions that integrated financial economic considerations (some based on our using the Markowitz [28] (first published in 1950 however) approach to portfolio design to that of production-reliability schedules and the Value at Risk Approach used in practice by financial institutions). By the same token, I have briefly introduced the financial markets' pricing approach to these problems. Although, the problems we treated are elementary and require extensive and further research, these problems point out the potentiality of such approaches in safety-reliability design from an economic and risk management perspective.

Financial economics deals extensively with pricing problems to allow an exchange (and thereby point out to a market price at which risk is exchanged) while hedging is pursued in order to reduce particular risks through exchanges. In such cases, hedging might consist of buyer–sellers agreement based on optional contracts or simply risk sharing agreements based on warranties and other financial services that producers provide to buyers. Thus, the prices of contractual agreements and risk exchanges may be used to infer the price of risk which can be used as an "objective" valuation of safety-reliable configuration (expressing in fact person's risk attitudes and latent counterparty risks which are not easily defined or priced). Through the willingness of buyers and sellers to buy and sell optional characteristics associated to a product, the market price of safety and reliability may be sought as it will provide a richer information base that prices essentially the same future states. Namely, since financial instruments are not costless and require that we determine their price so they may be used rationally, these prices are the implied indicators of what reliability and safety would a market be willing to pay for (and therefore to be designed appropriately). For example, options require that a premium (the option price) be known and paid to limit the size of losses just as product warranties are required to pay a premium (the price of warranty) to assure the buyer against products failure and their consequential costs. Both financial-based and "subjective" approaches (utility based, VaR techniques, etc.) are important, each contributing to a better assessment of parties' implied preferences for risk, their willingness to buy and sell and the observation of real prices—at which the exchange of risk and its derivatives occurs. It is for this reason that such approaches may be useful in defining the implied safety and reliability observed in real and current prices.

References

1. Ale B (2002) Risk assessment practices in the Netherlands. Saf Sci 40:105–126
2. Ale B, Smith E, Pitblado R (2000) Safety around airport—developments in 1990s and future directions. Det Norske Veritas, London
3. Antle JM (1996) Efficient food safety regulation in the food manufacturing sector. Am J Agric Econ 78(4):1242–1247
4. Antle JM (2000) The cost of quality in the meat industry: implications for HACCP regulation. In: Unnevehr LJ (ed) The economics of HACCP-costs and benefits, Chap. 6. Eagan Press, Saint Paul, pp 81–96
5. Bahr N (1997) System safety engineering and risk assessment: a practical approach. Taylor and Francis, London
6. Bakker GJ, Blom HAP (1993) Air traffic collision risk modeling. In: Proceedings of the 32nd IEEE Conference on Decision and Control, San Antonio
7. Barlow R, Proschan F (1965) Mathematical theory of reliability. Wiley, New York
8. Barnett A (2000) Free-flight and en route air safety: a first-order analysis. Oper Res 48:833–845
9. Blom HAP, Klompstra MB, Bakker GJ (2003) Accident risk assessment of simultaneous converging instrument approaches. National Airspace Laboratory (Report NLR-TP-2003-557), Amsterdam
10. Blom HAP, Corker KM, Stroeve SH (2005) Study on the integration of human performance and accident risk assessment models: AIR-MIDAS & TOPAZ. In: Proceedings of the sixth USA/Europe Air Traffic Management R&D Seminar, Baltimore
11. Blom HAP, Stroeve SH, de Jong HH (2006) Safety risk assessment by monte carlo simulation of complex safety critical operations. In: Proceedings of the 14th safety critical systems symposium, Bristol
12. Blom HAP, Bakker GJ, Blanker PJG, Daams J, Everdij MHC, Klompstra MB (1998) Accident risk assessment for advanced ATM. In: Proceedings of the second USA/Europe Air Traffic Management R&D Seminar, Orlando
13. Boeing Commercial Airplanes (2006) Statistical summary of commercial jet airplane accidents: worldwide operations 1959–2005. Boeing Commercial Airplanes, Seattle
14. Bo-Hyun Cho L, Neal Hooker H (2009) Comparing food safety standards. Food Control 20:40–47
15. Cox JJ, Tait N (1991) Reliability, safety and risk management, Butterworth-Heinemann, January
16. Formal Safety Assessment (FSA) (2005) Passenger ship safety: effective voyage planning for passenger ships, Formal Safety Assessment—large passenger ships navigation, Sub Committee on Safety of Navigation, 50th session, 2005, NAV50/11/1, http://research.dnv.com/skj/FSALPS/FSA-LPS-NAV.htm
17. Formal Safety Assessment (FSA) (2006a) Consideration on utilization of Bayesian network at step 3 of FSA, Maritime Safety Committee, 81st session, MSC 81/18/1
18. Formal Safety Assessment (FSA) (2006b) FSA study on ECDIS/ENCs. Maritime Safety Committee, 81st session. MSC81/24/5, http://research.dnv.com/skj/FSA-ECDIS/ECDIS.htm
19. Federal Aviation Administration (2005) European Organization for Safety of Air Navigation (2005) ATM safety techniques and toolbox, safety action plan-15. Washington, DC
20. GAIN (2003) Guide to methods & tools for safety analysis in air traffic management. Global Aviation Information Network. www.gainweb.org

21. Ha JS, Seong PH (2003) A method for risk-informed safety significance categorization using the analytic hierarchy process and Bayesian belief networks. Department of Nuclear and Quantum Engineering, Korea Advanced Institute of Science and Technology, 373-1, Guseong-Dong, Yuseong-Gu, Daejeon 305-701, South Korea, 2003
22. Hale A (2002) Risk contours and risk management criteria for safety at major airports, with particular reference to the case of Schiphol. Safety Sci 40:299–323
23. Huang H-W, Shih C, Yih S, Chen M-H (2008) Integrated software safety analysis method for digital I&C systems. Ann Nucl Energy 35:1471–1483
24. International Maritime Organization (IMO) (2002) Guidelines for the application of Formal Safety Assessment (FSA) for use in the IMO rule-making process. http://www.imo.org
25. Janic M (2000) An assessment of risk and safety in civil aviation. J Air Trans Manag 6:43–50
26. Luxhøj JT (2003) Probabilistic causal analysis for safety risk assessments in commercial air transport, Workshop on investigating and reporting of incidents and accidents (IRIA), September 16–19, 2003, Williamsburg
27. Luxhoj J, Coit D (2006) Modeling low probability/high consequence events: an aviation safety risk model. In: Proceedings of the 2006 Reliability and Maintainability Symposium (RAMS), Newport Beach
28. Markowitz HM (1959) Portfolio selection; efficient diversification of investments. Wiley, New York
29. Netjasov F, Janic M (2008) A review of research on risk and safety modelling in civil aviation. J Air Trans Manag 14:213–220
30. Pedrali M, Cagno E, Trucco P, Ruggeri F (2004) Towards the integration of human and organizational factors in risk assessment. A case study for the marine industry, Second international Asranet colloquium, Barcelona
31. Phadke MS (1986) Quality engineering using robust design. Prentice Hall, Englewood Cliffs
32. Pikaar AJ, Piers MA, Ale B (2000) External risk around airports—a model update. National Airspace Laboratory (Report NLR-TP-2000-400), Amsterdam
33. Rouvroye JL, van den Bliek EG (2002) Comparing safety analysis techniques. Reliab Eng Syst Saf 75:289–294
34. Taguchi G, Elsayed EA, Hsiang T (1989) Quality engineering in production systems. McGraw Hill, New York
35. Tapiero CS (1996) The management of quality and its control. Chapman and Hall, London
36. Tapiero S (1997) The economic effects of reliable and unreliable testing technologies, ICJM
37. Tapiero CS (2004) Risk and financial management: mathematical and computational concepts. Wiley, London
38. Tapiero CS (2004) Risk management. In: Teugels J, Bjorn S (eds) John Wiley Encyclopedia on Actuarial
39. Tapiero S (2005) Value at risk and inventory control. Eur J Oper Res 163(3):769–775
40. Tapiero CS (2006) Risks and assets pricing. In: Huang P (eds) Handbook of engineering statistics. Springer
41. Tapiero S (2007) Consumers risk and quality control in a collaborative supply chain. Eur J Oper Res 182:683–694
42. Tapiero Charles S (2005) Reliability design and RVaR. Int J Reliab Qual Safety Eng (IJRQSE) 2(4):347–353
43. Tapiero CS (2010) Finance and risk assets pricing. Wiley, Hoboken
44. Transportation Safety Board of Canada (TSB) (1998) Safety study of the operational relationship between ship master/watchkeeping officers and marine pilots. http://www.bst.gc.ca
45. Truccoa P, Di Giulio A, Randazzo G, Pedrali M (2003) Towards a systematic organisational analysis for improving safety assessment of the maritime transport system, in safety and reliability. In: Bedford T, Van Gelder PHAJM (eds) Swets & Zeitlinger, Lisse, ESREL'03, pp 513–21
46. Truccoa P, Cagnoa E, Ruggerib F, Grandea O (2008) A Bayesian belief network modelling of organisational factors in risk analysis: a case study in maritime transportation. Reliab Eng Syst Saf 93:823–834

Reliability Analysis of Structures Under Hybrid Uncertainty

Subrata Chakraborty and Palash Chandra Sam

1 Introduction

In engineering applications it is important to model and adequately treat all the available information during the analysis and design phase. Typically, the information are originated from different sources: field measurements, experts' judgments, objective and subjective considerations. Over these features, the influences originated from the human errors, imperfections in the construction techniques, influence of the boundary and environmental conditions are added. All these aspects can be brought back to one common denominator: *presence of uncertainty*. The uncertainty can be viewed as a part or class of imperfection in the information that attempts to model a system behavior in the real world (see Fig. 1). It is the gradual assessment of the truth content of postulation e.g., in relation to the occurrence of a defined event. Normally, the uncertainty is viewed in two categories, namely aleatory and epistemic. The aleatory uncertainty is classified as objective and irreducible uncertainty with sufficient information on the input uncertain data. These are inherently connected to the problem at hand and cannot be influenced by the designer. The epistemic uncertainty is classified as subjective and reducible uncertainty that stems from the lack of knowledge about the input uncertain data. It arises from the cognitive sources involving the definition of certain parameters, human errors, inaccuracies, manufacturing and measurement tolerances, etc. In brief, the objective uncertainty is concerned with the tendency of

S. Chakraborty (✉) · P. C. Sam
Department of Civil Engineering, Bengal Engineering and Science University, Shibpur, Howrah 711103, India
e-mail: schak@civil.becs.ac.in; schak@mailcity.com

H. Pham (ed.), *Safety and Risk Modeling and Its Applications*,
Springer Series in Reliability Engineering, DOI: 10.1007/978-0-85729-470-8_3,

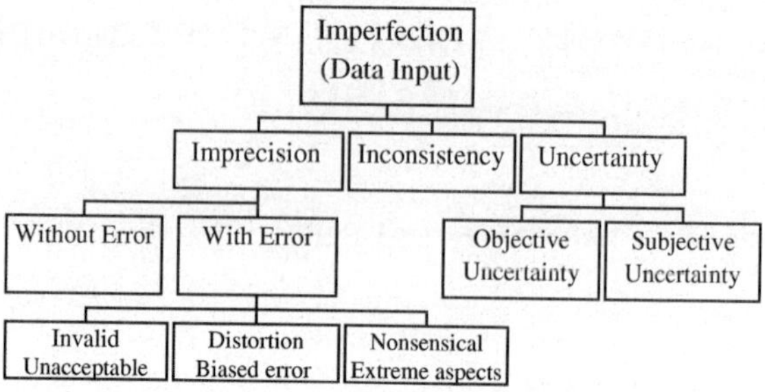

Fig. 1 Various forms of imperfection in the input information

an event to occur, and the subjective uncertainty is concerned with the ability to occur. In real-life application it is classified in such a way that a mathematically founded and realistic description is ensured in the structural analysis and safety evaluation [1].

The profession has accepted the fact that the existence of uncertainty cannot be avoided in the analysis and design of engineering system. It is now well recognized that when the existence of uncertainty is taken into account it leads to a more cost-effective design rather than when it is planned to eliminate or greatly reduce it for final design that will be safe, reliable and robust against uncertainty. Thus, the consideration of uncertainty in engineering analysis and design process is gaining increasing importance in the profession. The uncertainty quantification in a typical engineering decision-making process requires: (1) Characterization of the uncertainty involved in various system parameters and external environment; (2) Propagation of this uncertainty through engineering models and computational tools. The first step of characterization of the uncertainty involves the development of methodologies to model the uncertainty of both the aleatoric and the epistemic type. Regardless of the type being considered, the characterization process depends on the Experimental Research and Expert Judgment. The outcome of the process is in the form of probability distribution function (*pdf*), membership function (*mf*), interval bounds, etc., depending on the quantity and quality of data feasible to acquire. The propagation of uncertainty mainly involves two aspects i.e., the response analysis of system considering the uncertain input parameters and the associated safety analysis compatible with the decision-making process. The response of structural model under excitation can be mathematically expressed by a set of equations with suitably prescribed boundary and initial conditions as:

$$\Omega[Y(\mathbf{x},t)] = F(\mathbf{x},t) \tag{1}$$

where $Y(\mathbf{x}, t)$ is the output of the system due to input $F(\mathbf{x}, t)$ and Ω is the differential operator, representing the mechanical system. The model as defined by Eq. 1 involves two independent variables i.e., the position vector $\mathbf{x}$ and the time variable t. Depending on the nature of the variables involved in a typical

Table 1 Classifications of structural analysis problem

Level	System Operator (Ω)	Input (F)	Output vector (Y)	Remark
1	Deterministic	Deterministic	Deterministic	Conventional deterministic structural analysis
2	Deterministic	Deterministic	Uncertain	Random vibration
3	Uncertain	Uncertain	Deterministic	Stochastic system (SFEM, SFDM, SBEM, FFEM, etc.)
4	Uncertain	Uncertain	Uncertain	Most general models

problem, various types of analysis problems may arise. Those problems can be broadly classified as depicted in Table 1. Nowadays, the response analysis under uncertainty is generally performed through numerical model and various methods like the stochastic finite difference method (*SFDM*), the stochastic finite element method (*SFEM*), the stochastic boundary element method (*SBEM*), the fuzzy finite element method (*FFEM*), etc., emerged from this. The response surface method (*RSM*) based various metamodel strategies are also very common to approximate expensive computer simulations [2, 3]. The safety assessment is performed through reliability analysis either in the probabilistic or in the possibilistic approach depending on the quality and quantity of information available about the input uncertain parameters.

The present chapter deals with the safety evaluation of structure when it is characterized by both the probabilistic and the possibilistic uncertain parameters and referred to as hybrid uncertain systems. In doing so, various methods of safety evaluation of structural system under uncertainty are summarized first. Then, relying on the fundamental concept of various emerging methods of transformation of the possibilistic variables to equivalent probabilistic variables or vice versa, the feasibility of reliability analysis of such hybrid uncertain systems either in the probabilistic or in the possibilistic format is demonstrated. Finally, a numerical example illustrates the capability, consistency and suitability of the transformation approaches to evaluate the safety of hybrid uncertain system.

2 Safety Analyses of Structures Under Uncertainty

In last 20 years or so, many new methods have been developed to deal with the imperfection in input data. The large number of models reflects that there exist many aspects of imperfection and the probability theory is not the unique normative model that can be coped with all of them. In fact, numerous methods are used to deal with the uncertainty in natural sciences and engineering. These include from the probability theory and its variants (Bayesian theory, reliability theory), to multi-valued logic, fuzzy and related possibility theory, interval analysis, etc. This section briefly discusses such various developments of safety evaluation of structures and a critical comparison is also made with emphasis on the safety evaluation of hybrid uncertain system.

2.1 The Probabilistic Approach

For a long time, the probability theory was the only approach used to quantify the uncertainty of input parameters necessary to model the engineering system. Even now, the probabilistic methods are almost exclusively used in the industry. The applications vary from the structural design to biomechanics, to water quality control, to estimate the output of drainage systems, designing integrated circuits, traffic control and traffic flow estimation, and so on. In probabilistic approach, the assurance of performance is referred to as the reliability. Mathematically, this performance function is described as,

$$Z = g(\mathbf{X}) \tag{2}$$

where **X** is the vector consisting of the uncertain parameters. The probability of failure due to a single failure mode can be mathematically expressed by the following multi-dimensional integral,

$$p_f(Z<0) = \int \cdots \int_{g(\mathbf{X})<0} f_{\mathbf{X}}(\mathbf{X})\mathbf{dX} \tag{3}$$

where $f_{\mathbf{X}}(\mathbf{X})$ is the n-dimensional joint *pdf* of the basic random variables. The computation of probability of failure (p_f) by the above equation is the fundamental of safety evaluation in probabilistic approach known as the full distribution approach. In general, the joint *pdf* of random variables is almost impossible; moreover, evaluating the multiple integral is a formidable task. The Monte Carlo simulation technique is used as the robust alternative to compute the p_f for both the explicit and implicit limit state function with known *pdf* of the random variables. But it requires a number of deterministic analyses ranging between few hundreds to tens of thousands depending on the magnitude of p_f. Normally, the second moment based approximation methods i.e., the first-order reliability methods (FORM) and the second-order reliability methods (SORM) are proposed in the literature to obtain a reasonable estimate of p_f with significantly lower computational effort. In the present study, the FORM is used to obtain the safety in probabilistic format for structural system under hybrid uncertainties. A brief description of this method is provided in Appendix A. Further details may be seen elsewhere [4–6].

The second moment methods require fewer deterministic analyses provided that the closed form expressions of the sensitivity derivatives of the performance function with respect to the random variables are available. Otherwise, one may need to opt for the *RSM* to approximate the failure surface. The second moment methods may yield erroneous estimate of p_f for which there are multiple most probable points as in the case of structural dynamic problems. In this regard, the most significant criticism on the widespread use of classical reliability methods is that the information input in the analysis has to be in a precise probabilistic format and the limit state function through which this information is propagated is a precise model. The statistical distributions of the parameters, good information on correlations, etc., are seldom known for all random variables in real-life design problems. Moreover, the probabilistic approach is based on the accuracy of the probabilistic model of the random parameters. It has been shown [7] that even small errors in the statistical parameters may have large effects on the computed p_f, especially when these probabilities are very small.

2.2 The Possibilistic Approach

The reliability evaluation of structural system is carried out based on the classical probability theory in which some of the variables are considered to be random and rests are assumed as deterministic. The limit of applicability of such an approach is attained when insufficient reliable data are only available to describe the real-life systems with the aid of *pdf*. The effectiveness of probabilistic approach is thus lost with non-probabilistic uncertainty in the system input parameters. Moreover, the real-world problems are more complex than their corresponding mathematical model and to compensate this gap, some linguistic explanation occasionally adds to the results obtained through the models. In order to quantify such information, it is possible to apply the uncertainty measure on the basis of existing available data with the additional information from expert knowledge and experience. This has led to the development of various possibilistic methods. The related developments are summarized in the following.

2.2.1 Interval Analysis and Convex Modeling

In many cases, for example in preliminary design phases, even though some experimental data are available, it is not enough to construct the *pdf*, reliably. The available data can be used, particularly in combination with the engineering experience, to set some tolerances or bounds only. A typical example is the uncertainty in the parameters arising from the manufacturing tolerances, materials defects and variation in the operating conditions or errors in the observations. In such cases, the uncertain design variables can be well modeled using the non-probabilistic convex models of uncertainty. Examples of convex models include intervals, ellipse or any convex sets. This kind of mathematical model is called uncertain but bounded (*UBB*) model. BenHaim and Elishakoff [7] proposed a version of worst-case design based on the convex models for design problems with

UBB variables. A quantitative non-probabilistic reliability measure based on the convex model was first introduced by BenHaim [8] and the subsequent developments on the structural reliability analysis in which the bounds on the magnitude of uncertainty that is only required are notable [9–11].

The interval method of analysis seems to be a logical alternative when the parameters required to create the probabilistic models cannot be precisely determined due to lack of data. Interval analysis considers rectangular model that encloses all the possible combinations of the uncertain variables i.e., consists of all the possible probabilities that are consistent with the available information. It is basically a worst-scenario method since all the *UBB* variables vary independently and thus may reach their extreme values simultaneously leading to overly conservative design. The ellipsoid model considers all the variables to be correlated with each other, which excludes the extreme combination of the uncertain parameters and thus avoids overconservative designs. However, in reality, only part of these *UBB* variables are actually correlated while some others vary independently. Therefore, a more realistic option is to divide all the *UBB* quantities into groups and treat them with a multi-ellipsoid convex model [12].

2.2.2 The Possibility/Fuzzy Set Theory Based Approaches

The possibility is an alternative approach to the probability, initially introduced to model the uncertainties when the available information is linguistic. Various methods have been developed to deal with such subjective uncertainty. It is based on the possibility distribution defined by the *mf* obtained from the numerical data along with the expert knowledge and experience. A brief note on the fundamentals of possibility theory and fuzzy uncertainty are described in Appendix B. The fuzzy set theory based structural reliability analysis developments are summarized here.

The possibilistic safety evaluation algorithms mainly tried to explore the entire range of the uncertain variables to estimate the p_f. In a typical interval analysis or α-cut in fuzzy set approach, repeated standard probabilistic reliability analyses are performed to obtain the worst failure distribution pattern. Brown [13] first applied the fuzzy measure concept with classical structural reliability theory to obtain more reliable failure modes. Subsequently, Shiraishi and Furuta [14], Yao and Furuta [15] used fuzzy logic in structural reliability application. The outline of fuzzy theory and its contribution to structural reliability assessment is presented in Furuta [16] demonstrating the applicability of fuzzy set theory in various structural engineering applications. Yubin et al. [17] analyzed the fuzziness between the structural reliable state and the failure state. They proposed an intermediary transition between the two states and for a given confidence level α, the structural fuzzy random reliabilities were evaluated. Cremona and Gao [18] presented an alternative to the probabilistic theory based on the possibility theory that uses a new confidence measure: the measure of possibility to estimate the distribution of possibility of failure. Linear and nonlinear limit state function involving non-interactive generalized fuzzy variables are transformed to Gaussian fuzzy number and a possibilistic reliability index computation procedure is developed analogous to the probabilistic reliability index computation. This approach is used in the present numerical study for possibilistic safety assessment and briefly described in Appendix C. Möller et al. [1] developed a safety assessment method by transferring the uncertain input variables described by the fuzzy *mf* to fuzzy output response through α-cut using efficient optimum vortex method. Möller et al. [19] further introduced the fuzzy FORM (*FFORM*) of analysis by transferring the fuzzy random variables having known fuzzy *mf* to standard normalized variables. Bing et al. [20] proposed a method for fuzzy reliability analysis where the fuzzy linear regression model was used in conjunction with the finite element method. Similar to the stress-strength inference model in the classical reliability theory, the fuzzy stress-random strength is proposed to evaluate the fuzzy random reliability of mechanical structure. Jiang and Chen [21] developed a computational model to obtain the fuzzy reliability of mechanical component in which the fuzzy reliability was obtained using the conventional probability and mathematical transition. Starting from the fundamentals of fuzzy mathematics to fuzzy structural analysis and subsequently fuzzy reliability analysis is well covered in the text on fuzzy randomness [22].

2.3 The Safety Assessment Approaches and Hybrid Uncertainties

The probabilistic and the possibilistic methods of reliability analysis of structures as discussed above have been developed independently i.e., the former is to consider the random uncertainty and the latter is for nonstochastic uncertainty. Langley [23] has shown that the same numerical algorithm could be used for finding both the probability and the possibility of failure, modifying only the function to be minimized. Chakraborty [24] has reviewed the various safety assessment alternatives under uncertainty. Though, the valuable comparisons between these two methods are available in the literature [25, 26], no work seems to be available to find the relationship between the two approaches. Moreover, as such there is no consensus about what method to be used in many real-life problems. The general understanding is that if the uncertainties are modeled accurately, the probabilistic methods are better than their counterpart for efficient design. But, the scarcity of data available may need to make strong assumptions which can be sensitive enough in terms of safety of the system. It has been repeatedly invoked in the literature that the possibility is a better choice in such a case as the approach is not only safe but also simpler to apply. Moreover, it is difficult to make the probabilistic approach more conservative to protect the design from inaccuracies made due to lack of information. But, it is easy to increase the degree of conservatism in the possibilistic approach. In general, the possibility of failure effectively imposes a factor of safety on the probability of failure [25]. However, the issue of switching from the probabilistic to the possibilistic approach by justifying whether the information is little is not very clear. It is further noted that the possibility theories are of little use in the design of system with large number of failure modes that are known to be independent [26]. On the contrary, the probabilistic safety analysis is computationally expensive and there may not be enough information at early stage of design cycle. Thus, the possibilistic methods requiring less information yet can provide a measure of reliability that is attractive at preliminary design stage. At early design phase, the subjective uncertainty representing the design imprecision and inexactness in choosing among design alternatives usually dominates the preliminary design configuration. This conservatism would certainly ensure performance but could adversely affect the optimum cost. With the progress of iterative design process, these types of uncertainty reduced gradually, but the objective uncertainty remains throughout the design process. The general consensus is to opt for probabilistic approach when the numbers of possibilistic variables present in the system are greatly reduced [26].

Interestingly, most of these observations are based on the fundamental assumption that the input system parameters information are either all possibilistic or all probabilistic. This poses serious restriction on the necessary flexibility to the designer to start with both the probabilistic and the possibilistic description of the variables depending on the nature of available data so that one needs less assumption at early stage of modeling. Various transformation methods have emerged in the literature to transform the possibilistic variables to equivalent probabilistic variables or vice versa. The concept of establishing the equivalence between the fuzzy and random variables through entropy principle is used in structural analysis [13, 27, 28–30]. Chakraborty and Sam [31] applied the various transformation approaches to carry out the safety assessment in probabilistic format for hybrid uncertain system. Marano et al. presented [32] a fuzzy time-dependent reliability analysis procedure using information entropy for reinforced concrete beam subject to pitting corrosion considering random and fuzzy system parameters. Rao et al. [33] proposed a procedure of unified solution to tackle the hybrid uncertainties by combining the *SFEM* solution with the *FFEM* solution. Ferrari and Savoia [34], Savoia [35] defined an equivalent class of *pdfs* compatible with the corresponding fuzzy descriptions. Smith et al. [36] proposed to scale the fuzzy *mf* of the system response to a *pdf*. The implicit limit state functions are modeled using high-quality approximations to estimate the reliability of system with mixed parameters [37]. Du [38] proposed a unified uncertainty analysis considering the effect of mixed uncertainty and it is shown that the belief and the plausibility measures can be converted to obtain the lower bound and the upper bound probabilities in the context of evidence theory. A definition of structural reliability under mixed model representation is proposed by Luo et al. [39] employing the multi-ellipsoid to represent the bounded uncertainties. In this regard, the works on design optimization of system under random and fuzzy variables are worth mentioning to deal with the mixed uncertainties [40, 41].

It is generally realized that either the probabilistic or the possibilistic approach is not compatible to tackle the hybrid uncertainties. But, the concept of various transformation methods can be readily used to unify the mixed variables so that the existing probabilistic or the possibilistic reliability analysis format will be compatible. Relying on the fundamental concept of such transformations, the feasibility of reliability analysis of hybrid uncertain system either in the probabilistic or in the possibilistic format is demonstrated in the present chapter. This will provide the necessary flexibility to the designers to model the structural parameter uncertainty either probabilistically or possibilistically depending on the nature and quality of the input data. A numerical example illustrates the capability and consistency of the various strategies of transformation to evaluate the safety of hybrid uncertain system.

3 Safety Evaluation Under Hybrid Uncertainty

The focus of present study is on the reliability analysis of structures characterized by both the probabilistic and the possibilistic variables. The probabilistic variables are described by the associated *pdf* and the possibilistic variables are described by the associated fuzzy *mf*. The limit state function of the related reliability analysis problem involves the probabilistic and also the possibilistic variables. Thus, to make the analysis compatible with the reliability analysis in the probabilistic format or in the possibilistic format as described in appendices A and C, respectively, one needs to express the performance function either in terms of the random variables or in terms of the fuzzy variables only depending on the approach of analysis desired to apply. It is thus, generally realized that the fuzzy variables need to be transformed to equivalent random variables or vice versa for safety analysis under mixed uncertain variables. However, the transformation should be such as to satisfy the consistency conditions. One condition is that the possibility of an event should be greater than or equal to its probability. A more restrictive condition is that the possibility of any event that has non-zero probability must be 1. This condition leads to overly conservative design. In most cases, the transformations are obtained by satisfying the first condition only. In the subsequent sections, potential alternatives to transfer the fuzzy variables to equivalent random variables or vice versa are presented so that the limit state function can be expressed in terms of the random variables or in terms of the fuzzy variables only so that one can choose the method of analysis conveniently either in the probabilistic or in the possibilistic format.

3.1 Equivalent Entropy-Based Transformation

Entropy is a measure of the uncertainty of a random variable. It can also be looked upon as a measure of the imprecision for a fuzzy variable. Using the basic concept of entropy, the fuzzy imprecision can be transformed to random uncertainty or vice versa. The basis of this transformation is that the measurement is invariant under transformation. The principle allows one to use all the available information without unwittingly adding any information that is not contained in the evidence. The concept has been successfully applied to update the uncertain parameters. The probabilistic entropy i.e., Shannon's entropy, H_x, of an equivalent random variable, x, is defined as,

$$H_x = -\int_x p(x) \ln p(x)\, dx \tag{4}$$

where $p(x)$ is the *pdf* of x. The entropy of a non-probabilistic variable as defined by DeLuca and Termini [42] is given by,

$$G_x^{'} = -\int_x [\mu(x)(\ln \mu(x)) + (1 - \mu(x)) \ln (1 - \mu(x))] dx \tag{5}$$

The second term in Eq. 5 corresponds to the complementary events. Alternatively, the non-probabilistic entropy, $G_x^{'}$, can be defined as,

$$G_x^{'} = -\int_x \mu(x) \ln \mu'(x) dx, \quad \text{where } \mu'(x) = \frac{\mu(x)}{\int_x \mu(x) dx} \tag{6}$$

where $\mu(x)$ is the *mf* of x. The contribution due to the complementary term is dropped from Eq. 5 as the *mf* is standardized. It may be noted that the use of Eq. 5 leads to overly conservative design.

The equivalent entropy can be evaluated for normal random variable as,

$$\begin{aligned} H_x &= -\int_x p(x) \ln p(x) dx = -\int_{-\infty}^{+\infty} p(x)[-\ln(\sqrt{2\pi}\sigma) - \frac{1}{2}\left(\frac{x - \bar{x}}{\sigma_x}\right)^2] dx \\ &= \ln\left(\sqrt{2\pi e}\sigma\right) \end{aligned} \tag{7}$$

where $\bar{x}$ and σ_x are the mean and standard deviation of the uncertain variable. During the transformation of uncertain variable, the entropy of the transformed variable should have the same entropy of the original variables i.e.,

$$H_x = G_x^{'} \tag{8}$$

Substituting Eqs. 7 in 8, the parameter of the equivalent normal variable can be obtained as,

$$\sigma = \frac{1}{\sqrt{2\pi}} e^{G_x^{'} - 0.5} \tag{9}$$

The mean of the equivalent random variable is assumed to be same as that of the core of the fuzzy variable i.e., the value corresponding to *mf* value of 1.0.

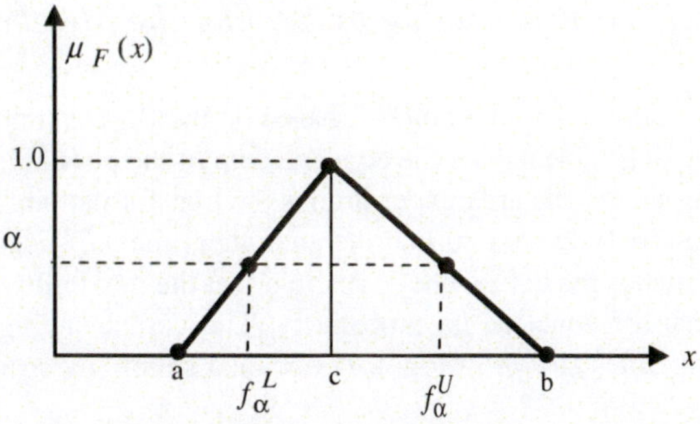

Fig. 2 The Fuzzy distribution and α-cut description

In general, the application is not restricted to the normal random variables only. Any random variable having distribution function defined by two parameters can be obtained by the transformation. However, the choice of the distribution should be such as to minimize the loss of information. The maximum entropy principle is mathematically an optimization problem where one seeks a *pdf* which maximizes the entropy function. If the distribution is defined by two parameters, the Gaussian distribution is obtained by this criterion [43]. Thus, the normal distribution satisfies most conservatively the consistency condition that the probability must be less than or equal to possibility as it maximizes the entropy.

Similar to above, to perform the transformation from random variables to fuzzy variables, all non-normal variables are first transformed to equivalent normal variables. Subsequently, the support of the fuzzy variables can be obtained by using Eq. 8. As earlier, the core value is assumed to be same as the mean of the equivalent normal variable. The application is not restricted to symmetric triangular fuzzy distribution only. Nonsymmetric triangular or trapezoidal fuzzy distribution property can also be readily obtained by using the associated expression of the fuzzy entropy corresponding to the desired *mf* of the fuzzy variables. For a triangular fuzzy variable as shown in Fig. 2, the *mf* can be mathematically expressed as,

$$\begin{aligned}\mu_F(x) &= \frac{(x-a)}{(c-a)} \quad \text{for}\, a \leq x \leq c \\ &= \frac{(b-x)}{(b-c)} \quad \text{for}\, c \leq x \leq b\end{aligned} \tag{10}$$

And, the fuzzy entropy can be derived using Eq. 10 in 6 as,

$$G_x' = 0.5 - \ln\frac{2}{b-a}. \tag{11}$$

Now, Eqs. 8 and 11 can be used to obtain the support width of the transform fuzzy variable i.e. $(b - a)$ in terms of standard deviation of the normal random variable.

3.2 Transformation by Scaling the Membership Function

A transformation approach based on the Bayesian approach is proposed by Smith et al. [36] to reduce the conservatism of the possibility theory. Scaling the *mf* with respect to the area under it does the transformation. The scaling factor is obtained to satisfy the axiom that the area under the *pdf* should be unity. It also intuitively satisfies the consistency principle that the possibility of an event should be greater than or equal to its probability. The corresponding *pdf* of the triangular fuzzy variables shown in Fig. 2 is obtained simply by scaling the *mf* i.e.,

$$\begin{aligned} p(x) &= \frac{k(x-a)}{(c-a)} \quad \text{for } a \le x \le c \\ &= \frac{k(b-x)}{(b-a)} \quad \text{for } c \le x \le b, \text{ where, } k = \frac{2}{b-a} \end{aligned} \tag{12}$$

The cumulative distribution function (*CDF*) can be obtained by integrating above equation.

3.3 The Reliability Bounds

The probability of an event is a direct measure of the confidence one can attribute to the event. The possibility and necessity measure do not satisfy the additive axiom rather defined the limiting cases of inequalities. The knowledge of possibility (or necessity) of the event alone is not sufficient to estimate the confidence on it, nevertheless, the equality relation is assumed so that, once the possibility of an event is known, the necessity of the complementary event is determined or vice versa. Following the interpretation of the possibility distribution based on the evidence theory, Dubious and Prade [44] defined an equivalent class of probability distribution where the lower bound and the upper bound of the probability are shown to be the possibility and the necessity. Based on this, Ferrari and Savoia [34] describe the compatible *CDF* in which the left boundary coincides with the increasing branch of the fuzzy variables and the right side boundary coincides with its complement i.e., the decreasing branch. For a fuzzy variable having a *mf*, $\mu(x)$, the conservative estimate of the upper bound and the lower bound of the *CDF* is then defined as,

$$F_x^*(x) = \begin{cases} 0, & x < x_L \\ \mu_x(x), & x_L \le x \le x_C \\ 1, & x > x_C \end{cases} \quad F_{*x}(x) = \begin{cases} 0, & x < x_C \\ 1-\mu_x(x), & x_C \le x \le x_R \\ 1, & x > x_R \end{cases} \tag{13}$$

The upper bound and the lower bound *pdf* can be obtained by differentiating the associated *CDF*. The upper bound and the lower bound *CDF* and *pdf* corresponding to any fuzzy distribution are conceptually clarified in Fig. 3. The *pdf*

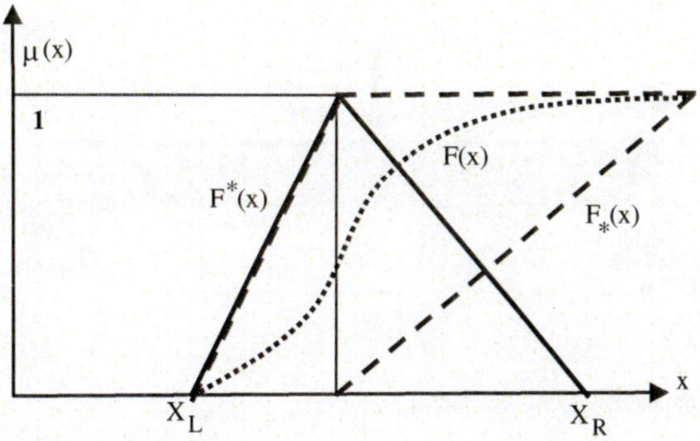

Fig. 3 The conservative bounds of the probability distribution

compatible with the fuzzy description lies in between these two distributions and infinite number of distributions may exist.

For a symmetric triangular fuzzy number as described in Fig. 2, the lower and the upper branches of the *mf* can be written as,

$$\mu_L(x) = \frac{x-a}{c-a}, \quad \mu_R(x) = \frac{x-c}{b-c} \tag{14}$$

Following Eq. 13, the upper and the lower *CDF*s are obtained as,

$$F_x^*(x) = \begin{cases} 0, & x<a \\ \frac{x-a}{c-a}, & a \le x \le c \\ 1, & x>c \end{cases} \quad F_{*x}(x) = \begin{cases} 0, & x<c \\ \frac{x-c}{b-c}, & c \le x \le b \\ 1, & x>b \end{cases} \tag{15}$$

And the associated *pdfs* will be,

$$f_x^*(x) = \begin{cases} 0, & x<a \\ \frac{1}{c-a}, & a \le x \le c \\ 1, & x>c \end{cases} \quad f_{*x}(x) = \begin{cases} 0, & x<c \\ \frac{1}{b-c}, & c \le x \le b \\ 1, & x>b \end{cases} \tag{16}$$

3.4 Reliability Analysis Under Hybrid Uncertainty

It is apparent that to perform the safety evaluation of structures characterized by mixed uncertain variables, all the parameters must be either random in case of probabilistic approach or all must be fuzzy type for possibilistic approach of safety evaluation. A structural system in which some variables are random, defined by their respective *pdf*, and some are fuzzy defined by the associated *mf*, one can easily derive the properties of the equivalent random variables following the fundamental concept of transformation as described in Sects. 3.1 and 3.2. If the transformed variables are not normal, those can be further transformed to equivalent normal variables by applying suitable transformation like the Rackwitz–Fiessler algorithm. Now, the standard second moment format as described in

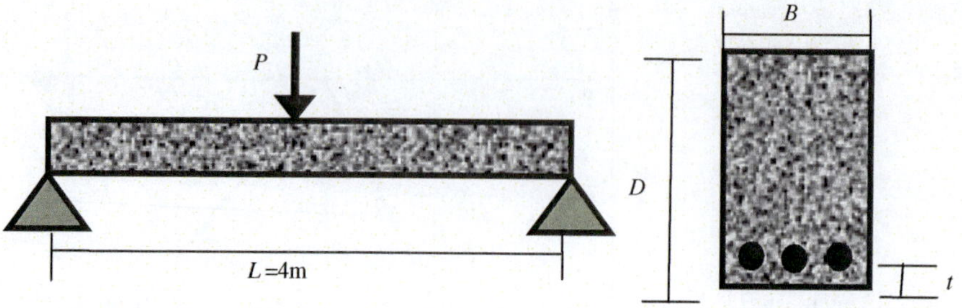

Fig. 4 The concrete beam

Appendix A can be readily applied to obtain the reliability index and the corresponding probability of failure. For a comparative study between these two approaches, a conservative upper bound and lower bound of p_f can be computed following the equivalent class of probability distribution boundaries as described in Sect. 3.3. The lower *CDF* and the *pdf* for the fuzzy variables are obtained from the *mf* and those are transformed to equivalent normal variables using Eq. 6. To obtain the lower bound p_f, the same procedure as that used to obtain the upper bound value can be followed. Similarly, to perform the possibilistic reliability analysis, the random variables defined by the associated *pdf* can be transformed to equivalent fuzzy variables following the transformation concept. In the present possibilistic reliability analysis, the entropy-based transformation is used to obtain the properties of the symmetrical fuzzy distribution. Finally, the possibility of failure is obtained by using the procedure described in *Appendix C*.

4 Numerical Study

A reinforced concrete beam as shown in Fig. 4 is taken up to elucidate the proposed safety evaluation procedure of hybrid uncertain system. Specifically, the purpose of the example problem is to demonstrate the capability of the transformation approaches to tackle the presence of mixed uncertain parameters to perform the probabilistic or the possibilistic safety analysis of structure.

The beam is reinforced with steel bars of area $(A_s) = 1{,}400\ \text{mm}^2$, having a length of 4.0 m subjected to an external load P at its center (Fig. 4). The characteristic strength of concrete (f_{ck}) and the yield stress of steel (F_y) are considered as random variables and assumed to be normally distributed. The load P, width (b), depth (D) and the cover of reinforcements (t) of the beam are modeled as fuzzy variables. The known property values of the variables of the concrete beam are depicted in Table 2.

In the present study, the reliability is defined with respect to the flexural mode of failure. Considering the ultimate flexural strength of the beam, the equation of failure surface is,

Table 2 The property values of the variables

Variables	F_y (N/mm^2)	f_{ck} (N/mm^2)	B (mm)	D (mm)	t (mm)
Mean value	320	20	300	500	50

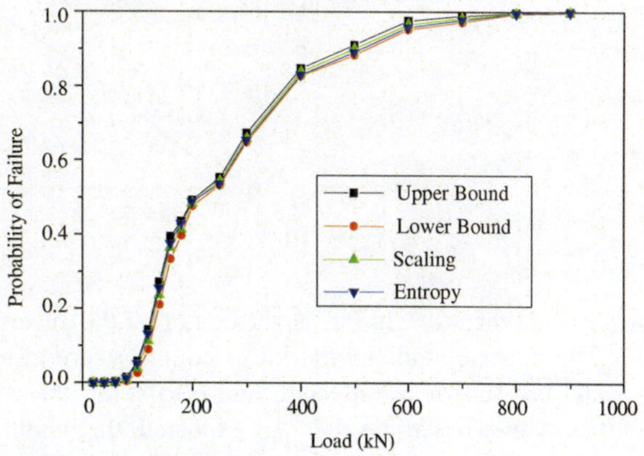

Fig. 5 The variation of probability of failure with increasing nominal value of the concentrated load with symmetric triangular fuzzy variables

$$g = F_y \cdot A_s \cdot (D - t)\left[1 - \frac{0.77F_y.A_s}{f_{ck} \cdot b \cdot (D - t)}\right] - M_e \tag{17}$$

where $M_e = PL/4$ is the moment due to external load. Denoting, $P = X_1$, $b = X_2$, $t = X_3$, $D = X_4$, $F_y = X_5$ and $f_{ck} = X_6$ and substituting these in Eq. 17, the failure surface equation becomes as,

$$g(x) = 1400X_4 \cdot X_5 - 1400X_5 \cdot X_3 - \frac{1.5092 \times 10^6 X_5^2}{X_6.X_2} - 1000X_1 \tag{18}$$

For symmetric triangular fuzzy distribution of the fuzzy variables *P, b, D* and t as shown in Fig. 2, $a = \bar{x}_i - w\sigma_{xi}$, $c = \bar{x}_i$ and $b = \overline{x_i} + w\sigma_{xi}$, where, $\overline{x_i}$ and σ_{xi} are the mean and standard deviation of the *i*th uncertain variable, respectively, and w defines the support width of the *mf* of the fuzzy variables.

4.1 Probabilistic Reliability Analysis

The reliability analysis in probabilistic format is performed by the *FORM* as described in *Appendix A*. The probability of failure of the beam considering the limit state function as described by Eq. 18 is computed by transferring the fuzzy variables to equivalent random variables following Sects. 3.1 and 3.2. The reliability results are presented in Fig. 5 with increasing nominal value of the concentrated load. The support width (w) of the symmetrical triangular fuzzy variables described as a function of the standard deviation (σ) is taken as 3, unless

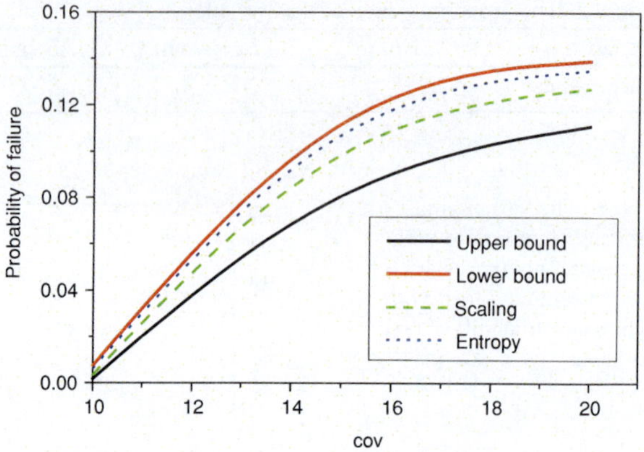

Fig. 6 The variation of probability of failure with varying cov

mentioned otherwise. In Fig. 5, the cov of all the uncertain parameters are taken as 0.2. The upper bound and the lower bound p_f are also computed following Sect. 3.3 and are shown in the same figure to study the consistency of the proposed transformation-based results. As expected the width of the bounded solution increases as the level of uncertainty increases. However, the p_f are within the bounded solution as obtained based on the conservative evidence theory.

The variation of p_f with increasing cov (representing more uncertainty) of all the uncertain variables is shown in Fig. 6 considering the same support width i.e., $w = 3.0$. The nominal value of the load, P, is taken as 10 kN to develop this figure. It is generally seen that the failure probability obtained by the equivalent entropy-based formulation gives a higher value of p_f. The results are obvious due to the normal distribution assumption, which maximizes the entropy. Any other distribution includes less uncertainty. Since, any entropy that is less than its maximum implies unwarranted use of additional information, it will be biased in some manner.

The probability of failure is also computed considering unsymmetrical triangular variation of the fuzzy variables as shown in Fig. 7. Note that for ith such fuzzy variable, $x_L = a = \bar{x}_i - w_1\sigma_{xi}$, $c = \bar{x}_i$ and $x_U = b = \bar{x}_i + w_2\sigma_{xi}$. The left and right end supports are now defined by two unequal parameters w_1 and w_2. For an unsymmetrical triangular fuzzy distribution with $w_1 = 1.0$ and $w_2 = 2.0$, the variation of p_f with increasing mean load is shown in Fig. 8.

4.2 Possibilistic Reliability Analysis

To perform the safety analysis in possibilistic format, the random variables are now transformed to equivalent symmetric fuzzy variables by applying the entropy based transformation as described in Sect. 3.1. Subsequently, the possibility of failure is computed following the procedure described in Appendix C. The possibility of failure results are presented in Fig. 9 with increasing nominal value of

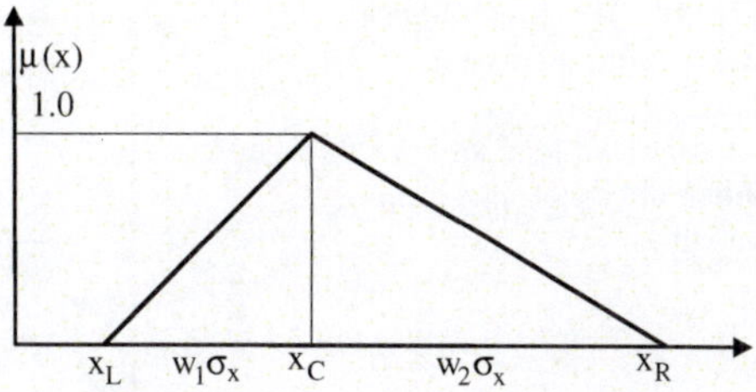

Fig. 7 The unsymmetrical triangular fuzzy distribution

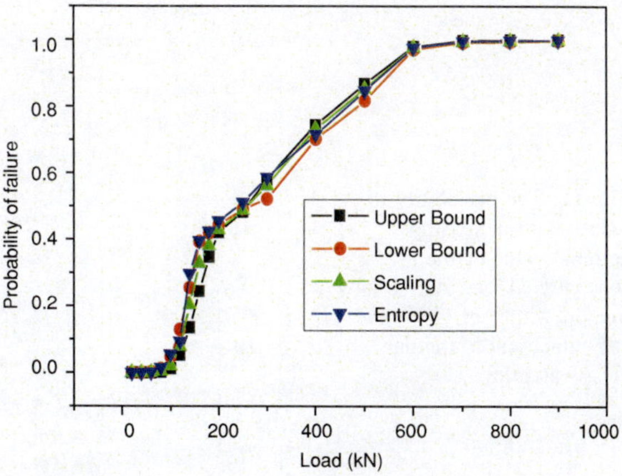

Fig. 8 The variation of probability of failure for unsymmetrical fuzzy distribution

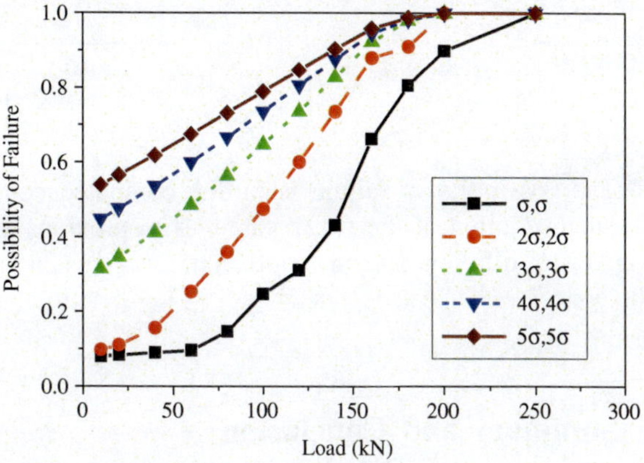

Fig. 9 The possibility of failure with increasing nominal value of the concentrated load and different support widths of the symmetric triangular fuzzy variables

the concentrated load and various support widths of the symmetric triangular fuzzy variables described as a function of standard deviation (σ). The cov of uncertain variables are taken as 0.2 to develop these figures.

The variation of possibility of failure for different cov of uncertain parameters is shown in Fig. 10 with increasing nominal value of the concentrated load. The value of w to define the support widths of the symmetric triangular fuzzy variables is taken as 3.

Fig. 10 The possibility of failure with increasing nominal value of the concentrated load and different cov of uncertain variables

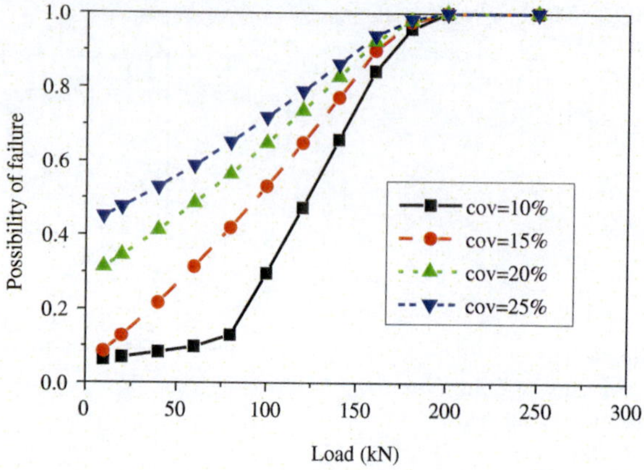

Fig. 11 The possibility of failure with increasing nominal value of the concentrated load and various support widths of the unsymmetrical triangular fuzzy variables

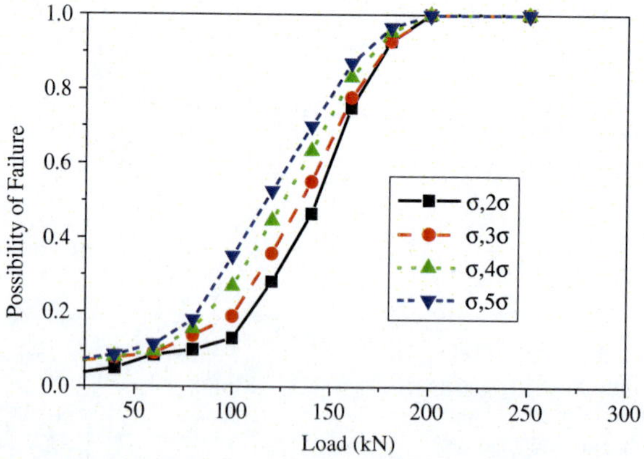

The possibility of failure is further computed considering unsymmetrical triangular variation of the fuzzy variables as previous and the results are shown in Fig. 11 for different support width configurations and increasing nominal value of the concentrated load.

5 Summary and Conclusion

The concept of establishing the equivalence between the fuzzy and the random variables is found to be potential to perform the safety evaluation either in the probabilistic or the possibilistic framework under hybrid uncertainties. Relying on such emerging transformation concepts, the safety analysis of structural system characterized by hybrid uncertain parameters is demonstrated in the present chapter. The transformation approaches satisfy the consistency condition that the possibility of an event should be greater than or equal to its probability.

The entropy-based transformation directly provides the equivalent normally distributed parameters, whereas in scaling approach there is an additional requirement that the probabilistic parameters need to be transformed to equivalent normal variables. The probability of failure obtained by the proposed transformation approaches is in conformity with the failure probability bounds derived based on the evidence theory. From the numerical study it is seen that the trend of failure probability obtained by both the scaling and the entropy-based approach is the same. However, there is a definite difference in the reliability index obtained based on the two transformation approaches. The bounds on failure probability based on the evidence theory are too far apart and too conservative which is obvious as the evidence theory gives conservative estimate of the upper bound and the lower bound of the *CDF*. These bounds are computed to serve as a check whether the transformation-based algorithm produces reliability results that are within the bounds obtained from a different approach following the consistency principle. The authors are of the opinion that the entropy-based transformation hinges on sound mathematical basis and the theory of expressing the uncertainty information is well established and applied in various fields of engineering. It is safe to apply entropy based transformation approach when no specific judgment is available regarding the distribution pattern of the uncertain variables. The conventional safety analysis algorithms normally assume all the variables to be of single type i.e., either all probabilistic or all possibilistic. As such this approach introduces gross assumptions at the very beginning during modeling of the system, whereas in the proposed approach there is always some chance to lose some of the information during transformation stage. But, the present approach offers more flexibility to the designer in realistic modeling of the system. However, it is not very clear in open literature, which is more justified and it is felt to need more study on this aspect.

Acknowledgments The first author gratefully acknowledges the financial support from the Alexander von Humboldt Foundation during the preparation of this manuscript. He is also grateful to Professor R. S. Langley, University of Cambridge, UK for introducing him to the research topic.

Appendix A: A Brief Note on the Second Moment Reliability Analysis

In the reliability analysis of structures by the FORM and SORM, typically a most probable point (MPP) is found on the failure surface in a standard normal space, $\boldsymbol{U}$. Any set of continuous basic random variables, $\boldsymbol{X}$, is transformed to $\boldsymbol{U}$-space using a one-to-one transformation i.e., $\boldsymbol{U} = T(\boldsymbol{X})$. The MPP, $\boldsymbol{u}^*$, lies on the failure surface, $g(\boldsymbol{X}) = G(\boldsymbol{U}) = 0$, and it is the closest point on the limit state surface to the origin in U-space. The MPP, $\boldsymbol{u}^*$, can be obtained by solving the optimization problem,

$$\text{minimize}, \beta = \sqrt{\mathbf{U}^{\mathrm{T}}\mathbf{U}}; \quad \text{such that}, G(\mathbf{U}) = \mathbf{0} \tag{A.1}$$

Various algorithms are available to solve the problem i.e., the Hasofer–Lind method, the Rackwitz–Fiessler algorithm using the Newton–Raphson root solving approach, the sequential quadratic programming, etc. Once the MPP is obtained, the p_f can be estimated using the FORM or SORM. The FORM is based on a linear approximation of the limit state built at the MPP in U-space and the p_f is obtained as, $p_f = \Phi(-\beta)$, where Φ is the standard Gaussian *CDF* and β is the reliability index defined by,

$$\beta = \frac{-\nabla_{\mathbf{u}^*}(G)}{\|\nabla_{\mathbf{u}^*}G\|}\mathbf{u}^* \tag{A.2}$$

In SORM, the second-order approximations are fitted to the limit state surface and the failure probability is computed using an asymptotic formula [45]. In this regard it is to be noted that the above format assumes that all the random variables involved are normally distributed. The non-normal variables can be transformed to equivalent Gaussian random variables by applying suitable transformation. For example, the Rackwitz–Fiessler method can be applied to obtain the equivalent mean and standard deviation of the transformed normal variables as below,

$$\sigma' = \frac{\varphi\{\Phi^{-1}[(\mathrm{F}_{\mathbf{X}}(\mathbf{x}^*))]\}}{\mathrm{f}_{\mathbf{X}}(\mathbf{x}^*)}, \quad \mu = \mathbf{x}^* - \Phi^{-1}[(\mathrm{F}_{\mathbf{X}}(\mathbf{x}^*))]\sigma' \tag{A.3}$$

where $F_{\mathbf{X}}$ and $f_{\mathbf{X}}$ are the *CDF* and *pdf* of the random variable, respectively.

Appendix B: A Brief Note on the Possibility Theory

The possibility is an alternate approach to the probability, initially used to model the uncertainties when the available information is linguistic. Zadeh [46] introduced the theory of fuzzy sets as a basis for possibility. It is based on the possibility distribution defined by the *mf* obtained from the numerical data along with the expert knowledge and experience. A fuzzy set is a set containing the elements that have varying degrees of membership in the set. The *mf* describes the grade of membership to the fuzzy set for each element in the domain, varying from zero to one. For a fuzzy set $\tilde{x}$, the *mf* is defined as $\mu_{\tilde{x}}(x)$ for all x that belongs to the domain X i.e.,

$$\tilde{x} = \{(x, \mu_{\tilde{x}}(x)) | x \in X, \mu_{\tilde{x}}(x) \in (0,1)\} \tag{B.1}$$

If, $\mu_{\tilde{x}}(x) > 0$, x is definitely a member of the subset $\tilde{x}$. If $\mu_{\tilde{x}}(x) = 0$, x is definitely not a member of the subset $\tilde{x}$.

A *mf* for a fuzzy variable F is described in Fig. 12. The core comprises those elements x of the universe such that $\mu_F(x) = 1$, the support comprises those elements $\boldsymbol{x}$ of the universe such that $\mu_F(\boldsymbol{x}) > 0$ and the boundaries comprise those elements $\boldsymbol{x}$ of the universe such that $1 > \mu_F(\boldsymbol{x}) > 0$.

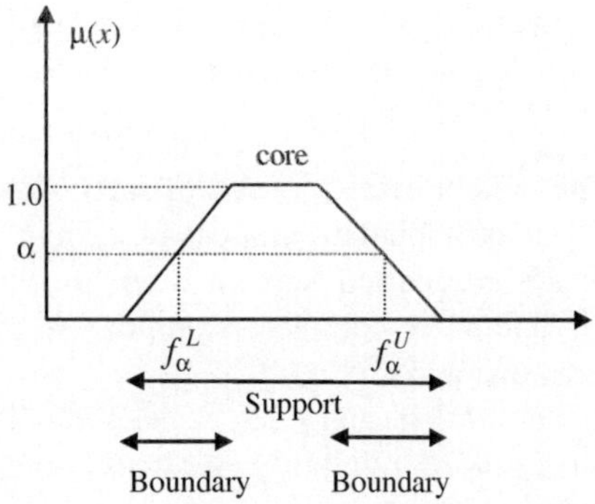

Fig. 12 The Fuzzy distribution and α-cut description

A convenient way to represent the fuzzy variables is the α-cut method of its possibility distribution as described by the *mf*. The α-cut subsets of F are defined as,

$$F_\alpha = \{x \in X, \mu_F(x) \geq \alpha\} \quad 0 \leq \alpha \leq 1. \tag{B.2}$$

For *i*th α-cut level, the upper bound and the lower bound f_α^L and f_α^U as depicted in Fig. 12 are given by,

$$f_\alpha^L = \min\{f : f \in F_\alpha\} \text{ and} f_\alpha^U = \min\{f : f \in F_\alpha\} \tag{B.3}$$

For a given fuzzy variable, if n sets of α-cuts are established for all the fuzzy quantities of the variable, an array of size 2 by n can be used to represent all of the quantities as the following:

$$F = \{(f^L, f^U)1, (f^L, f^U)2, \ldots, (f^L, f^U)n\} \tag{B.6}$$

The possibility is also defined as a special case of the plausibility measure, which is used in the evidence theory. According to this definition, if the body of evidence about a set of events are nested, then the plausibility of each event reduces to the possibility. On the other hand, if the body of the evidence consists of singletons, then the plausibility reduces to probability. The properties of the possibility, when it is defined as a special case of plausibility and the properties of Zadeh's possibility are almost same. Based on this interpretation of possibility, Dubious and Prade [44] has shown that a normalized possibility distribution $\pi(\omega)$ can be effectively represented by a fuzzy number Q from its *mf*, $\mu_Q(\omega)$ as,

$$\pi(\omega) = \mu_Q(\omega) \tag{B.7}$$

And it is shown that the fuzzy number can be used to select a class of probability measures P for which the possibility and the necessity represent the upper bound and the lower bound of the probability of each event.

Appendix C: A Brief Note on Possibilistic Reliability Evaluation

The possibilistic approach of safety evaluation used in the present study is based on the development proposed by Cremona and Gao [18]. The essential concept of which is reported here in order to outline the procedure of possibilistic safety evaluation. A complete examination of the theory of interests can be found in Cremona and Gao [18].

For a safety margin Z = g(X) as described by Eq. 2, the possibilistic approach for assessing reliability will try to evaluate the safety in terms of possibility of failure:

$$\prod_f = \prod(g(X) \leq 0). \tag{C.1}$$

The set of fuzzy intervals $\{\boldsymbol{X}\}$ is first transformed into a set of Gaussian fuzzy numbers $\{\boldsymbol{U}\}$ by applying τ-transform like the Rosenblatt transformation in classical reliability theory. The details of the forward and inverse of the transformation are provided in Cremona and Gao [18]. Once, the different values are transformed, the possibility of failure is directly expressed in terms of Gaussian fuzzy number:

$$\Pi_f = \Pi(g(\{X\})) = \Pi(g(\tau^{-1}(\{X\}))) = \Pi(g_U(U)) \tag{C.2}$$

The possibilistic reliability index and the possibilistic design point are then obtained as the solutions of a minimization problem expressed as follows:

$$\begin{aligned} &\lambda = \min\left(\|U\|_\infty\right) \\ &\text{according to } g_U(U) = 0 \quad |U_i| = \left|U_j\right|, \forall i,j. \end{aligned} \tag{C.3}$$

For failure possibility <1.0, the degree of failure possibility can be obtained by two approaches: (1) the possibilistic reliability index λ is first sought and the possibility of failure is determined subsequently, and (2) the possibility of failure is first sought, the possibilistic reliability index λ is then obtained. Both the approaches can be implemented by standard numerical optimization technique.

References

1. Möller B, Beer M, Graf W, Hoffman A (1999) Possibility theory based safety assessment. Comput Aided Civil Infrastruct Eng Special Issue on Fuzzy Model 14:81–91
2. Box GEP, Draper NR (1987) Empirical model building and response surface. Wiley, New York
3. Jin R, Chen W, Simpson T (2001) Comparative studies of metamodelling techniques under multiple modeling criteria. Struct Multidis Optim 23:1–13
4. Ditlevsen O, Madsen HO (1996) Structural reliability methods. Wiley, West Sussex

5. Melchers RE (1999) Structural reliability analysis and prediction. Wiley, West Sussex
6. Haldar A, Mahadevan S (2000) Reliability assessment using stochastic finite element analysis. Wiley, USA
7. BenHaim Y, Elishakoff I (1990) Convex models of uncertainty in applied mechanics. Elsevier Science, Amsterdam
8. BenHaim Y (1995) A non-probabilistic measure of reliability of linear systems based on expansion of convex models. Struct Saf 17:91–109
9. Penmetsa RC, Grandhi RV (2002) Efficient estimation of structural reliability for problems with uncertain intervals. Comput Struct 80:1103–1112
10. Qiu Z, Yang D, Elishakoff I (2008) Probabilistic interval reliability of structural systems. Int J Solids Struct 45:2850–2860
11. Karanki DR, Kushwaha HS, Verma AK, Ajit S (2009) Uncertainty analysis based on probability bounds (p-box) approach in probabilistic safety assessment. Risk Anal 29(5):662–675
12. Luo Y, Kang Z, Luo Z, Li A (2008) Continuum topology optimization with non-probabilistic reliability constraints based on multi-ellipsoid convex model. Struct Multidis Optim 37(2):107–119
13. Brown CB (1979) A fuzzy safety measure, entropy constructed probabilities. J Eng Mech ASCE 105(5):855–871
14. Shiraishi N, Furuta H (1983) Reliability analysis based on fuzzy probability. J Eng Mech ASCE 109(6):1445–1459
15. Yao J, Furuta H (1986) Probabilistic treatment of fuzzy events in civil engineering. Prob Eng Mech 1(1):58–61
16. Furuta H (1995) Reliability and optimization of structural systems. In: Rackwitz R, Augusti G, Borri A (eds) Proceedings of the VI IFIP WG 7.5 working Congress. Chapman and Hall, London
17. Yubin L, Qiao Z, Wang G (1997) Fuzzy random reliability of structures based on fuzzy random variables. Fuzzy Sets Syst 86:345–355
18. Cremona C, Gao Y (1997) The possibilistic theory: theoretical aspects and applications. Struct Saf 19(2):173–201
19. Möller B, Graf W, Beer M (2003) Safety assessment of structures in view of fuzzy randomness. Comput Struct 81:1567–1582
20. Bing Li, Zhu M, Xu K (2000) A practical method for fuzzy reliability analysis of mechanical structures. Reliab Eng Syst Safety 67:311–315
21. Jiang Q, Chen CH (2003) A numerical algorithm of fuzzy reliability. Reliab Eng Syst Safety 80:299–307
22. Möller B, Beer M (2005) Fuzzy randomness uncertainty in civil engineering and computational mechanics. Springer-Verlag, London
23. Langley RS (2000) Unified approach to probabilistic and possibilistic analysis of uncertain systems. J Eng Mech ASCE 126(11):1163–1172
24. Chakraborty S (2003) Safety assessment of structures under hybrid uncertainty, Technical Report, CUED/C-MECH/TR-86 (ISSN 0309-7420), Cambridge University
25. Sophie QC (2000) Comparing probabilistic and fuzzy set approaches for design in the presence of uncertainty, Ph.D. Thesis, Virginia Polytechnic Institute, State University
26. Nikolaidis E, Chen S, Cudney H, Raphael TH, Rosca RT (2004) Comparison of probability and possibility for design against catastrophic failure under uncertainty. J Mech Des ASME 126:386–394
27. Kam TY, Brown CB (1983) Updating parameters with fuzzy entropies. J Eng Mech ASCE 198(6):1334–1343
28. Haldar A, Reddy RK (1992) A random fuzzy analysis of existing structures. Fuzzy Sets Syst 48:201–210
29. Rahman MS, Khalid M, Zahaby El (1997) Probabilistic liquefaction risk analysis using fuzzy variables. Soil Dyn Earthq Eng 16:63–79
30. Zhenyu L, Chen Q (2002) A new approach to fuzzy finite element analysis. Comput Meth Appl Mech 191:5113–5118

31. Chakraborty S, Sam PC (2007) Probabilistic safety analysis of structures under hybrid uncertainty. Int J Numer Meth Eng 70(4):405–422
32. Marano GC, Quaranta G, Mezzina M (2008) Fuzzy time-dependent reliability analysis of RC beams subject to pitting corrosion. J Mater Civil Eng ASCE 20(9):578–587
33. Rao SS, Chen L, Mulkay E (1998) Unified finite element method for engineering systems with hybrid uncertainties. AIAA J 36(7):1291–1299
34. Ferrari P, Savoia M (1998) Fuzzy number theory to obtain conservative results with respect to probability. Comput Meth Appl Mech Eng 160:205–222
35. Savoia M (2002) Structural reliability analysis through fuzzy number approach, with application to stability. Comput Struct 80:1087–1102
36. Smith SA, Krisnamurthy T, Mason BH (2002) Optimized vertex method and hybrid reliability, 43rd AIAA/ASME/ASCE/AHS/ASC Structures, Structural Dynamics and Materials Conference, *AIAA*-2002-1465
37. Adduri PR, Penmetsa RC (2008) Confidence bounds on component reliability in the presence of mixed uncertain variables. Int J Mech Sci 50:481–489
38. Du X (2008) Unified uncertainty analysis by the first order reliability method. J Mech Design ASME 130:091401–091410
39. Luo Y, Zhan Kang, Alex Li(2009) Structural reliability assessment based on probability and convex set mixed model. Comput Struct 87:1408–1415
40. Du L, Choi KK (2008) An inverse analysis method for design optimization with both statistical and fuzzy uncertainties. Struct Multidis Optim 37:107–119
41. Huang H-Z, Zhang X (2009) Design optimization with discrete and continuous variables of aleatory and epistemic uncertainties. J Mech Design ASME 131:0310061–0310068
42. DeLuca A, Termini S (1970) A definition of non probabilistic entropy in the setting of fuzzy set theory. Inform Control 20:301–312
43. Puig B, Akian J (2004) Non-Gaussian simulation using Hermite polynomials expansion and maximum entropy principle. Prob Eng Mech 19:293–305
44. Dubois D, Prade H (1991) Random sets and fuzzy interval analysis. Fuzzy Sets Syst 42:87–101
45. Kiureghian AD, Lin HZ, Hwang SJ (1987) Second-order reliability approximations. J Eng Mech ASCE 113(8):1208–1225
46. Zadeh L (1978) Fuzzy sets as a basis for a theory of possibility. Fuzzy Sets Syst 1:3–28

Maintenance and Warranty Concepts

Minjae Park and Hoang Pham

1 Introduction

In general, a warranty is an obligation attached to products that require the manufacturer to provide compensation for customer (buyer) according to the warranty terms when the warranted products fail to perform their intended functions. A warranty is important to the manufacturer as well as the customer of any commercial product since it provides protection to both parties. As for the customer, a warranty provides a resource for dealing with items that fail due to the uncertainty of the product's performance and unreliable products. For the manufacturer, it provides protection since the warranty terms explicitly limit the responsibility of a manufacturer in terms of both time and type of product failure. Because of the role of the warranty, manufacturers have developed various types of warranty policies to grab the interest of the customers. However, manufacturers cannot extend the warranty period without limit and maximize warranty benefits because of the cost related to it. Many researchers have studied in the last several decades on various warranty modeling and policies along with its maintenance policies. This chapter focuses on the developments of warranty modeling with various maintenance policies as well as the methodologies with various aspects that can be used to derive the mathematical warranty modeling. The concepts of warranty and review of the overall information about the warranty policy such as warranty's role, concept and different types will be discussed. The basic mathematical maintenance concepts

H. Pham (✉)
Department of Industrial and Systems Engineering, Rutgers University,
96 Frelinghuysen Road, Piscataway, NJ 08854-8018, USA
e-mail: hopham@rci.rutgers.edu

M. Park
College of Business Administration, Hongik University, Sangsu-dong,
Mapogu, Seoul 121-791, Korea

H. Pham (ed.), *Safety and Risk Modeling and Its Applications*,
Springer Series in Reliability Engineering, DOI: 10.1007/978-0-85729-470-8_4,

including counting processes such as renewal process, quasi-renewal process, non-homogeneous Poisson process, compound and marked Poisson process and bivariate exponential distribution also will be provided.

As the market becomes competitive and diversified, it is hard for the manufacturers to differentiate its product to consumers with only quality and an eye catching design. Also with the massive information available to consumers regarding the product manufacturers need to find a better way to communicate with its customers to differentiate and to inform its product. In order to achieve this goal, many companies promote the warranty policies through maintenance aspects as an effective tool to attract consumers. This chapter discusses on the developments of various maintenance and warranty policies that can be used to derive the mathematical warranty modeling.

2 Concept and Role of the Warranty Policy

Warranty policy is a guarantee or an obligation to repair or replace a defective product or parts when the product does not perform its expected function during a given time period. This is a contract between the customer and the manufacturer upon the point when the policy is sold.

Warranty benefits both the consumer and the manufacturer as it is set to protect both parties. The consumer is protected as it guarantees a resource to deal with any defects or errors while using the product. Similarly, the manufacturer is protected because the warranty terms explicitly limit the responsibility in terms of both time and type of product failure. The warranty policy is an obligation attached to products that require the manufacturer to provide compensation for consumers according to the warranty terms when the warranted products fail to perform their intended functions [1].

As for a manufacturer, with the increase in demand for better quality warranty, it tries to develop an appealing policy and strategically use it as a promotional/marketing tool. Companies often emphasize on the benefits received under the policy such as details of the compensation for the defects, the charge or the period of the warranty. However, given that any service under the warranty policy is a potential cost item for a company, drafting a policy which is economically optimal so that it minimizes the cost but maximizes the satisfaction of the consumer is critical.

In summary, the warranty policy concept is to protect both the consumer and the manufacturer. The consumer is provided a resource for dealing with items that fail to function properly, i.e., unreliable products. Whereas the manufacturer is provided protection because warranty terms explicitly limit the responsibility in terms of both time and type of product failure. When products are getting more complicated, it would be difficult for customers to make a purchasing decision. So, the warranty policy would provide one of the role models for products' quality and reliability. And the longer warranty period costs more expenses for the sellers. When a manufacturer wants to provide better warranty condition than their other competitive sellers, they are supposed to provide better quality of products. Otherwise, they could not save their warranty cost. Such trade-offs would make the warranty policy be a strong marketing tool to increase the sales rate and to advertise the quality of products.

2.1 Warranty Policies

There are various characteristics which categorize the warranty policy separately. These characteristics include the number of warranty dimensions, the renewability of a warranty and the warranty compensation methods.

2.1.1 Dimensional Policies

First, consider the number of warranty dimensions. Most warranties in practice are one dimensional for which the warranty terms are based on product age or product usage, but not both. Compared to one-dimensional warranty, two-dimensional warranties are more complex since the warranty obligation depends on both product age and product usage as well as the potential interaction between them. Two-dimensional warranties are often seen in automobile industry. The Hyundai Motor Company is currently offering 10 years with 100,000 miles warranty on the power train for most of their new models. Several researchers [2, 4] have studied the warranty policy based on the automobile industry's data.

2.1.2 Renewing Warranty and Non-Renewing Warranty

One of the basic characteristics of warranties is whether they are renewable or not. For a regular renewable policy with warranty period, whenever a product fails in the warranty period, a customer is compensated according to the terms of the warranty contract and the warranty policy is renewed for another period. As a result, a warranty cycle starting from the point of sale, ending at the warranty expiration date, is a random variable whose value depends on the warranty period, the total number of failures under the warranty and the actual failure inter-arrival times. This topic is one of the future research topics. The majority of warranties in the market are non-renewable for which the warranty cycle, which is the same as the warranty period, is not random, but pre-determined since the warranty obligation will be terminated as soon as warranty period unit of time passes after the sale. These types of policies are also known as fixed period warranties.

2.1.3 Free Replacement Warranty, Pro-Rata Warranty and Combination Warranty

According to the methods of compensation specified in a warranty contract upon premature failures, there are three basic types of warranties: free replacement/

repair warranty (FRW), pro-rata warranty (PRW) and combination warranty (CMW). Under FRW, a failed item is replaced/repaired at no cost to the buyer if the failure occurs in the warranty period. On the other hand, under PRW, warranty services are not provided free of charge, but are provided at a pro-rated cost with the proration depending on the amount of usage or service time provided by the item prior to its failure [2]. Combination warranty contains both features of FRW and PRW, which often contains two warranty periods, a free replacement period followed by a pro-rata period. Full-service warranty, also known as preventive maintenance warranty, is a policy that may be offered for expensive deteriorating complex products such as automobiles. Under these type of policies, consumers not only receive free repairs upon premature failures, but also free preventive maintenance.

3 Warranty Cost Analysis

This session discusses about the researches on warranty policies and related topics that many researchers [2–130] have done in the literature by several different categorized groups.

3.1 One-Dimensional Warranty and Two-Dimensional Warranty

One-dimensional warranty is characterized by the warranty period, which is defined in terms of a single variable. Single variable could be time, age or usage. In the case of two-dimensional warranties, there are two dimensions to express warranty polices. One is representing time and the other representing item usage. As a result, many different types of warranties may be defined based on the characteristics of warranty policies [2]. And many researchers have studied the cost analysis based on two-dimensional warranty [12, 20–22, 31, 32, 39, 68, 71, 72]. Yun and Kang [68] examine new warranty servicing strategy, considering imperfect repair with a two-dimensional warranty. Baik et al. [72] study two-dimensional failure modeling for a system where degradation is due to age and usage with minimal repair. Most of the products have one of the two attributes with some exceptions, for example, a vehicle. Several researchers [2, 4] have studied the warranty policy based on the automobile industry's data. Compared to one-attribute warranties, two-attribute warranties are more complex [4–8]. Chun and Tang [23] propose several decision models that estimate the expected total cost incurred under various types of two-attribute warranty policies. Kim and Rao [41] consider two-attribute warranty policies for non-repairable items and the item failures are described in terms of a bivariate exponential distribution. Jiang and Ji [73] study a multiple attribute value model based on four attributes such as cost, availability, reliability and lifetime.

3.2 Renewing Warranty and Non-Renewing Warranty

Under a renewing warranty, the product which fails during its warranty period is replaced by a new one at a cost to the manufacturer or at a pro-rated cost to the user and the warranty is renewed. Under a non-renewing warranty, the manufacturer guarantees a satisfactory service only during the original warranty period. Renewable warranties are usually given to the non-repairable and inexpensive products such as home appliances and so on. Compared to the renewable warranties, the period of non-renewable warranties is relatively longer. So this might be one of the possible reasons why such policies are not as popular as non-renewable ones for warranty issuers [74]. Jung et al. [38] investigate the optimal replacement policies following the expiration of warranty such as renewing warranty and non-renewing warranty. Chukova and Hayakawa [16, 17] evaluate the warranty costs over the warranty period under non-renewing and renewing warranty policies over the life cycle of the product. Sahin and Polatoglu [54] prove that the cost rate function is pseudo-convex under a fixed-maintenance period policy under non-renewing and renewing warranty policies. Chen and Chien [10] investigate a model to study the effect of PM carried out by the buyer on items sold under a renewing FRW.

3.3 Warranty Period and Post-Warranty Period

During warranty period, as mentioned above, there are several kinds of warranty polices such as FRW, PRW or CMW. However, during post-warranty period, customers have to repair or replace the failure product at their own expense. Jung and Park [37] consider two types of warranty policies such as renewing warranty and non-renewing warranty with warranty period and post-warranty period. They derive the expressions for the expected maintenance costs for the periodic preventive maintenance during post-warranty period. Jung et al. [38] study the optimal replacement policies during post-warranty period considering the expected downtime per unit time and the expected cost rate per unit time. Jung [36] considers the optimal period for the periodic PM during the post warranty period which minimizes the expected long-run maintenance cost per unit time.

3.4 Warranty Reserve

Warranty reserve is one of the important factors which would be considered for the warranty policies. Therefore, several researchers [9, 33, 34, 51, 58, 63, 75] have considered the warranty reserve for the cost analysis. Patankar and Mitra [51] investigate the effect of warranty execution on the expected warranty reserves of a linear pro-rata rebate plan. Ja et al. [33, 34] consider a policy where warranty is not renewed on product failure within the warranty period but the product is minimally repaired by the manufacturer with the warranty reserves.

4 Reliability and Warranty

The relationship between warranty policies and products' reliability is very closely related. If the product's reliability is good, then the product's warranty could be extended. Otherwise, the product's warranty should be considered again. However, there are some exceptions. To increase a product's sales, some providers extend the product's warranty period. They use the warranty policy as a marketing tool. The reliability of product is determined by several important factors such as product's design, development, manufacturing stages and so on. It depends on the selection of suppliers and their cooperation in quality efforts as well. This implies that several important factors must take into account the interaction between warranty and reliability. A company either gives a warranty that is far shorter than the expected life of their item or increases the cost to a very high level to cover expected warranty costs. Therefore, a product's reliability is one of the important measures to investigate the warranty cost analysis [46]. On the other hand, Percy [76] presents some new ideas for improving a product's reliability by adopting the Bayesian methodology.

4.1 Maintenance Policies and Warranty

The maintenance objectives are to minimize the maintenance related operating costs, to maximize equipment availability and reliability or prolong equipment lifetime [73]. For deteriorating complex products, it is essential to perform preventive maintenance to achieve satisfactory reliability performance. Maintenance involves planned and unplanned actions carried out to retain a system at or restore it to an acceptable operating condition. Planned maintenance is usually referred to as preventive maintenance while unplanned maintenance is labeled as corrective maintenance or repair [1]. Two well-known preventive maintenance policies are block replacement policy and age replacement policy. Barlow and Hunter [77] suggested these two types of preventive maintenance. Since then, a lot of researches have been done regarding maintenance polices. Jhang and Sheu [78] derive the expected long-run cost per unit time for each policy. Sheu [79] considers a two-typed failures system which is subject to shocks that arrive by an NHPP with age and block replacement policy. Wang [80] summarized, classified and compared various existing maintenance policies for both single-unit and multi-unit systems. Also, Pham and Wang [81] summarize various treatment methods and optimal policies on the imperfect maintenance. Jung and Park [37] develop the optimal periodic preventive maintenance policies following the expiration of

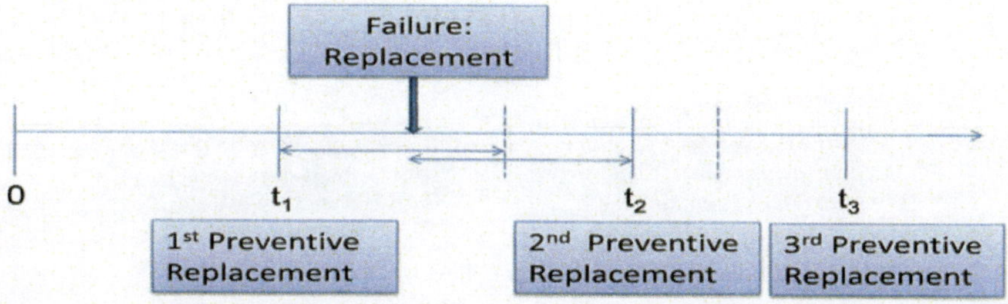

Fig. 1 Age replacement policies

warranty. Garbatov and Soares [82] plan the maintenance from an economic point of view so as to minimize maintenance costs but satisfying a minimum reliability level. Also, several researchers [11, 83–85] investigate the maintenance policies based on the Bayesian approach. Chen and Popova [11] propose two kinds of Bayesian maintenance polices. Additionally, a set of maintenance policies which consist of minimal repair and preventive maintenance is analyzed for the case of known and unknown failure parameters of the item's lifetime distribution. Sheu et al. [85] and Juang and Anderson [84] consider a Bayesian theoretic approach to determine an optimal adaptive preventive maintenance policy with minimal repair. A Bayesian approach is established to formally express and update the uncertain parameters for determining an optimal adaptive preventive maintenance policy. Stephens and Crowder [60] analyze the discrete time warranty data based on the MCMC model.

4.2 Age Replacement Polices

In the age replacement policy, a preventive replacement is performed after a given continuous operation time T without failure, and a failure replacement is performed if the system fails before T [73]. This model has been generalized by many researchers [14, 73, 78, 79, 86–91]. In Fig. 1, a product is replaced at a certain age t, or upon failure, whichever occurs first. And if the failure replacement happened then the next preventive replacement is rescheduled from the time of failure replacement. Sheu and Chien [89] consider a generalized age-replacement policy of a system subject to shocks, which arrived by NHPP, with random leadtime. Bai and Yun [86] proposed a generalized replacement policy based on the system age and the minimal repair which is similar to the age replacement policy. Kumar and Westberg [88] develop maintenance model under age replacement policy using proportional hazards model and TTT-plotting. Yeh et al. [91] investigate the effects of a renewing FRW on the age replacement policy for a non-repairable product and compare maintenance polices with warranty and without warranty.

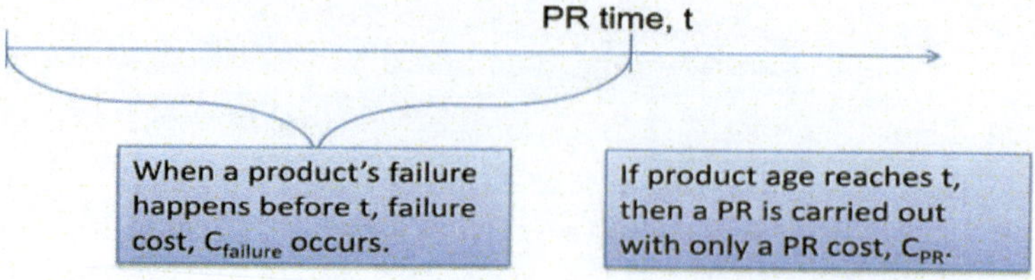

Fig. 2 Cost model based on the age replacement policy

Chien [14] investigates the effects of an imperfect renewing FRW on the age replacement policy for a product with an increasing failure rate. Sheu et al. [92] propose a generalized replacement policy where a system has two types of failures and is replaced at the minor failure or catastrophic failure or at age T, whichever occurs first.

4.2.1 Cost Models

Figure 2 presents the basic cost model based on the age replacement policy [77, 93]. Let PR be preventive replacement and C_{PR} and $C_{failure}$ stand for preventive replacement cost and failure cost, respectively. If a random variable x is a failure time, a cost coefficient is defined as

$$C(t) = \begin{cases} C_{failure} & \text{if} \quad x < t \\ C_{PR} & \text{if} \quad x \geq t \end{cases}$$

$E(T(t))$ is the expected duration and the expected cost rate is given by

$$\text{Expected cost rate} = \frac{E(C(t))}{E(T(t))} = \frac{C_{PR}R(t) + C_{failure}F(t)}{\int_0^t R(x)dx}$$

4.2.2 Availability Model

In a similar way as deriving the cost model [73, 94], the availability model based on the age replacement policy is given by

$$A(t) = \frac{\text{MTBF}}{\text{MTBF} + \text{MTTR}} = \frac{1}{1 + \frac{\text{MTTR}}{\text{MTBF}}}$$
$$= \frac{1}{1 + \frac{T_f F(t) + T_p R(t)}{\int_0^T R(t)\mathrm{d}t}},$$

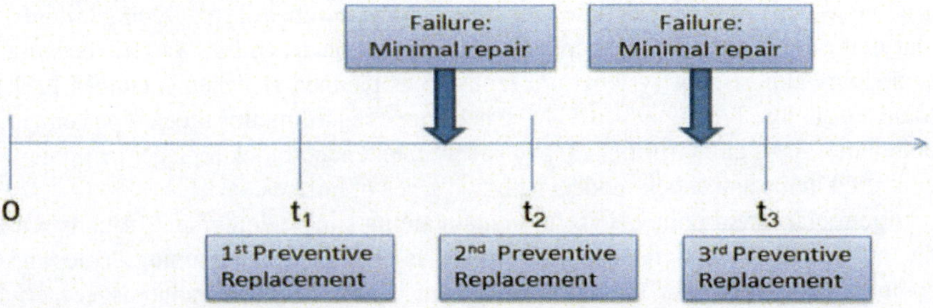

Fig. 3 Block replacement policies

where MTBF stands for a mean time between failure and MTTR stands for a mean time to replacement. T_f is a time of performing a failure replacement and T_p is a time of performing a preventive replacement.

4.2.3 Reliability Model

There are several reliability models [73]. One of them is explained here. The PR occurrence rate is just the number of PR over total replacement by time *t*. And higher occurrence rate is more reliable. The reliability model based on the age replacement policy is given by

$$\begin{aligned}\text{Occurrence rate for PR} &= \frac{\#\,\text{of PR by } T}{\#\,\text{of FR by } T + \#\,\text{of PR by } T}\\ &= \frac{\#\,\text{of PR by } T}{\#\,\text{of total replacement}}\end{aligned}$$

4.2.4 Lifetime Model

Lastly, we consider the lifetime model [94]. This is a well-known model based on the age replacement policy. The mean residual life function, $m(t)$, is given by

$$m(t) = E(X - t|X > t) = \frac{1}{R(t)}\int_t^{\infty} xf(x)\mathrm{d}x - t$$

4.3 Block Replacement Policies

In the block replacement policy, an operating system is replaced by a new one at times $kT, k = 1, 2, \ldots$ and at failures. In Fig. 3, they perform preventive maintenance after it has been operating at time t regardless of the number of intervening failures. One of the drawbacks of block replacement policy is that it is rather wasteful because sometimes almost-new systems are replaced at planned replacement times [95]. Many researches [78, 79, 95–101] have been done regarding this block replacement policy too. Sheu and Griffith [101] consider an extended block replacement policy with used items and shock models with two types of failures.

Age replacement policy is useful in maintaining simple equipments usually. On the other hand, block replacement policy is useful in maintaining large and complex equipment. For the age replacement policy, between maintenance periods, a failed component/system is replaced at the moment. However, in the block replacement policy, between maintenance periods, a failed component/system is repaired minimally.

4.3.1 Cost Model

In a similar way of cost model in age replacement policy [102, 103], let C_{PR} and C_{CR} stand for preventive replacement cost and corrective replacement cost, respectively. Consider a single component system. The system is replaced on failure and preventively at times T, $2T$,..., etc. Let $H(t)$ denote the mean number of replacements in the interval $(0, t)$ of a unit (system).

$E(T(t))$ is the expected duration and the expected cost rate is given by

$$\text{Expected cost rate} = \frac{C_{PR} + C_{CR}H(t)}{T}$$

4.3.2 Modified Cost Model 1

Park and Yoo [104] propose the modified block replacement policy where a block replacement is performed at failure k, counting after the pre-determined individual failure-replacement interval $(0, \tau]$. They called this policy as the block replacement policy based on idle count. C_d is downtime cost per unit. Additionally, $M(t)$ represents the mean number of failure replacements during $(0, \tau]$ and $R^{(i)}(\tau)$ is the time-to-failure i from τ for the fleet. C_dD is the mean downtime cost per unit. Let $G_\tau(t)$ be the cdf of the residual life at τ.

$$\text{Expected cost rate} = \frac{C_{PR} + C_{CR}M(\tau) + C_dD}{\tau + E\{R^{(k)}(\tau)\}},$$

$$\text{where } D = \sum_{j=1}^{k-1} \frac{j}{N} \binom{N}{j} \int_0^\infty [G_\tau(t)]^i [1 - G_\tau(t)]^{N-i} dt$$

4.3.3 Modified Cost Model 2

Nakagawa [105] proposes another modified block replacement policy with an idle period, units are replaced at failure until a fixed time T and then follows an idle period d, during which failed units are left idle. $I(d)$ is the mean downtime per unit during d.

$$\text{Expected cost rate} = \frac{C_{\mathrm{PR}} + C_{\mathrm{CR}} M(T) + C_d I(d)}{\tau + d}, \quad \text{where } I(d) = \int_0^d G_T(t)\mathrm{d}t$$

4.4 Maintenance Cost Analysis

Boland and Proschan [106] investigate a model for the minimal repair-periodic replacement policy and consider the problem of determining the period which minimizes the total expected cost of repair and replacement. Park et al. [107] consider the situation where each PM relieves stress temporarily and hence slows the rate of system degradation, while the hazard rate of the system remains monotonically increasing. Canfield [108] obtains the cost optimization of the PM intervention interval by determining the average cost-rate of system operation. Wang and Pham [109] investigate availability, maintenance cost and optimal maintenance polices of the series system with n constituting components under the general assumption that each component is subject to correlated failure and repair, imperfect repair, shut-off rule and arbitrary distributions of times to failure and repair.

4.5 Maintenance Policies and Warranty

Maintenance polices and warranties are very interesting topics to study together. Several researchers [15, 36, 42, 44, 50, 67, 110, 111] have studied them at the same time. Monga and Zuo [44] present a study on reliability-based design of a series–parallel system considering burn-in, warranty and maintenance and they use genetic algorithms to obtain optimal values of system design, burn-in period, preventive maintenance intervals and replacement time. Lin et al. [42] present a cost minimization model for an optimal design of a mixed series–parallel system with deteriorating components. Pascual and Ortega [50] develop a model to help a maintenance decision-making situation of a given equipment. It means that it determines optimal life-cycle duration and intervals between overhauls by minimizing global maintenance costs. Yeh and Lo [67] investigate preventive maintenance warranty polices for repairable products. They determine the optimal number of preventive maintenance actions and the maintenance schedule when the length of a warranty period is pre-specified. Djamaludin et al. [110] develop a framework to study preventive maintenance actions when items are sold under warranty and review the models that have appeared in the literature.

5 Other Topics

In order to set up the warranty policy, a policy maker should have some information about a product's failure. For example, there are past failure data, experimental data regarding the product's failure, intuition of the product's failure. The Bayesian decision approach is a way to incorporate this information into the decision-making process [11]. Jung and Han [35] determine an optimal replacement policy for a repairable system with warranty period based on the Bayesian approach in case of renewing FRW and renewing PRW. Huang and Zhuo [30] proposed a Bayesian decision model for determining the optimal warranty policy for repairable products. Fang and Huang [28] present an approach along with Bayesian process to tackle a complex decision problem and based on that approach, the optimal prior and posterior decisions of pricing scheme, production plan and warranty policy can be determined simultaneously. Gutierrez-Pulido et al. [29] provide an approach for the determination of warranty length that takes into account the following aspects: choice of a good estimate of the failure-time model of the product and the use of a utility function that incorporates different considerations of costs, marketing and quality. Chukova et al. [19] design a procedure for estimating the degree of repair as well as other modeling parameters by Markov chain Monte Carlo (MCMC) methods.

5.1 Burn-in Process and Warranty

The burn-in process is a part of the production process whereby manufactured products are operated for a short period of time before release [112]. Burn-in is used to improve product quality pre-sale. Particularly for products with an initially high failure rate sold under warranty, burn-in can be used to reduce the warranty cost [56]. Several researchers [13, 56, 57, 59, 64, 69, 70, 112–116] have investigated the warranty policy using the burn-in process. Wu et al. [64] develop a cost model to determine the optimal burn-in time and warranty length for non-repairable products under the fully renewing FRW and PRW policy. In Chang's paper [113], the optimal burn-in decision has to take both the critical time and its post-burn-in mean residual life into consideration for improving reliability due to the features of unimodal failure rate function and its upside down unimodal mean residual life. Rangan and Khajoui [117] construct a new stochastic model which treats burn-in, warranty and maintenance strategies together in order to define coordinated strategies for system design and management. Wu and Clements-

Croome [116] consider a product with a long time dormant period and investigate two burn-in policies, which incur different burn-in costs and different burn-in effects on the products. Sheu and Chien [56] consider a general repairable product sold under warranty and determine the burn-in time required before the product is put on sale. Burn-in time is optimized to minimize the expected total cost under various warranty policies. In Yun et al.'s papers [69, 70], optimal burn-in time to minimize the total mean cost, which is the sum of manufacturing cost with burn-in and cumulative warranty cost, is studied under cumulative FRW and PRW.

5.2 *Software Reliability and Warranty*

On the other hand, based on various software systems, many researchers [25, 27, 52, 53, 55, 62, 118–125] have investigated and studied the warranty policy considering several factors such as maintenance and upgrade of software models. Using software reliability, Pham and Zhang [121] develop cost models with warranty cost, time to remove each error detected in the software system and risk cost due to software failure. Sahin and Zahedi [53, 55] present a framework and develop a Markov decision model to analyze warranty, maintenance and upgrade decisions for software packages under different market conditions. Voas [125] presents several methodologies according to the specific needs of the organization requesting assurances about the software's integrity and the peculiarities of that type of software. Williams [62] suggests an approach to calculating the delivery cost of a software product when warranty is to be provided with an imperfect debugging phenomenon.

5.3 *Bayesian Approach and Warranty*

The Bayesian decision method is another approach for the warranty analysis. In this section, we investigate many papers [11, 19, 28–30, 35, 43, 45, 60, 61, 76, 112, 126–128] which cover the warranty policy and the maintenance policy based on the Bayesian decision method.

6 Mathematical Approach

In this subsection, we investigate several backgrounds to study warranty analysis mathematically. Several processes have been considered to stand for failure intervals. Among them, two types of stochastic processes, renewal processes and non-homogeneous Poisson processes [4, 79, 129–131], are very useful for warranty cost modeling. We study renewal process [132–135], quasi-renewal process [136, 137] and its extensions. When Poisson process parameter λ is constant, it is Poisson process. However, when the parameter is not constant, it is non-homogeneous Poisson process. And there are two more applications such as combined Poisson process and marked Poisson process [134].

6.1 Renewal Processes

Consider a counting process for which the times between successive events are independent and identically distributed with an arbitrary distribution. Such a counting process is called a renewal process [138, 139]. Let $\{N(t), t \geq 0\}$ be a counting process and let X_n denote the time between the $(n - 1)$st and the nth events of this process, $n \geq 1$. If the sequence of non-negative r.v. $\{X_1, X_2,\ldots\}$ is independent and identically distributed, then the counting process $\{N(t), t \geq 0\}$ is said to be a renewal process.

6.2 Quasi-Renewal Processes

Let X_n be the inter-occurrence time between the $(n - 1)$th and nth events of the process. Let $f_i(x)$, $F_i(x)$ and $h_i(x)$ be the pdf, cdf, and failure rate of random variable X_i, respectively. We say $\{N(t), t > 0\}$ is a *quasi-renewal process* (QRP) [140, 141] associated with the distribution F and the parameter α, $\alpha > 0$ a constant, if $X_n = \alpha^{n-1} \cdot Z_n$, $n = 1, 2,\ldots$ where Z_ns are iid and $Z_n \sim F$, where $\{N(t), t > 0\}$ is a counting process. The pdf, cdf and failure rate, respectively, for $n = 2, 3, 4,\ldots$ are given by

$$f_n(x) = \frac{1}{\alpha^{n-1}} f_1\left(\frac{1}{\alpha^{n-1}}x\right)$$

$$F_n(x) = F_1\left(\frac{1}{\alpha^{n-1}}x\right)$$

$$h_n(x) = \frac{1}{\alpha^{n-1}} h_1\left(\frac{1}{\alpha^{n-1}}x\right)$$

QRP is described in Fig. 4. W stands for warranty period and X_i is the ith inter-failure interval. Then $X_2 = \alpha \cdot X_1$ and $X_3 = \alpha^2 \cdot X_1$. Eventually, X_m is equal to $\alpha^{m-1} \cdot X_1$

6.3 Extensions of Quasi-Renewal processes

Bai and Pham [6] suggested two extensions of QRP such as truncated quasi-renewal process and censored quasi-renewal process. They omit a certain range of possible values for the truncated QRP. After rescaling of the pmf makes it possible

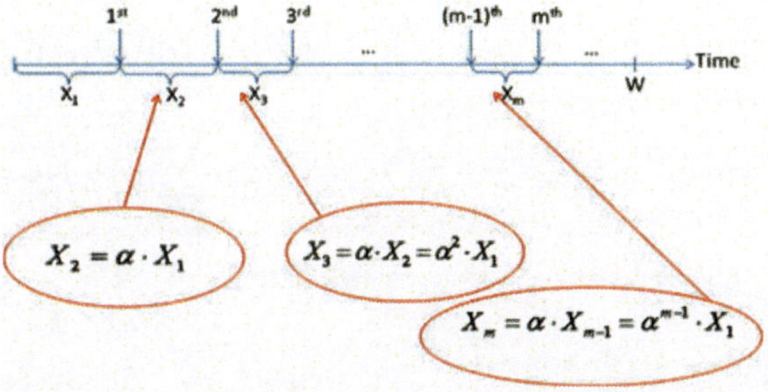

Fig. 4 Quasi-renewal process

to satisfy the necessary condition of distribution which summation of probability is equal to 1. The truncated QRP above m means that for a given t, $N(t)$ can only take values of 0,1,..., m. For such $N(t)$, pmf is given by

$$P\{N(t) = i\} = \frac{G^{(i)}(t) - G^{(i+1)}(t)}{1 - G^{(m+1)}(t)}, \quad i = 0, 1, \ldots, m$$

where $G^{(i)}(t)$ is the convolution of the inter-occurrence times $X_1, X_1, \ldots, X_i$ and $G^{(0)}(t) = 1$. So truncated QRP's first moment and second moment are obtained by

$$\begin{aligned} E(N(t)) &= \sum_{i=0}^{m} i \cdot P^T\{N(t) = i\} \\ &= \sum_{i=0}^{m} i\left(\frac{G^{(i)}(t) - G^{(i+1)}(t)}{1 - G^{(m+1)}(t)}\right) \\ &= \frac{\sum_{i=1}^{m} G^{(i)}(t) - mG^{(m+1)}(t)}{1 - G^{(m+1)}(t)} \end{aligned}$$

and

$$\begin{aligned} E\left(N^2(t)\right) &= \sum_{i=0}^{m} i^2 \cdot P^T\{N(t) = i\} \\ &= \sum_{i=0}^{m} i^2\left(\frac{G^{(i)}(t) - G^{(i+1)}(t)}{1 - G^{(m+1)}(t)}\right) \\ &= \frac{\sum_{i=1}^{m+1} (2i - 1)G^{(i)}(t) - m^2 G^{(m+1)}(t)}{1 - G^{(m+1)}(t)} \end{aligned}$$

If we use these two moments, we obtain the variance of $N(t)$. There is another extension of QRP, censored QRP. It is similar to truncated QRP but it is slightly different. If there are above a certain number of failures, they would be transformed into the last number of failures. It means that any observed failure above a certain number, m, is transformed into a single value m. Censored QRP's first moment and second moment are given by

$$\begin{aligned} E(N(t)) &= \sum_{i=0}^{m-1} i\left(G^{(i)}(t) - G^{(i+1)}(t)\right) + mG^{(m)}(t) \\ &= \sum_{i=1}^{m-1} iG^{(i)}(t) - \sum_{j=2}^{m} (j-1)G^{(j)}(t) + mG^{(m)}(t) \\ &= \sum_{i=1}^{m} G^{(i)}(t) \end{aligned}$$

and

$$\begin{aligned} E\left(N^2(t)\right) &= \sum_{i=0}^{m-1} i^2\left(G^{(i)}(t) - G^{(i+1)}(t)\right) + m^2G^{(m)}(t) \\ &= \sum_{i=1}^{m-1} i^2G^{(i)}(t) - \sum_{j=2}^{m} (j-1)^2G^{(j)}(t) + m^2G^{(m)}(t) \\ &= \sum_{i=1}^{m} (2i-1)G^{(i)}(t) \end{aligned}$$

6.4 Non-Homogeneous Poisson Processes

$\{N(t),\ t \geq 0\}$ is said to be a non-homogeneous Poisson process [4, 9, 119, 129] with intensity function $\lambda(t)$ if it satisfies

$N(T) = 0$
$\{N(t),\ t \geq 0\}$ has independent increments
Pr (exactly 1 event in $(t, t + h)) = \lambda(t)h + o(h)$
Pr (more than 1 event in $(t, t + h)) = o(h)$
Then $\Pr(N(t) = n) = e^{-m(t)}\frac{m(t)^n}{n!}, n \geq 0$ where $m(t) = \int_0^t \lambda(s)ds$

$N(t)$ has a Poisson distribution with mean $m(t)$ which is the mean value function of the process.

6.5 Compound Poisson Processes and Marked Poisson Processes

Both compound Poisson and marked Poisson processes [134] appear often as models of physical phenomena. Given a Poisson process $N(t)$ of rate $\lambda > 0$, suppose that each event has associated with it a random variable, possibly representing a value, an interval. The successive values X_1, X_2, X_3,... are assumed to be independent random variables. Then, a compound Poisson process is the cumulative value process defined by

$$Z(t) = \sum_{k=1}^{N(t)} X_k \quad \text{for } t \geq 0$$

If $\lambda > 0$ is the rate for the process $N(t)$ and $\mu = E(X_1)$ and $\sigma^2 = \mathrm{Var}(X_1)$ are the common mean and variance for $X_1, X_2, X_3, \ldots,$ then the moments of $Z(t)$ can be determined as follows:

$$E(N(t)) = \lambda\mu t; \quad \mathrm{Var}(N(t)) = \lambda(\sigma^2 + \mu^2)t$$

A marked Poisson process is the sequence of pairs $(W_1, X_1), (W_2, X_2), \ldots,$ where $W_1, W_2, \ldots$ are the waiting times or event times in the Poisson process $N(t)$.

Theorem 2.1 [134] Let $(W_1, X_1), (W_2, X_2), \ldots$ be a marked Poisson process where $X_1, X_2, X_3, \ldots$ are the waiting times in a Poisson process of rate λ and $X_1, X_2, X_3, \ldots$ are independent identically distributed continuous random variables having probability density function $f(x)$. Then $(W_1, X_1), (W_2, X_2), \ldots$ form a two-dimensional non-homogeneous Poisson point process where the mean number of points in a region A is given by

$$\mu(A) = \iint_A \lambda f(x)\mathrm{d}x\mathrm{d}t$$

6.6 Bivariate Exponential Distribution

The bivariate distributions have been investigated by many researchers [22, 23, 62, 65, 103, 110–113] for the reliability applications. Specifically, some researchers have studied the cases of bivariate gamma distributions [110, 112], bivariate exponential distributions [22, 23] and bivariate logistic distributions [28, 62].

Among various bivariate distributions, bivariate exponential distributions (BED) are one of the most common distributions applied in reliability engineering. The BEDs have also attracted many practical applications in reliability problems. However, unfortunately, there is no clear and explicit form for the BED unlike bivariate normal distributions. Therefore, a lot of researchers [22, 23, 52, 61, 65, 97, 113] have tried to develop various types of BEDs.

References

1. Wang H, Pham H (2006) Reliability and optimal maintenance. Springer, London
2. Blischke W (1994) Warranty cost analysis. CRC Press, USA
3. Blischke W, Murthy D (1996) Product warranty handbook. CRC Press, USA
4. Majeske K (2007) A non-homogeneous Poisson process predictive model for automobile warranty claims. Reliab Eng Syst Saf 92:243–251
5. Bai J, Pham H (2004) Discounted warranty cost of minimally repaired series systems. IEEE Trans Reliab 53:37–42
6. Bai J, Pham H (2005) Repair-limit risk-free warranty policies with imperfect repair. IEEE Trans Syst Man Cybern Part A 35:765–772

7. Balachandran K, Maschmeyer R, Livingstone J (1981) Product warranty period: a Markovian approach to estimation and analysis of repair and replacement costs. Acc Rev 1:115–124
8. Balcer Y, Sahin I (1986) Replacement costs under warranty: cost moments and time variability. Oper Res 34:554–559
9. Buczkowski P, Kulkarni V (2006) Funding a warranty reserve with contributions after each sale. Prob Eng Inform Sci 20:497–515
10. Chen J, YH C (2007) Renewing warranty and preventive maintenance for products with failure penalty post-warranty. Qual Reliab Eng Int 23:107–121
11. Chen T, Popova E (2000) Bayesian maintenance policies during a warranty period. Stoch Models 16:121–142
12. Chen T, Popova E (2002) Maintenance policies with two-dimensional warranty. Reliab Eng Syst Saf 77:61–69
13. Chien Y (2005) Optimal burn-in time for general repairable products sold under failure-free renewing warranty. Int J Qual Reliab Manag 22:651–666
14. Chien Y (2008) Optimal age-replacement policy under an imperfect renewing free-replacement warranty. IEEE Trans Reliab 57:125–133
15. Chukova S, Arnold R, Wang D (2004) Warranty analysis: an approach to modeling imperfect repairs. Int J Prod Econ 89:57–68
16. Chukova S, Hayakawa Y (2004) Warranty cost analysis: non-renewing warranty with repair time. Appl Stoch Models Bus Ind 20:59–72
17. Chukova S, Hayakawa Y (2004) Warranty cost analysis: renewing warranty with non-zero repair time. Int J Reliab Qual Safety Eng 11:93–112
18. Chukova S, Hayakawa Y (2005) Warranty cost analysis: quasi-renewal inter-repair times. Int J Qual Reliab Manag 22:687
19. Chukova S, Hayakawa Y, Arnold R (2007) Warranty analysis: estimation of the degree of imperfect repair via a Bayesian approach. Recent Adv Stoch Oper Res, pages 3–22, World Scientific
20. Chukova S, Hayakawa Y, Johnston M (2006) Two-dimensional warranty: minimal/complete repair strategy
21. Chukova S, Hayakawa Y, Johnston M (2007) Optimal two-dimensional warranty repair strategy. Proc Inst Mech Eng Part O J Risk Reliab 221:265–273
22. Chukova S, Johnston M (2006) Two-dimensional warranty repair strategy based on minimal and complete repairs. Math Comput Model 44:1133–1143
23. Chun Y, Tang K (1999) Cost analysis of two-attribute warranty policies based on the product usage rate. IEEE Trans Eng Manag 46:201–209
24. Djamaludin I, Murthy D, Kim C (2001) Warranty and preventive maintenance. Int J Reliab Saf Eng 8:89–108
25. Dohi T, Okamura H, Kaio N, Osaki S (2001) Age-dependent optimal warranty policy and its application to software maintenance contract. Front Sci Ser 4:2547–2552
26. Duchesne T, Marri F (2009) General distributional properties of discounted warranty costs under minimal repair and risk adjusted warranty costs. IEEE Trans Reliab 58:143–151
27. Evol J, Pract R (2001) Policy analysis for warranty, maintenance, and upgrade of software systems. J Softw Maint Evol Res Pract 13:469–493
28. Fang C, Huang Y (2008) A Bayesian decision analysis in determining the optimal policy for pricing, production, and warranty of repairable products. Expert Syst Appl 35:1858–1872
29. Gutierrez-Pulido H, Aguirre-Torres V, Christen J (2006) A Bayesian approach for the determination of warranty length. J Qual Technol 38:180
30. Huang Y, Zhuo Y (2004) Estimation of future breakdowns to determine optimal warranty policies for products with deterioration. Reliab Eng Syst Saf 84:163–168
31. Iskandar B, Murthy D (2003) Repair-replace strategies for two-dimensional warranty policies. Math Comput Model 38:1233–1241
32. Iskandar B, Murthy D, Jack N (2005) A new repair-replace strategy for items sold with a two-dimensional warranty. Comput Oper Res 32:669–682

33. Ja S, Kulkarni V, Mitra A, Patankar J (2002) Warranty reserves for non-stationary sales processes. Naval Res Logist 49:499–513
34. Ja S, Kulkarni V, Mitra A, Patankar J, Inc A, Southlake T (2001) A nonrenewable minimal-repair warranty policy with time-dependent costs. IEEE Trans Reliab 50:346–352
35. Jung G, Han S (2002) A Bayesian approach to optimal replacement policy for a repairable system with warranty period. Korean Commun Stat 9:21–31
36. Jung G, Lee C, Park D (2000) Periodic preventive maintenance policies following the expiration of warranty. Asia Pac J Oper Res 17:17–26
37. Jung G, Park D (2003) Optimal maintenance policies during the post-warranty period. Reliab Eng Syst Saf 82:173–185
38. Jung K, Han S, Park D (2008) Optimization of cost and downtime for replacement model following the expiration of warranty. Reliab Eng Syst Saf 93:995–1003
39. Jung M, Bai D (2007) Analysis of field data under two-dimensional warranty. Reliab Eng Syst Saf 92:135–143
40. Kim C, Djamaludin I, Murthy D (2004) Warranty and discrete preventive maintenance. Reliab Eng Syst Saf 84:301–309
41. Kim H, Rao B (2000) Expected warranty cost of two-attribute free-replacement warranties based on a bivariate exponential distribution. Comput Ind Eng 38:425–434
42. Lin D, Zuo M, Yam R, Meng M (2000) Optimal system design considering warranty, periodic preventive maintenance, and minimal repair. J Oper Res Soc 51
43. Lu Y (2005) Bayesian sampling plan for Weibull lifetime data under warranty policy with limited size on test equipment. Tamkang University, Department of Statistics
44. Monga A, Zuo M (1998) Optimal system design considering maintenance and warranty. Comput Oper Res 25:691–705
45. Moskowitz H, Chun Y, University P, Krannert E (1988) Graduate School of Management Institute for Research in the Behavioral, and M. Sciences, A Bayesian Model for the Two-attribute Warranty Policy: Institute for Research in the Behavioral, Economic, and Management Sciences, Krannert Graduate School of Management, Purdue University
46. Murthy D (2006) Product warranty and reliability. Ann Oper Res 143:133–146
47. Murthy D, Djamaludin I (2002) New product warranty: a literature review. Int J Prod Econ 79:231–260
48. Park M, Pham H (2009) Warranty system-cost analysis using quasi-renewal processes. Opsearch J Oper Res Soc India 45:263–274
49. Park M, Pham H (2008) Renewable warranty models using quasi-renewal processes. 14th ISSAT international conference on reliability and quality in design, Aug 2008
50. Pascual R, Ortega J (2006) Optimal replacement and overhaul decisions with imperfect maintenance and warranty contracts. Reliab Eng Syst Saf 91:241–248
51. Patankar J, Mitra A (1995) Effects of warranty execution on warranty reserve costs. Manag Sci 41:395–400
52. Rinsaka K, Dohi T (2005) Determining the optimal software warranty period under various operational circumstances. Int J Qual Reliab Manag 22:715
53. Sahin I (2001) Policy analysis for warranty, maintenance, and upgrade of software systems. J Softw Maint Res Pract 13:469–493
54. Sahin I, Polatoglu H (1996) Maintenance strategies following the expiration of warranty. IEEE Trans Reliab 45:220–228
55. Sahin I, Zahedi F (2001) Control limit policies for warranty, maintenance and upgrade of software systems. IIE Trans 33:729–745
56. Sheu S, Chien Y (2005) Optimal burn-in time to minimize the cost for general repairable products sold under warranty. Eur J Oper Res 163:445–461
57. Sheu S, Lin C (2005) Optimal burn-in time to minimize the cost for repairable assembly products under warranty. Int J Pure Appl Math 22:367
58. Singpurwalla N, Wilson S (1993) The warranty problem: its statistical and game theoretic aspects. SIAM Rev 35:17–42

59. Song SI, Cho YC, Park HK (2001) A study on optimal cost model of combined ESS and burn-in under warranty policy. J Soc Korea Ind Syst Eng 24:1–10
60. Stephens D, Crowder M (2004) Bayesian analysis of discrete time warranty data. J Roy Stat Soc Ser C 53:195–217
61. Wang G (1989) A Bayesian theoretical approach to the evaluation of renewal functions with application to warranty analysis. University of Southern California
62. Williams D (2007) Study of the warranty cost model for software reliability with an imperfect debugging phenomenon. Turk J Elec Eng 15:200
63. Wortman M, Elkins D (2005) Stochastic modeling for computational warranty analysis. Naval Res Logist 52:224–231
64. Wu C, Chou C, Huang C (2007) Optimal burn-in time and warranty length under fully renewing combination free replacement and pro-rata warranty. Reliab Eng Syst Saf 92:914–920
65. Wu S, Li H (2007) Warranty cost analysis for products with a dormant state. Eur J Oper Res 182:1285–1293
66. Yeh R, Chen G, Chen M (2005) Optimal age-replacement policy for nonrepairable products under renewing free-replacement warranty. IEEE Trans Reliab 54:92–97
67. Yeh R, Lo H (2001) Optimal preventive-maintenance warranty policy for repairable products. Eur J Oper Res 134:59–69
68. Yun W, Kang K (2007) Imperfect repair policies under two-dimensional warranty. Proc Inst Mech Eng Part O J Risk Reliab 221:239–247
69. Yun W, Lee Y, Chung I, Ferreira L (2001) Optimal burn-in time under cumulative pro-rata replacement warranty. Int J Reliab Appl 2:241–252
70. Yun W, Lee Y, Ferreira L (2002) Optimal burn-in time under cumulative free replacement warranty. Reliab Eng Syst Saf 78:93–100
71. Manna D, Pal S, Sinha S (2008) A note on calculating cost of two-dimensional warranty policy. Comput Ind Eng 54:1071–1077
72. Baik J, Murthy D, Jack N (2004) Two-dimensional failure modeling with minimal repair. Naval Res Logist 51:345–362
73. Jiang R, Ji P (2002) Age replacement policy: a multi-attribute value model. Reliab Eng Syst Saf 76:311–318
74. Bai J (2004) On the study of warranties for repairable complex systems. In: Department of Industrial and System Engineering, Ph.D. Dissertation, Rutgers University
75. Manna D, Pal S, Sinha S (2006) Optimal determination of warranty region for 2D policy: a customers' perspective. Comput Ind Eng 50:161–174
76. Percy D (2002) Bayesian enhanced strategic decision making for reliability. Eur J Oper Res 139:133–145
77. Barlow R, Hunter L (1960) Optimum preventive maintenance policies. Oper Res 8:90–100
78. Jhang J, Sheu S (2000) Optimal age and block replacement policies for a multi-component system with failure interaction. Int J Syst Sci 31:593–603
79. Sheu S (1998) A generalized age and block replacement of a system subject to shocks. Eur J Oper Res 108:345–362
80. Wang H (2002) A survey of maintenance policies of deteriorating systems. Eur J Oper Res 139:469–489
81. Pham H, Wang H (1996) Imperfect maintenance. Eur J Oper Res 94:425–438
82. Garbatov Y, Guedes Soares C (2001) Cost and reliability based strategies for fatigue maintenance planning of floating structures. Reliab Eng Syst Saf 73:293–301
83. van Noortwijk J (2000) Optimal maintenance decisions on the basis of uncertain failure probabilities. J Qual Maint Eng 6:113–122
84. Juang M, Anderson G (2004) A Bayesian method on adaptive preventive maintenance problem. Eur J Oper Res 155:455–473
85. Sheu S, Yeh R, Lin Y, Juang M (2001) A Bayesian approach to an adaptive preventive maintenance model. Reliab Eng Syst Saf 71:33–44

86. Bai D, Yun W (1986) An age replacement policy with minimal repair cost limit. IEEE Trans Reliab 35:452–454
87. Chien Y, Sheu S (2006) Extended optimal age-replacement policy with minimal repair of a system subject to shocks. Eur J Oper Res 174:169–181
88. Kumar D, Westberg U (1997) Maintenance scheduling under age replacement policy using proportional hazards model and TTT-plotting. Eur J Oper Res 99:507–515
89. Sheu S, Chien Y (2004) Optimal age-replacement policy of a system subject to shocks with random lead-time. Eur J Oper Res 159:132–144
90. Sheu S, Griffith W (2001) Optimal age-replacement policy with age-dependent minimal-repair and random-lead time. IEEE Trans Reliab 50:302–309
91. Yeh R, Chen G, Chen M (2005) Optimal age-replacement policy for nonrepairable products under renewing free-replacement warranty. IEEE Trans Reliab 54:92–97
92. Sheu S, Griffith W, Nakagawa T (1995) Extended optimal replacement model with random minimal repair costs. Eur J Oper Res 85:636–649
93. Pham H, Wang H (2000) Optimal (t, T) opportunistic maintenance of a k-out-of-n: G system with imperfect PM and partial failure. Naval Res Logist 47:223–239
94. Elsayed E (1996) Reliability engineering. Addison Wesley Longman, England
95. Sheu S (1997) Extended block replacement policy of a system subject to shocks. IEEE Trans Reliab 46:375–382
96. Acharya D, Nagabhushanam G, Alam S (1986) Jointly optimal block-replacement and spare provisioning policy. IEEE Trans Reliab 35:447–451
97. Archibald Y, Dekker R (1996) Modified block-replacement for multiple-component systems. IEEE Trans Reliab 45:75–83
98. Berg M, Cleroux R (1982) The block replacement problem with minimal repair and random repair costs. J Stat Comput Simul 15:1–7
99. Sheu S (1991) A generalized block replacement policy with minimal repair and general random repair costs for a multi-unit system. J Oper Res Soc 42:331–341
100. Sheu S (1996) A modified block replacement policy with two variables and general random minimal repair cost. J Appl Prob 33:557–572
101. Sheu S, Griffith W (2002) Extended block replacement policy with shock models and used items. Eur J Oper Res 140:50–60
102. Acharya D, Nagabhushanam G, Alam S (1986) Jointly optimal block-replacement and spare provisioning policy. IEEE Trans Reliab 35:447–451
103. Barlow R, Proschan F (1965) Mathematical theory of reliability. SIAM, Wiley & Sons
104. Park K, Yoo Y (1993) (t, k) block replacement policy with idle count. IEEE Trans Reliab 42:561–565
105. Nakagawa T (1982) Modified block replacement with two variables. IEEE Trans Reliab 31:398–400
106. Boland P, Proschan F (1982) Periodic replacement with increasing minimal repair costs at failure. Oper Res 30:1183–1189
107. Park D, Jung G, Yum J (2000) Cost minimization for periodic maintenance policy of a system subject to slow degradation. Reliab Eng Syst Saf 68:105–112
108. Canfield R (1986) Cost optimization of periodic preventive maintenance. IEEE Trans Reliab 35:78–81
109. Wang H, Pham H (2006) Availability and maintenance of series systems subject to imperfect repair and correlated failure and repair. Eur J Oper Res 174:1706–1722
110. Djamaludin I, Murthy D, Kim C (2001) Warranty and preventive maintenance. Int J Reliab Saf Eng 8:89–108
111. Jung K, Park M, Park D (2010) System maintenance cost dependent on life cycle under renewing warranty policy. Reliab Eng Syst Saf 95(7):816–821
112. Perlstein D, Jarvis W, Mazzuchi T (2001) Bayesian calculation of cost optimal burn-in test durations for mixed exponential populations. Reliab Eng Syst Saf 72:265–273
113. Chang D (2000) Optimal burn-in decision for products with an unimodal failure rate function. Eur J Oper Res 126:534–540

114. Tseng S, Tang J, Ku I (2003) Determination of burn-in parameters and residual life for highly reliable products. Naval Res Logist 50:1–14
115. Wu C, Su C (2002) Determination of the optimal burn-in time and cost using an environmental stress approach: a case study in switch mode rectifier. Reliab Eng Syst Saf 76:53–61
116. Wu S, Clements-Croome D (2007) Burn-in policies for products having dormant states. Reliab Eng Syst Saf 92:278–285
117. Rangan A, Khajoui S (2007) Optimal system design based on burn in, warranty and maintenance. RFID Eurasia, 2007 1st Annual 1–5
118. Pham H (2000) Software reliability. Springer, London
119. Pham H (2003) Software reliability and cost models: perspectives, comparison, and practice. Eur J Oper Res 149:475–489
120. Pham H, Wang H (2001) A quasi-renewal process for software reliability and testing costs. IEEE Trans Syst Man Cybern Part A 31:623–631
121. Pham H, Zhang X (1999) A software cost model with warranty and risk costs. IEEE Trans Comput 48:71–75
122. Rinsaka K, Dohi T (2004) Determination of optimal warranty period in a software development project
123. Stytz M (2003) The case for software warranties. IEEE Secur Priv 1:80–82
124. Voas J (2000) Limited software warranties, pp 3–7
125. Voas J (2000) Limited software warranties. In: Proceedings of Seventh IEEE international conference and workshop on the engineering of computer based systems, pp 56–61
126. Bolstad W (2004) Introduction to Bayesian statistics. Wiley-Interscience, Hoboken
127. Hulting F, Robinson J (1988) The reliability of a series system of repairable subsystems: a Bayesian approach. Technometrics 30:143–154
128. Marquez D, Neil M, Fenton N (2007) A new Bayesian network approach to reliability modelling. Math Meth Reliab (MMR07)
129. Krivtsov V (2007) Practical extensions to NHPP application in repairable system reliability analysis. Reliab Eng Syst Saf 92:560–562
130. Pham H, Zhang X (2003) NHPP software reliability and cost models with testing coverage. Eur J Oper Res 145:443–454
131. Jain M, Maheshwari S (2006) Generalized renewal process (GRP) for the analysis of software reliability growth model. Asia Pac J Oper Res 23:215
132. Ross S (2000) Introduction to probability models. San Diego
133. Kulkarni V (1995) Modeling and analysis of stochastic systems. Chapman & Hall/CRC, Boca Raton
134. Taylor H, Karlin S (1984) An introduction to stochastic modeling. Academic Press, San Diego
135. Ross S (2000) Introduction to probability models. Academic press, San Diego
136. Wang H (1997) Reliability and maintenance modeling for systems with imperfect maintenance and dependence, Ph.D. Dissertation: Rutgers University
137. Wang H, Pham H (1996) A quasi-renewal process and its applications in imperfect maintenance. Int J Syst Sci 27:1055–1062
138. Heyman D, Sobel M (2004) Stochastic models in operations research. Courier Dover Publications, USA
139. Cinlar E (1975) Introduction to stochastic processes. Prentice-Hall, Englewood Cliffs
140. Rehmert I, Nachlas J (2009) Availability analysis for the quasi-renewal process. IEEE Trans Syst Man Cybern 39:272–280
141. Rehmert IJ, Nachlas JA (2009) Availability analysis for the quasi-renewal process. IEEE Trans Syst Man Cybern Part A 39:272–280

Part II
Risk Modeling

Risk Analysis

Terje Aven

1 Introduction

A risk analysis is a methodology to determine the nature and extent of risk. It comprises the following three main steps:

1. Identification of relevant threats/hazards
2. Cause and consequence analysis, including analysis of exposures and vulnerabilities
3. Risk description

The basis of the risk analysis is the systematic use of analytical—largely probability-based—methods which have been constantly improved over the past years. Probabilistic risk assessments for large technological systems, for instance, include tools such as fault and event trees. The processing of data is often guided by inferential statistics and organised in line with decision analytic procedures. These tools have been developed to generate knowledge about cause–effect relationships, express the strength of these relationships and characterise remaining uncertainties. In short, risk analysis specify what is at stake, assess uncertainties and calculate probabilities for (un)wanted consequences, to produce a risk picture.

A number of approaches and methods exist for analysing risks. We distinguish between two main categories:

a. *Statistical methods*. Data are available to predict the future performance of the activity or system analysed. These methods can be based on data extrapolation or probabilistic modelling.

T. Aven (✉)
University of Stavanger, Stavanger, Norway
e-mail: terje.aven@uis.no

H. Pham (ed.), *Safety and Risk Modeling and Its Applications*,
Springer Series in Reliability Engineering, DOI: 10.1007/978-0-85729-470-8_5,

b. *System analysis methods*. These methods are used to analyse systems where there is a lack of data to accurately predict the future performance of the system. Insights are obtained by decomposing the system into subsystems/components for which more information is available. Overall probabilities and risk are a function of the system's architecture and of the probabilities on the subsystems/component level [43]. Examples of such methods are FMEA (failure mode and effect analysis), FTA (fault tree analysis), ETA (event tree analysis), QRA (Quantitative risk analysis) and PRA (probabilistic risk assessment).

This approach also includes analysis of specific scenarios, which show different plausible pathways from release of an agent to the final outcome (loss).

In this chapter we restrict attention to the system analysis methods, and the QRAs. The other system analysis methods mentioned above can be viewed as tools being used in a QRA.

Apostolakis [4] presents an excellent summary of the benefits of QRA, mainly based on experience from nuclear power, but they are relevant also for other industries:

1. Considers a number of scenarios that involve multiple failures, thus providing an in-depth understanding of system failure modes.
2. Increases the probability that complex interactions between events/systems/operators will be identified.
3. Provides a common understanding of the problem, thus facilitating communication among various stakeholder groups.
4. Is an integrated approach, thus identifying the needs for contributions from diverse disciplines such as engineering and the social and behavioral sciences.
5. Focuses on uncertainty quantification and creates a better picture of what the community of experts knows or does not know about a particular issue, thus providing valuable input to decisions regarding need for research in diverse disciplines, e.g., physical phenomena and human errors.
6. Facilitates risk management by identifying the dominant accident scenarios so that resources are not wasted on items that are insignificant contributions to risk.

Nonetheless, despite these benefits, QRAs are subject to strong criticism. For example, Reid [46] argues there is a common tendency of underestimation of the uncertainties in QRAs. The disguised subjectivity of risk assessments is potentially dangerous and open to abuse if it is not recognised. According to Stirling [54], using risk assessment when there does not exist strong knowledge about the probabilities and outcomes is irrational, unscientific and potentially misleading. Renn [47] summarises critique drawn from the social sciences over many years and concludes that technical risk analyses represent a narrow framework that should not be the single criterion for risk identification, evaluation and management.

A substantial part of the research and development within risk assessment is motivated by the need for meeting this critique and developing improved methods and models. The right move forward is not to reject QRA, but to improve the tool and its use. The challenge is how decision-making on risk can be informed by the best available technical and scientific knowledge (refer Stirling [53]: 100). The aim of this paper is to discuss to what extent the trends seen are in fact meeting the challenges raised.

The ambition of the discussion is of course not to cover all important issues of QRAs. The scope is limited to some critical issues, and these are mainly of foundational character.

This chapter reviews basic principles and methods in risk analysis. Some key challenges related to conduction and use of the analyses are identified and discussed. These relate to inter alia the treatment of uncertainties and the incorporation of human and organisational factors.

2 Basic Features of a QRA

To explain the main ideas of a QRA we will look into the three main steps (1–3) of the analysis mentioned in the previous section: (1) Identification of relevant threats/hazards, (2) Cause and consequence analysis and (3) Risk description.

The level of depth of the various analysis steps depends of course on the purpose and scope of the analysis.

A simple example will be used to illustrate the basic steps of the analysis and some of the common methods used in the analysis. The example is to large extent based on Garrick et al. [25] and Aven [8]. We refer to the former reference for the technical details. Various aspects of the approach adopted by Garrick et al. [25] are discussed.

The system is a hypothetical electrical grid. The electrical infrastructure is critical to the nations's well-being. The analysis covers analyses related to cyber attacks and physical attack on a hypothetical electric power grid.

The system has many functions, but the main function is the delivery of electric power to the consumers. The following damage levels are defined:

0 no damage
1 transient outage (4–24 h) to network 1
2 transient outage to the region and network 1
3 long-term outage (more than 24 h) to network 1
4 long-term outage to network 1 and transient outage to the region (a transient outage is an outage which is automatically restored with no human intervention)
5 long-term outage to the region and network 1.

In addition the analysis focuses on (I) The number of attacks (suitably defined) and (II) The proportion of attacks being "successful".

2.1 Step 1: Identify Relevant Threats and Hazards

As a basis for this activity an analysis of the system is carried out, to understand how the system works, so departures from normal, successful operation can be easily identified. Once the system is understood, vulnerabilities that require special analysis can be identified. The electrical grid has four main elements; substations, transmission lines, supervisory control and data acquisition (SCADA) systems and energy management systems (EMSs).

A first list of hazards/threats is normally identified based on this system analysis, experience from similar type of analyses, statistics, brainstorming activities and specific tools such as Failure mode and failure effect analysis (FMEA) and Hazards and operability studies (HAZOP).

The FMEA was developed in the 1950s and was one of the first systematic methods that was used to analyse failures in technical systems. If we describe or rank the criticality of the various failures in the failure mode and effect analysis, the analysis is often referred to as an FMECA (Failure Modes, Effects and Criticality Analysis). The criticality is a function of the failure effect and the frequency/probability.

FMEA is a simple analysis method to reveal possible failures and to predict the failure effects on the system as a whole. The method is inductive; for each component of the system we investigate what would happen if this component fails. The method represents a systematic analysis of the components of the system to identify all significant failure modes and to see how important they are for the system performance. Only one component is considered a time, the other components are then assumed to function perfectly. FMEA is therefore not suitable for revealing critical combinations of component failures.

A HAZOP is a systematic analysis of how deviation from the design specifications in a system can arise, and an analysis of the risk potential of these deviations. Based on a set of guidewords, scenarios that may result in a hazard or an operational problem are is identified. The following guidewords are commonly used: NO/NOT, MORE OF/LESS OF, AS WELL AS, PART OF, REVERSE, AND OTHER THAN. The guidewords are related to process conditions, activities, materials, time and place. For example, when analysing a pipe from one unit to another in a process plant, we define the deviation "no throughput" based on the guideword NO/NOT, and the deviation "higher pressure than the design pressure" based on the guideword MORE OF. Then causes and consequences of the deviation are studied. This is done by asking questions, for example for the first mentioned deviation in the pipe example above:

- What must happen to ensure the occurrence of the deviation "no throughput" (cause)?
- Is such an event possible (relevance/probability)?
- What are the consequences of no throughput (consequence)?

Fig. 1 A simple example of a fault tree

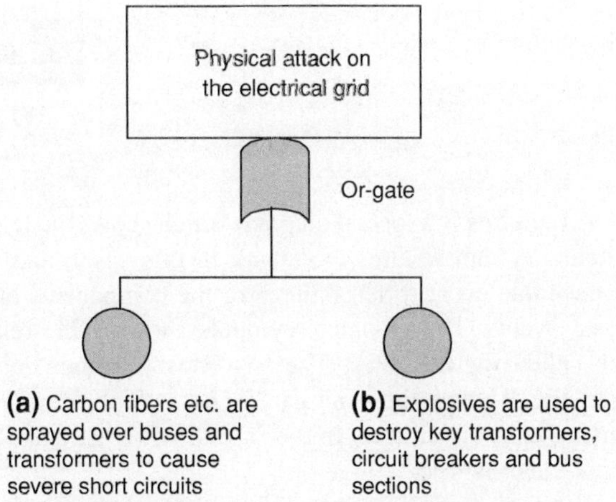

As a support in the work of formulating meaningful questions based on the guidewords, special forms have been developed.

Normally the source identification is strongly integrated with step 2, as source identification quickly leads to discussions of scenarios, causes, uncertainties and likelihoods.

For the case, we restrict attention to potential terrorist attacks. Further detailing is carried out in Step 2.

2.2 *Step 2: Cause and Consequence Analysis*

We focus on the following threats:

i. a physical attack on the electrical grid
ii. a complementary simultaneous cyber attack on the electrical grid.

Now starting from the events (i) and (ii), a standard analysis can be performed using event trees and fault trees to identify the possible consequences of the initiating events (i) or (ii) and possible scenarios leading to the events based on a backward approach. This process leads to a number of specified scenarios.

We first look at the backward analysis, asking how the initiating events (i) and (ii) may occur.

Numerous physical methods could be used to damage equipment at each substation with varying degrees of damage to the network. For example, carbon fibres could be sprayed over buses and transformers to cause severe short circuits, and explosives could be used to destroy key transformers, circuit breakers and bus sections. Without going into more details, we see the contours of an analysis establishing possible scenarios that can lead to the event (i). This analysis could be structured by using fault trees, see the simple example in Fig. 1. The attack occurs if either the basic event A occurs or B (or both).

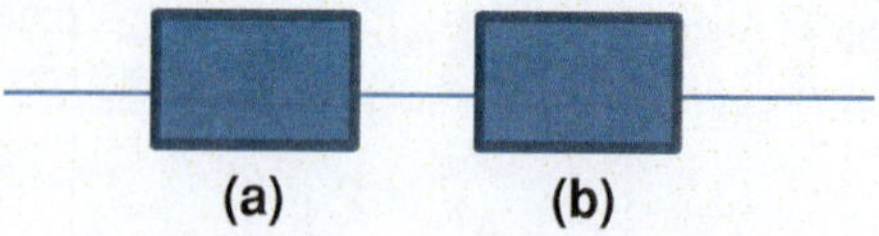

Fig. 2 A reliability block diagram (Series system).

A fault tree is a logical diagram which shows the relation between a top event—often a system failure (the attack in this case), and events that may cause this undesirable event (often failures of the components of the system), referred to as basic events. The graphical symbols showing the relation between these events are called logical gates. The two most common logical gates are the Or-gate (see Fig. 1) and the And-gate. The output from an And-gate is true if the input-events are all true. In the case of an Or-gate, the output is true if at least one input-event is true.

The fault tree (FTA) method was developed by Bell Telephone Laboratories in 1962 when they performed a safety evaluation of the Minuteman Launch Control System. The Boeing company further developed the technique and made use of computer programs for both quantitative and qualitative fault tree analysis. Since the 1970s fault tree analysis has become very widespread and is today one of the most used risk analysis methods. The applications of the method include most industries. The space industry and the nuclear power industry have perhaps been the two industries that have used fault tree analysis the most.

A fault tree that comprises only And- and Or-gates can alternatively be represented by a reliability block diagram. This is a logical diagram which shows the functional ability of a system. Each component in the system is illustrated by a rectangle as shown in Fig. 2. If there is connection from a to b in Fig. 2, this means that the system is functioning. Usually "functioning" means absence from one or more failure modes.

The diagrams in Figs. 1 and 2 represent alternative ways of representing the same activity or system.

The next stage in the analysis would be to perform a resource attacker's analysis, which addresses the following points:

- The need for resources to carry out the attack
- Possible attackers
- Their motivation, competence, resource basis and ability in general to perform the attack
- Knowledge, access and target vulnerabilities, as well as non-retribution and the ability to assess the success of an attack (cf. Anton et al. [2]).

A comprehensive analysis would also address factors that would influence the performance of the attackers, and the systems and barriers to withstand the attacks and reduce their consequences. Examples of such performance influencing factors are political conditions, intelligence, surveillance and reconnaissance, and self-awareness.

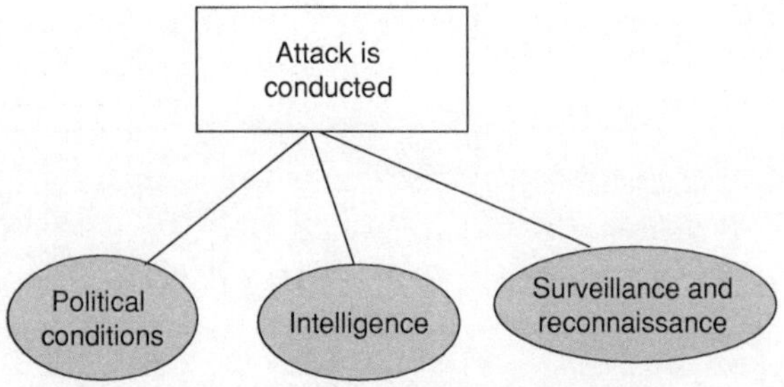

Fig. 3 Factors influencing the possible occurrence of an attack

An influence diagram (Bayesian belief network) could be used to show the influencing factors. A simple example is shown in Fig. 3.

Now, if an attack occurs, what are the consequences? To analyse these, event trees are used. They produce scenarios starting from the initiating events, in our case (i) and (ii). A simple event tree based on (ii) is shown in Fig. 4.

The Network 1 event represents the event that attackers damaging sufficient equipment in Network 1 cause a long-term power outage. The horizontal path from the Network 1 event occurs if the attackers do not disable enough equipment to cause a network power outage. The failure path from the Network 1 top event (the vertical path in the event tree) occurs if the attack results in long-term damage to network power supplies.

The SCADA event represents the event that intruders initiating a cyber attack that causes short-term power failures throughout the regional grid. The horizontal path from the SCADA event occurs if the intruders do not disable the regional grid. The failure path from the SCADA event (the vertical path in the event tree) occurs if the intruders cause a regional power outage.

Based on the various scenarios, a damage category is determined, as shown in Fig. 4. This event tree provides a basis for analysing uncertainties and probabilities related to the various attack scenarios. As noted by Garrick et al. [25], in practice, "it is often necessary to increase the level of detail in the supporting analyses to examine the threats, vulnerabilities, and causes that may contribute to each undesired condition. The increased detail facilitates a more systematic evaluation of each potential cause of failure and provides a logical framework for assessing the effectiveness of specific mitigation measures". The detailed analyses also often contribute to reduce the inherent uncertainties in the phenomena and processes studied, and identify the most important sources of uncertainty.

An example of a more detailed analysis is a vulnerability analysis of critical systems, such as the SCADA system. A vulnerability analysis studies the possible consequences (specifically addressing the system weaknesses) of possible threats and hazards. The vulnerability study can be conducted in many different ways. An example is the approach by Anton et al. [2], which is based on an identification of vulnerabilities using a checklist covering a rather comprehensive taxonomy applied to physical, cyber, human/social and infrastructure objects. It covers attributes related to

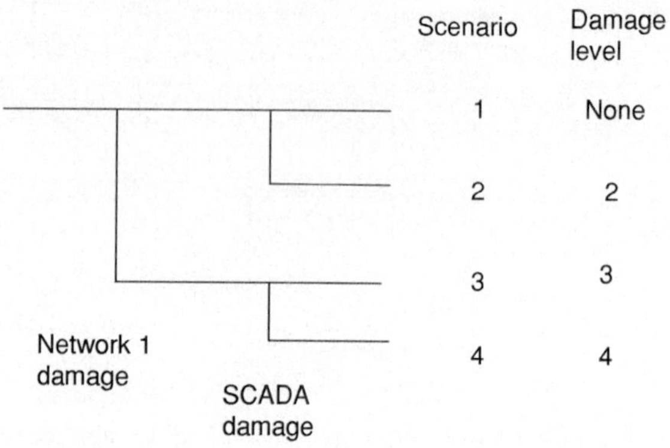

Fig. 4 Simple event tree for the initiating event (ii)

Design/architecture (Singularity—uniqueness, centrality, homogeneity; Separability; Logic/implementation errors—fallibility; Design sensitivity—fragility, limits, finiteness; Unrecoverability),

Behavioral (Behavioral sensitivity/fragility; Malevolence; Rigidity; Malleability; Gullibility—deceivability, naivete; Complacency; Corruptibility—controllability) and

General (Accessible—detectable, identifiable, transparent, interceptable; Hard to manage or control; Self-unawareness and unpredictability; Predictability).

By a systematic review of these attributes, the aim is to identify vulnerabilities which are beyond the standard well-known vulnerabilities.

2.3 Step 3: Risk Description

Up to now we have performed a qualitative analysis with the purpose of gaining insights. The analysis structures the analysts' knowledge on the threats and vulnerabilities and what the consequences of a threat are. Normally there would be a number of scenarios developing from a specific initiating event. There are uncertainties present, and these uncertainties we need to assess and describe.

The key question to address are:

- How likely is it that a physical attack on the electrical grid occurs in a specified period of time?
- Are there large uncertainties in the phenomena that influence the occurrence of this event?
- What are the essential factors creating the uncertainties?
- How likely are the different consequences, using the appropriate categorisations?
- Are there large uncertainties in the phenomena that influence these likelihoods?
- What are the essential factors creating the uncertainties?

Fig. 5 An example of a risk matrix. Here x corresponds to a scenario in damage category 2 and probability of 10–50%

>50%					
10-50%			x		
1-10%					
<1%					
Probability Damage category	0	1	2	3	4

Various kind of risk matrices can be informative to present the likelihoods. The traditional risk matrix show combinations of possible consequences (with some defined categories) and associated probabilities (See Fig. 5).

In Fig. 5 the x is to be interpreted as the assessor's probability that the scenario resulting in damage category 2 will occur. The assignments are based on relevant models and data. These models and data are a part of the background information for the assigned probabilities.

Focus in the analysis is on quantities like

X: the number of future attacks (properly defined)
Y_i: 1 if attack i is successful and 0 otherwise
Y: the proportion of the attacks being successful
Z: the number of successful attacks.

Then we assess uncertainties, using probabilities, and this leads to probability distributions of the above quantities. Given a number of attacks, say 10, we may, for example, assign a probability of 20% that one of these is successful. From the analysis we may establish a probability distribution for Z, the number of successful attacks.

Probabilities and expected values are used to express uncertainties and degrees of belief. However, we need to see beyond these values. The analyses are based on judgements, made by some experts, a number of assumptions and suppositions are made, and there could be large uncertainties associated with the phenomena being studied. The outcomes could be surprising relative to the assigned probabilities and expected values. In a risk analysis a number of such probability assignments are performed, and the hidden uncertainties could create surprising outcomes someplace. You do not know where it will come, but it certainly could happen.

An assumption could be that the "war of terror" is further escalating, but we could be wrong.

To identify and express such uncertainties different types of procedures are used [8, 50]. These relate to the background information (assumptions and suppositions) of the assigned probabilities, as well as factors such as

– Vulnerabilities
– Complexity in technology

- Complexity in organisations
- Available information
- Time horizon
- The thoroughness, etc., of the analysis. What is the experts' competence, seen in relation to the best available knowledge? Do we have available relevant experience data? To what extent would further analysis reduces the uncertainties about the potential consequences? Etc.
- IRGC, International Risk Governance Council [31] consequence features, such as
  - Delay effects—which describes the time of latency between an initial event and the actual damage.
  - Reversibility—which describe the possibility to restore the situation to the state before damage occurred.

For example, the feature "delay effects" could lead to a focus on activities or mechanisms that could initiate deteriorating processes causing future surprises.

Addressing the uncertainties also mean to consider the *manageability*; i.e., to what extent is it possible to control and reduce the uncertainties, and obtain desirable outcomes. Some risks are more manageable than others, meaning that the potential for reducing the risk is larger for some risks compared to others. By proper uncertainty management, we seek to obtain desirable consequences.

Risk should be described by addressing such issues along with the probabilities.

The risk analyses systemise the knowledge and uncertainties about the phenomena, processes, activities and systems being analysed. This knowledge and these uncertainties are described and discussed and this provides a basis for evaluating what is important (tolerable and acceptable) and for comparing options.

Expressing risk also means to perform sensitivity analyses. The purpose of these analyses is to show how sensitive the output risk indices are with respect to changes in basic input quantities, for example assumptions and suppositions.

The analysis is used to identify critical systems, and thus provide a basis for selecting appropriate measures. To illustrate this, let R be a risk index, for example expressing the expected number of fatalities (PLL) or the probability of a system failure, and let R_i be the risk index when subsystem i is in the functioning state. Then a common way of ranking the different subsystems is to compute the risk improvement potential (also referred to as the risk achievement worth) $I_i = R_i - R$, i.e., the maximum potential risk improvement that can be obtained by improving system i (Aven [6], Haimes [28]). The potential I_i is referred to as a *risk importance measure*.

3 Challenges

The section addresses some important challenges of risk analysis:

1. Treatment of uncertainties
2. The causal chains and event modelling

3. Incorporation of human and organisational factors
4. The decision context.

These relate in particular to the scientific foundation of the QRAs but also the applications of the QRA.

3.1 Treatment of Uncertainties

A proper treatment of uncertainties in risk assessments is one of the main challenges of QRA. The risk analyses are to describe the uncertainties (refer to item 5, Sect. 1) but many risk analyses ignore uncertainties beyond best estimates. Consider the following statement from an experienced risk analyst team about uncertainty in quantitative risk analysis [14]:

> The analyses are based on the "best estimates" obtained by using the company's standards for models and data. It is acknowledged that there are uncertainties associated with all elements in the analysis, from the hazard identification to the models and probability calculations. It is concluded that the precision of the analysis is limited, and that one must take this into considerations when comparing the results with the risk acceptance criteria and tolerability limits.

Based on such a statement one may question what uncertainty in QRA means. Everything is uncertain, but is not risk assessment performed to assess and describe the uncertainties? Again refer to item 5 above.

We have to acknowledge that there are different types of risk assessments (QRAs) and they treat uncertainties to varying degree. Traditionally six levels have been identified [41], and the best-estimate approach mentioned above is referred to as level 4. The most detailed level, level 6, is the probability of frequency approach [32]. Here second-order probabilities (subjective probabilities) P are used to describe the assessors' epistemic uncertainties about the relative frequency interpreted probabilities p_f. The latter probabilities describe stochastic or aleatory uncertainties (variations). The analysis may produce a 90% credibility interval for p_f, $[a, b]$, saying that the analyst is 90% confident that p_f lies in the interval $[a, b]$.

If we look at the prevailing practice concerning uncertainty treatment in QRAs today the first impression may indicate that not much has changed since the late 1970s and the beginning of the 1980s. However, these issue of uncertainties and treatment of uncertainties in QRAs has been continuously addressed by analysts and researchers since then. Some of the main challenges focused have been:

- The meaning and usefulness of the distinction between epistemic and aleatory uncertainties.
- The meaning of model uncertainty and approaches for treating model uncertainty.
- The need for seeing beyond expected values and probabilities when assessing uncertainties
- Alternative representations of uncertainty than probability.

In the following we will look closer at these challenges and give some reflections on the way ahead (Sect. 3.1.5).

3.1.1 The Distinction Between Epistemic and Aleatory Uncertainties

If we study the lifetimes of a type of light bulbs, the distinction between aleatory uncertainty and epistemic uncertainty is clear and easy to understand. The variation in lifetimes produces the aleatory uncertainty. If we had full knowledge about the generated distribution of lifetimes, there would be no epistemic uncertainties. However, in practice we do not know this underlying true distribution and hence we need to address epistemic uncertainties. In practice this is typically done by assuming that the distribution belongs to a parametric distribution class, for example the Weibull distribution with parameters α and β. This distribution is to be considered a model of the underlying true distribution, and given this model the epistemic uncertainties are reduced to lack of knowledge concerning the correct parameters α and β.

For mass produced units and other situations with large populations of units, this uncertainty structure makes sense. We would however prefer to refer to the aleatory uncertainty as variation and not uncertainty, as variation or population variation explains better what we would like to express.

In a risk analysis context, the situations are often unique, and the distinction between aleatory uncertainty and epistemic uncertainty is then more problematic. Consider as an example the probability of a terrorist attack (properly specified). To define the aleatory uncertainty in this case we need to construct an infinite population of similar attack situations. The variation in this population generated by "success" (attack) and "failure" (not attack) represents the aleatory uncertainty. The proportion of successes equals the probability (chance in a Bayesian context, see, e.g., Singpurwalla [51]) of an attack. But is such a construction meaningful? No, it makes no sense to define a large set of "identical", independent attack situations, where some aspects (for example related to the potential attackers and the political context) are fixed and others (for example the attackers' motivation) are subject to variation. Say that the attack success rate is 10%. Then in 1,000 situations, with the attackers and the political context specified, the attackers will attack in about 100 cases. In these situations the attackers are motivated, but not in the remaining ones. Motivation for an attack in one situation does not affect the motivation in another. For independent random situations (refer the light bulb example above) such "experiments" are meaningful, but not in unique cases like this.

This type of problem is surprisingly seldom addressed in the literature. It is common to define the underlying aleatory-based probabilities and distributions, but without clarifying their meaning. Researchers express that there is only one type of uncertainty, stemming from lack of knowledge (i.e., it is epistemic), seem to represent a small minority. See, e.g., Helton and Burmaster [28]. Yet, the Bayesian paradigm (as, for example, presented by Lindley [35]) is based on this idea. Probability is considered a measure of uncertainty about events and outcomes (consequences), seen through the eyes of the assessor and based on the available background information and knowledge. Probability is a subjective measure of uncertainty, conditional on the background knowledge. The reference is a certain standard such as drawing a ball from an urn. If we assign a probability of 0.4 for an event A, we compare our uncertainty of A to occur with drawing a red ball from an urn having 10 balls where 4 are red.

According to this paradigm, there is epistemic uncertainty associated with the occurrence of an attack. And the analyst assigns a probability expressing his/her uncertainty about this event. If relevant, knowledge about the variability is included in the background knowledge. The variability gives rise to uncertainty but is not defined as uncertainty in this context. A relative frequency generated by random variation is referred to as a chance, to distinguish it from a probability, which is reserved for expressions of epistemic uncertainty based on belief [35, 51]. Thus, we may use probability to describe uncertainty about the unknown value of a chance, whenever a chance is introduced.

Consider again the light bulb example. Say that the problem is to express the probability that a specific new light bulb fails before x units of time, and we adopt the Bayesian approach using subjective probability P. Let A be the failure event. To specify the probability of A given the background knowledge K, i.e., $P(A|K)$, the analyst may use the standard Bayesian approach conditioning on the parameters, which gives that

$$P(A|K) = \int G(x|\,\alpha,\, \beta)\, dF(\alpha,\, \beta|\, K), \tag{1}$$

where F is the prior (or posterior) distribution of α and β given the background knowledge, and $G(x|\alpha, \beta)$ is the Weibull lifetime distribution. If α and β are known, the probability that the light bulb fails before x units of time is set equal to the chance $G(x|\alpha, \beta)$. Here P is a subjective probability reflecting epistemic uncertainties. The aleatory uncertainty (the chance distribution of lifetimes) is expressed by the Weibull distribution G. We may also express epistemic uncertainties about the chance distribution, for example a probability of 0.10 that the chance at $x = 1$ is at least 0.2.

3.1.2 Model Uncertainty

Model uncertainty in a QRA context is a difficult concept. Its meaning and how to deal with it have been addressed by many researchers. See, e.g., Dewooght [21], Nilsen and Aven [39], Haimes [27] and the references therein.

The models are used as tools to obtain insight into risk, to express risk, and they form part of the conditions and the background knowledge on which the analysis is built. It is obviously important to reflect on how suitable the model is for its objective. In this regard, however, it is not only the approximation of the real world that is the point, but also the model's ability to reflect the essential aspects of the real world, and to simplify complicated features and conditions.

Let us return to the Weibull distribution introduced in Sect. 3.1.5. This distribution is a model (a simplified representation) of the real world. We may interpret model uncertainty as the accuracy of this model relative to the true distribution, and if the aim of the risk analysis is to determine this distribution the issue of model uncertainty has a clear meaning and it obviously needs due attention. Model uncertainty, or model accuracy, is also relevant in the Bayesian context—the reference is the underlying chance distribution. However model uncertainty or accuracy has no meaning with respect to the subjective probability P. For the probability P the model is merely a tool judged useful for expressing the uncertainties about A. The model is a part of the background knowledge. If we change the model, we change the background knowledge. If we have specified $P(A|K)$ using the Weibull distribution according to formula (1), the probability is conditional on the use of this distribution. If we knew the true distribution P would be different, but since this distribution is unknown we condition on this model. We have introduced the Weibull distribution to simplify the problem. If we had considered the space of all distribution functions, the assignment process would not be feasible in practice.

What is then a satisfactory or a good model? How accurate needs a model to be considered acceptable? Well, the ultimate requirement for a model is that any improvement in the model to make it more accurate, as judged by the analyst, should not lead to a change in the conclusions made. However, this requirement may be difficult to verify—the best the analyst can do in most cases is to use sensitivity analyses to see how changes in the model affect the results. And of course, an experienced risk analyst should have an idea of what is important for the risk results and what is not. It would also be a task for the risk analysis discipline (community) to contribute to the development of adequate models—to standardise what a good model is. Note that a crude model can be preferred instead of a more accurate model in some situations if the model is simpler and it is able to identify the essential features of the system performance. Achieving the appropriate balance between simplicity and accuracy is a main task of the risk analyst.

The natural sciences provide theories and laws describing physical phenomena such as ignition and explosion. Similarly, social science theories provide a basis for modelling human and organisational factors. In principle there is no difference compared to the natural sciences. The uncertainties are larger, but seeing risk assessment as a tool to describe the uncertainties (refer again to item 5 in Sect. 1), this should not undermine the risk assessments.

To take into account model uncertainties, different approaches are used. In structural reliability analysis (SRA) attempts are made to explicitly reflect the model uncertainties, see, e.g., Aven [7]. Let Z be the true capacity of the system at the time of interest. Using the model $G(X) = X_1 - X_2$, where X_1 represents a strength measurement and X_2 represents a load measurement, we have put $Z = G(X)$. This means a simplification, and the idea is then to introduce an error term X_0, say, such that we obtain a new model $G_0(X) = X_0(X_1 - X_2)$. Clearly, this may give a better model, a more accurate description of the world. However, it may not be chosen in a practical case as it may complicate the assessments. It may be much more difficult to specify a probability distribution for (X_0, X_1, X_2) than for (X_1, X_2). There might be lack of relevant data to support the uncertainty analysis of X_0 and there could be dependencies between X_0 and (X_1, X_2). We have to balance the need for accuracy and simplicity.

In the literature various other methods have been suggested to reflect model uncertainties. A typical procedure used is the following [3]: let M_1 and M_2 be two alternative models to be used for assigning the probability A. Conditional on M_i, we have an assignment $P(A|K_i)$. Unconditionally, this gives

$$P(A|K) = P(A|K_1)\,p_1 + P(A|K_2)\,p_2,$$

where p_i is the analyst's subjective probability that the ith model (i.e., the set of associated assumptions) is true.

Such a procedure is analogous to the Weibull case if "true" refers to a condition of the real world that is true or not. The model in itself is not true. We have introduced the model to simplify a complex world. The same procedure could also be used for the distribution class but as concluded above, at a certain stage we have to accept the model with its strengths and limitations.

3.1.3 The Need for Seeing Beyond Expected Values and Probabilities

It is common to define and describe risk using probabilities and expected values. However, these perspectives have been challenged. The probabilities and expected values could camouflage uncertainties (e.g., Rosa [48], Aven [10]). The assigned probabilities are conditioned on a number of assumptions and suppositions. They depend on the background knowledge. Uncertainties are often hidden in the background knowledge, and restricting attention to the assigned probabilities could camouflage factors that could produce surprising outcomes. By jumping directly into probabilities, important uncertainty aspects are easily truncated, meaning that potential surprises could be left unconsidered.

Let us look at an example. Consider the risk, seen through the eyes of a risk analyst in the 1970s, related to future health problems for divers working on offshore petroleum projects. An assignment is to be made for the probability that a diver would experience health problems (properly defined) during the coming 30 years due to the diving activities. Let us assume that an assignment of 1% is made. This number is based on the available knowledge at that time. There are not strong indications that the divers will experience health problems. However, we know today, that these probabilities led to poor predictions. Many divers have experienced severe health problems [15]. By restricting risk to the probability assignments alone, we see that aspects of uncertainty and risk are hidden. There is a lack of understanding about the underlying phenomena, but the probability assignments alone are not able to fully describe this status.

Several risk perspectives and definitions have been proposed in line with these acknowledgments. For example, Aven and Renn [14] defines risk associated with an activity as *uncertainty about and severity of the consequences of the activity*, where severity refers to intensity, size, extension, scope and other potential measures of magnitude, and is with respect to something that human beings value (lives, the environment, money, etc.). Losses and gains, for example expressed by money or the number of fatalities, are ways of defining the severity of the consequences.

In case of large uncertainties, risk analyses could support decision-making, but other principles, measures and instruments are also required, such as the cautionary/precautionary principles [15, 30] as well as robustness and resilience strategies [31]. An informative decision basis is needed, but it should be far more nuanced than can be obtained by probabilistic analysis alone. This is stressed by many researchers, e.g., Apostolakis [4] and Apostolakis and Lemon [5]: QRA results are never the sole basis for decision-making. Safety-related decision-making is risk-*informed*, not risk-*based*. This conclusion is however not only justified by referring to the need for addressing uncertainties beyond probabilities and expected values. A main issue here is the fact that risk needs to be balanced with other concerns (See Sect. 4).

3.1.4 Alternative Representations of Uncertainty Than Probability

There are different traditions and schools concerning uncertainty assessments. In a QRA, probability is by far the most common approach to represent uncertainties. However, recently we have seen an increasing number of contributions where alternative representations are suggested and used. It has been questioned whether uncertainty can be represented by a single probability, or if imprecise (interval) probabilities are needed for providing a more general representation of uncertainty [19, 20, 56]. also been questioned whether probability is limited to special cases of uncertainty regarding binary and precisely defined events only. Suggested alternatives for addressing these cases include fuzzy probability [26, 58] and the concept of possibility [55, 57]. Furthermore, probabilities have been criticised for not reflecting properly the weight of the evidence they are based on, as is done in evidence theory [52].

One of the major objections against using probability to represent uncertainty seems to be that probability is a measure of randomness only. The existence of alternative interpretations of probability is not always recognised and, as pointed out by Natvig [38], work on fuzzy-based representations of uncertainty often seems to be motivated by the inadequacy of the relative frequency interpretation of probability; thus not taking into account other interpretations [24].

The interpretations of many of these alternative representation are not clear. To avoid confusion about the concept of uncertainty, it may be better to separate uncertainty from lack of precision (ambiguity, vagueness, fuzziness). We refer to Flage et al. [24] and Aven [11].

3.1.5 The Way Ahead

The scientific basis of the QRAs needs to be strengthened. A risk assessment must have a solid scientific basis, which clarifies the axioms, interpretations and measurement procedures of the representations of uncertainties, probabilities and risk [17]. If a probability of an attack or a probability of an oil spill is introduced, its meaning must be explained. If the produced probability numbers are subject to uncertainties, this uncertainty must be assessed and discussed. The background knowledge that the probabilities are based on must be described. The role of models must be clarified. How is model uncertainty understood and dealt with in the analysis?

The risk analysts need to be professional in the field of risk and uncertainty. Now we often see risk analysts with a poor background in the fundamentals of risk assessment. It is not sufficient to be a statistician or an engineer to act as a professional risk analyst. Risk analysis is a discipline in its own and requires education and training in topics like risk and probability concepts, risk analysis methods, uncertainty analysis, risk characterisations and risk communication. The stakeholders', including the decision-makers' and third parties', understanding of risk assessments and their results depends very much on the professional risk analyst's ability to communicate the risk picture. Risk professionals need to be sharp on the scope, as well as the boundaries and limitations of the assessments [11].

The risk analyses need to provide a much broader risk picture than what is typically the case today. Separate uncertainty analyses should be carried out, extending the traditional probabilistic based analyses. It is not sufficient to produce a risk picture by just reporting some probability estimates or assignments. What is needed is also

- Sensitivities showing how the risk indices depend on the background knowledge (assumptions and suppositions)
- Uncertainty assessments
- Description of the background knowledge, including models used.

The uncertainty assessments should not be restricted to standard probabilistic analysis, as this analysis could hide important uncertainty factors. The search for quantitative, explicit approaches for expressing the uncertainties, even beyond the subjective probabilities, may seem to be a possible way forward. However such an approach should be used with care. Trying to be precise and accurately expressing what is extremely uncertain does not make sense. Instead we recommend a more open qualitative approach for revealing such uncertainties. Some would feel this as less attractive from a methodological and scientific point of view, perhaps it is, but it would be more suited for solving the problem at hand, which is about analysis and management of risk and uncertainties.

What we would like to see is a broad risk description covering risk numbers, sensitivities as well as uncertainty factors.

We are sceptical to the alternative representations of uncertainty as mentioned in Sect. 3.1.4. However, further research is required to clarify the role of these representations in QRA. See discussion in Aven [11] and Aven and Zio [16]

3.2 The Causal Chains and Event Modelling

The traditional risk analysis approach used in QRAs can be viewed as a special case of system engineering [27]. This approach, which to large extent is based on causal chains and event modelling, has been subject to strong criticism. Many researchers argue that some of the key methods used in risk analysis are not able to capture "systemic accidents". Hollnagel [29], for example, argues that to model systemic accidents it is necessary to go beyond the causal chains—we must describe system performance as a whole, where the steps and stages on the way to an accident are seen as parts of a whole rather than as distinct events. It is not only interesting to model the events that lead to the occurrence of an accident, which is done in, for example, event and fault trees, but also to capture the array of factors at different system levels that contribute to the occurrence of these events. Leveson [34] makes her points very clear:

> Traditional methods and tools for risk analysis and management have not been terribly successful in the new types of high-tech systems with distributed human and automated decision-making we are attempting to build today. The traditional approaches, mostly based on viewing causality in terms of chains of events with relatively simple cause-effect links, are based on assumptions that do not fit these new types of systems: these approaches to safety engineering were created in the world of primarily mechanical systems and then adapted for electro-mechanical systems, none of which begin to approach the level of complexity, non-linear dynamic interactions, and technological innovation in today's socio-technical systems. At the same time, today's complex engineered systems have become increasingly essential to our lives. In addition to traditional infrastructures (such as water, electrical, and ground transportation systems), there are increasingly complex communication systems, information systems, air transportation systems, new product/process development systems, production systems, distribution systems, and others.
>
> The limitations of the traditional models and approaches to managing and assessing risk in these systems make it difficult to include all factors contributing to risk, including human performance and organisational, management and social factors; to incorporate human error and complex decision-making; and to capture the non-linear dynamics of interactions among components, including the adaptation of social and technical structures over time.

Leveson argues for a paradigm-changing approach to safety engineering and risk management. She refers to a new alternative accident model, called STAMP (System-Theoretic Accident Modeling and Processes).

A critical review of the principles and methods being used is of course important, and the research by Hollnagel et al. [45] and others in this field adds valuable input to the further development of risk analysis as a discipline. Obviously we need a set of different approaches and methods for analysing risk. No approach is able to meet the expectations with respect to all aspects. The causal chains and event modelling approach has shown to work for a number of industries and settings, and the overall judgement of the approach is not as negative as Leveson expresses. Furthermore, the causal chains and event modelling approach is continuously improved, incorporating human, operational and organisational factors, see e.g., I-Risk [40], ARAMIS [22], the BORA project [13], the SAM approach [42] and the Hybrid Causal Logic Method [37, 49]. It is not difficult to point at limitations of these approaches, but it is important to acknowledge that the suitability of a model always has to be judged by reference to its ability to represent the real world, but also its ability to simplify the world, as discussed above.

We consider the causal chains and event modelling to provide a sound basis for the risk analyses in many cases. However, we acknowledge the limitations of this approach, as well as other aspects of the analyses, and add alternative qualitative tools to see beyond these limitations. Insights provided by this alternative research paradigm can be used to strengthen the risk picture obtained by the more traditional approach.

3.3 Incorporation of Human and Organisational Factors (HOFs)

Another important research issue is the development of appropriate problem decomposition methods for risk and vulnerability analysis (including extending the logic modelling techniques—such as fault tree and event tree to include influence diagrams), essential for capturing different dimensions of complex risk issues. The work must be seen in relation to the existing approaches mentioned above; such as the HCL method [37, 49] and related methodologies (see e.g., Léger et al. [33], Ale et al. [1] and Luxhøj et al. [36]), which develop methodology for operational risk analysis including analysis of the performance of safety barriers, with respect to technical systems as well as human, operational and organisational factors.

To simplify, the basic problem can be summarised as follows, using an example from maintenance in a process plant:

1. Identify events A that summarise essential barrier performance. An example is 'ignition' or 'avoid ignition' given a specific leakage scenario.
2. Establish a deterministic model that links A and events B_i and quantities X_i on a more detailed level. A fault tree is an example of such a model.
3. Specify a set of operational and management factors F_i that could influence the performance of the barriers, and which have not been included in the model developed in stage 2. Examples of such factors are the quality of the maintenance work, the level of competence and the adequacy of organisation.
4. Specify probabilities $P(B_i|F)$, where F is the vector of the F_is.
5. Use probability calculus to obtain $P(A|F)$.

To carry out such an analysis there are a number of challenges, of which the following are some of the more important:

- Determine which F factors that should be included in the basic stage 2 model. The F factors are fixed, meaning that the probability assignments are conditioned on these factors. If some of the F factors are to be considered unknown to

the analyst, these factors need to be included in the basic model, or the factors should be divided into two categories, reflecting unknown factors on the one hand and some given factors on the other. Such a distinction is made in the SAM-method [42].
- Finding adequate procedures for assigning scores to the F factors and specifying the probabilities $P(B_i|F)$. The probabilities $P(B_i|F)$ need to be based on models and methods used for barrier performance analyses, such as human reliability analysis.

For the purpose of the present chapter we will not go further into the analysis. Our concern is the research direction, where the ambition is to explicitly incorporate the human and organisational factors. Developing suitable methodology is not straightforward. A detailed analysis requires substantial input data, and the data must be relevant. Such analyses cannot be performed without extensive use of expert judgements. However, expert judgement is not to be seen as something negative. The risk analysis is a tool for summarising the information available (including uncertainties), and expert judgements constitute an important part of this information.

The developments in this field seem to be partly based on the causal chains and event modelling approach. A natural question then is to what extent these developments are meeting the critique raised in the previous section. Are, for example, the non-linear dynamics of interactions among components adequately reflected? Such issues need to be addressed, and for these developments to gain acceptance, validation processes are required incorporating reviews and discussions of thee issues. Is the analysis to be considered a number-crunching exercise characterised by arbitrariness. Would it not be better to adopt a more qualitative approach? Is it possible to obtain confidence in the assigned probabilities when they are to reflect aspects like management involvement and culture?

Validation of risk analysis methods is an important but difficult issue. To achieve results that are trusted by the stakeholders, it is important to pay attention to the process behind the risk calculation results. For example, subjective input to the risk analysis should to as large extent as possible be assigned by broad groups of experts, rather than by one single expert. A method for assigning probabilities cannot be validated in the sense that you can check that the results are correct. However, all stakeholders need to have confidence in the process of transforming the analysts' knowledge and lack of knowledge into probabilities.

We expect further developments in this area. Due attention needs to be paid to validation and foundational issues.

3.4 The Decision Process

Our main focus here is the search for mechanistic decision rules, linked to acceptance (tolerability) limits. For example, the Norwegian offshore petroleum regulations state that risk acceptance criteria (tolerability limits) expressed as upper limits of acceptable (tolerable) risk should be developed, and before the risk analyses are carried out (PSA [15, 44]). Such criteria are common and we see an increasing trend. But one may question the appropriateness of introducing such criteria. Consider the following criterion:

> The probability of getting an oil spill during 1 year of operation causing an environmental damage having a restitution period of more than z years, should not exceed 1×10^{-x}.

At the political level it is obvious that it would not be possible to establish consensus about such a limit. Different parties would have different preferences. But for the Government it should it be possible to establish such a number? Say that it would make an attempt to do this. And suppose that it considers two options, a weak limit, say 1×10^{-3} and a strong limit say 1×10^{-4}. What limit should it choose? The answer would be the weak limit, as the strong limit could mean lack of flexibility in choosing the overall best solution. If the benefits are sufficiently large, the level 1×10^{-3} could be acceptable. Following this line of arguments, the use of such limits leads to the formulation of weak limits, which are met in most situations. QRAs are then used to test whether the risks are acceptable in relation to these weak limits. It is to large extent waste of money, the conclusions are obvious.

At the operational level, the same type of arguments will apply. The oil company is to determine an acceptance criterion, and it faces the same type of dilemmas as above. Why should it specify strong limits? It would restrict the company from obtaining the overall best solutions. The result is that weak limits are specified and risk assessments play the role of verification, a role that adds not much value.

If a high level of safety or security is to be obtained, other mechanisms need to be implemented than risk acceptance (tolerability) limits. If such criteria are established, they give a focus on obtaining a minimum safety standard, instead of continuous improvement and risk reduction.

The ALARP principle represents such a mechanism. The ALARP principle expresses that the risk should be reduced to a level that is as low as reasonably practicable. A risk reducing measure should be implemented provided it cannot be demonstrated that the costs are grossly disproportionate relative to the gains obtained [30]. Risk assessments play an important role in ALARP processes, as risk reduction needs to be based on risk assessments. Risk must be described and the effect of risk reducing measures determined. Although the ALARP principle has been in use for many years, its interpretation and the implementation procedures are still being discussed. Implementing a broad risk perspective as outlined in Sect. 2.3 would require updated ALARP procedures, as risk is more than computed probabilities and expected values.

4 Conclusions

We make the following conclusions:

1. The scientific basis of the risk analyses needs to be strengthened.
2. The risk analyses need to provide a much broader risk description than what is typically the case today. More weight should be given to uncertainties.

Table 1 Stages in the development of risk communication [23, 53]

1: all we have to do is get the numbers right
2: all we have to do is tell them the numbers
3: all we have to do is explain what we mean by the numbers
4: all we have to do is show them that they've accepted similar risks in the past
5: all we have to do is show them that it's a good deal for them
6: all we have to do is treat them nice
7: all we have to do is make them partners

3. Further research is needed for improving the modelling and analysis of human and organisational factors.
4. The ALARP principle and implementation need to be adapted to such an extended risk description. Risk acceptance (tolerability) limits should be used with care.

Other researchers have presented similar recommendations, see e.g., Stirling [53, 54] and Renn [47]. In the social scientist risk research community there has been a strong recognition of the need for moving away from a search for an "analytical fix" and towards addressing the social and institutional aspects of the problem. IRGC, International Risk Governance Council [31] and UK Cabinet Office [18] are examples of risk frameworks based on this recognition. Fischhoff [23] have summarised seven stages in this retreat away from what might be seen as "naïve positivism" in the risk debate [53], see Table 1.

Nonetheless, the risk analyses consultants and the formal decision-making on the regulation of risk remain relatively unaffected by this recognition (refer Stirling [53]: 100). There are many reasons for this, but a main factor is certainly the risk assessment tool in itself. The assessments are in general lacking a proper foundation and the perspective on uncertainties is too narrow.

The relative frequency based approach and the Bayesian schools of thought have collided for many years over the definition of probability, sowing considerable confusion over the definition of risk and the limits of probabilistic risk analysis [41]. Attempts at reconciliation of these two perspectives have led to the probability of frequency approach, as briefly described and discussed in the beginning of Sect. 3.1. However, on both sides of this approach we have other approaches, the pure classical approach based on relative frequency interpreted probabilities only, and broad risk perspectives as discussed for example by Aven and Renn [14]. The broad perspectives are categorised by an acknowledgement of the need for seeing beyond the probabilistic analysis. The main pillar of risk is not probability but uncertainty, refer the Aven and Renn [14] definition of risk: uncertainty about and severity of the consequences (or outcomes) of an activity with respect to something that human beings value.

Classical decision theorists have often taken the stand that the distinction between aleatory and epistemic uncertainties as in the probability of frequency approach is unnecessary because, according to the axioms of expected-utility decision analysis, it is irrelevant in rational choices [41]. All one needs is their common measure, the Bayesian (subjective) probability, that can characterise both. The different probabilities can then be combined for decision-making purposes as if all uncertainties were of the same nature. Such a view is however disputed by many researchers and analysts. As argued by Paté-Cornell [41], rationality can be viewed as more complex than the simple maximisation of expected utility. Decision-makers may need and/or ask for a full display of the magnitudes and the sources of uncertainties before making an informed judgement.

Such a perspective provides the basis for the risk perspective adopted in many risk frameworks (e.g., IRGC [31], Aven [9]). It considers uncertainty as the main component of a risk description—probability is just a tool used to express the uncertainties. To some extent we describe the uncertainties, but one should acknowledge that the full scope of these uncertainties cannot be transformed to a mathematical formula, using probabilities or other measures of uncertainty.

References

1. Ale B, Bellamy LJ, van der Boom R, Cooper J, Cooke RM, Goossens LHJ, Hale AR, Kurowicka D, Morales O, Roelen ALC, Spouge J (2009) Further development of a causal model for air transport safety (CATS): building the mathematical heart, Reliab Eng Syst Saf 94(9):1433-1441.
2. Anton PS, Anderson R, Mesic R, Scheiern M (2003) The vulnerability assessment & mitigation methodology. Rand report. ISBN 0 8330-3434-0.
3. Apostolakis G (1990) The concept of probability in safety assessments of technological systems. Science 250:1359–1364
4. Apostolakis GE (2004) How useful is quantitative risk assessment? Risk Anal 24:515-520.
5. Apostolakis GE, Lemon DM (2005). A screening methodology for the identification and ranking of infrastructure vulnerabilities due to terrorism, Risk Anal 25(2):361-376.
6. Aven T (1992) Reliability and Risk analysis. Elsevier, London.
7. Aven T (2003) Foundations of risk analysis. Wiley, NJ
8. Aven T (2007) A unified framework for risk and vulnerability analysis and management covering both safety and security. Reliab Eng Syst Saf 92:745–754
9. Aven T (2008) Risk analysis. Wiley, NJ
10. Aven T (2008) A semi-quantitative approach to risk analysis, as an alternative to QRAs. Reliab Eng Syst Saf 93:790–797
11. Aven T (2011) Selective critique of risk assessments with recommendations for improving methodology and practice. Reliability Engineering and System Safety. 96, 509-514
12. Aven T (2009) Trends in risk analysis. International Journal of Performability Engineering. 5, 447-461.
13. Aven T, Hauge S, Sklet S, Vinnem JE (2006) Methodology for incorporating human and organizational factors in risk analyses for offshore installations. Int J Mater Struct Reliab 4:1–14
14. Aven T, Renn O (2009) On risk defined as an event where the outcome is uncertain. J. Risk Research, 12, 1-11.
15. Aven T, Vinnem JE (2007) Risk management, with applications from the offshore oil and gas industry. Springer Verlag, London
16. Aven T, Zio E (2011) Some considerations on the treatment of uncertainties in risk assessment for practical decision-making. Reliab Eng Syst Saf 96:64-74.
17. Bedford T, Cooke R (2001) Probabilistic Risk Analysis. Foundations and Methods. Cambridge University Publishing Ltd, Cambridge

18. Cabinet Office (2002) Risk: improving government's capability to handle risk and uncertainty. Strategy unit report, UK
19. Coolen FPA (2004) On the use of imprecise probabilities in reliability. Qual Reliab Eng Int 20:193–202
20. Coolen FPA, Utkin LV (2007) Imprecise probability: a concise overview. In: Aven T, Vinnem JE (eds) Risk, reliability and societal safet. Proceedings of the European Safety and Reliability Conference (ESREL), Stavanger, 25–27 June 2007. Taylor & Francis, London
21. Dewooght J (1998) Model uncertainty and model inaccuracy. Reliab Eng Syst Safety 59:171–185
22. Duijm NJ, Goossens L (2006) Quantifying the influence of safety management on the reliability of safety barriers. J Hazard Mater 130(3):284–292
22. Fischhoff B (1995) Risk perception and communication unplugged: twenty years of process. Risk Anal 15:501–527
24. Flage R, Aven T, Zio E (2008)Alternative representations of uncertainty in system reliability and risk analysis—review and discussion. ESREL 2008
25. Garrick JB et al (2004) Confronting the risks of terrorism: making the right decisions. Reliab Eng Syst Safety 86(2):129–176
26. Gudder S (2000) What is fuzzy probability theory? Found Phys 30(10):1663–1678
27. Haimes YY (2004) Risk modelling, assessment, and management, 2nd ed. Wiley, New Jersey
28. Helton JC, Burmaster DE (1996) On the treatment of aleatory and epistemic uncertainty in performance assessment of complex systems. Reliab Eng Syst Saf 54(2–3) (Special Issue on Aleatory and Epistemic Uncertainty)
29. Hollnagel E (2004) Barriers and accident prevention. Ashgate Publishers, Aldershot
30. HSE (2001) Reducing risk, protecting people. HES Books, ISBN 0717621510
31. International Risk Governance Council (IRGC) (2005) White paper on risk governance. Towards an integrative approach. Author: O. Renn with Annexes by P. Graham. International Risk Governance Council, Geneva
32. Kaplan S, Garrick BJ (1981) On the quantitative definition of risk. Risk Anal 1:11–27
33. Léger A., Duval C., Farret R., Weber P., Levrat E. and Iung, B., Modeling of Human and Organizational Impacts for System Risk Analyses, In Conference proceedings PSAM 9, Hong Kong,19-23/5-08, 2008.
34. Leveson N (2007) Modeling and analyzing risk in complex socio-technical systems. NeTWork workshop, Berlin, 27–29 Sep 2007
35. Lindley DV (2006) Understanding uncertainty. Wiley, Hoboken
36. Léger A, Duval C, Farret R, Weber P, Levrat E, Iung B (2008) Modeling of human and organizational impacts for system risk analyses, In: conference proceedings PSAM 9, Hong Kong, 19-23/5-08, 2008.
37. Mohaghegh Z, Kazemi R, Mosleh A (2009) Incorporating organizational factors into Probabilistic Risk Assessment (PRA) of complex socio-technical systems: A hybrid technique formalization. Reliability Engineering and System Safety, 94, 1000-1018.
38. Natvig B (1983) Possibility versus probability. Fuzzy Sets Syst 10:31–36
39. Nilsen T, Aven T (2003) Models and model uncertainty in the context of risk analysis. Reliab Eng Syst Safety 79:309–317
40. Papazoglou IA, Bellamy LJ, Hale AR, Aneziris ON, Post JG, Oh JIH (2003) I-Risk: development of an integrated technical and Management risk methodology for chemical installations. J Loss Prev Process Ind 16:575–591
41. Paté-Cornell ME (1996) Uncertainties in risk analysis: six levels of treatment. Reliab Eng Syst Saf 54(2–3):95–111
42. Paté-Cornell EM, Murphy DM (1996) Human and management factors in probabilistic risk analysis: the SAM approach and observations from recent applications. Reliab Eng Syst Saf 53:115–126
43. Paté-Cornell E, Dillon R (2001) Probabilistic risk analysis for the NASA space shuttle: a brief history and current work. Reliab Eng Syst Saf 74:345–352
44. PSA, Regulations Petroleum Safety Authority, Norway, 2001.

45. Rasmussen J (1997) Risk management in a dynamic society: a modelling problem. Saf Sci 27(2/3):183–213
46. Reid SG (1992) Acceptable risk. In: Blockley DI (ed) Engineering safety. McGraw-Hill, New York, pp 138–166
47. Renn O (1998) Three decades of risk research: accomplishments and new challenges. J Risk Res 1(1):49–71 (1998)
48. Rosa EA (1998) Metatheoretical foundations for post-normal risk. J Risk Res 1:15–44
49. Røed W, Mosleh A, Vinnem JE, Aven T (2008) On the use of hybrid causal logic method in offshore risk analysis. Reliab Eng Syst Saf (To appear)
50. Sandøy M, Sandøy M, Aven T, Ford D (2005) On integrating risk perspectives in project management. Risk Manag Int J 7:7–21
51. Singpurwalla N (2006) Reliability and Risk. A Bayesian Perspective. Wiley, NJ
52. Shafer G (1976) A mathematical theory of evidence. Princeton University Press, Princeton
53. Stirling A (1998) Risk at a turning point?'. J Risk Res 1:97–109
54. Stirling A (2007) Science, precaution and risk assessment: towards more measured and constructive policy debate. Eur Mol Biol Org Rep 8:309–315
55. Unwin SD (1986) A fuzzy set theoretic foundation for vagueness in uncertainty analysis. Risk Anal 6(1):27–34
56. Utkin LV, Coolen FPA (2007) Imprecise reliability: an introductory overview. In: Levitin G (ed) Computational intelligence in reliability engineering—new metaheuristics, neural and fuzzy techniques in reliability. Springer, London
57. Zadeh LA (1965) Fuzzy sets. Inform Control 8:338–353
58. Zadeh LA (1968) Probability measures of fuzzy events. J Math Anal Appl 23:421–427

Risk Importance Measures

Enrico Zio

1 Introduction

Quantitative information about the role that the components of a system play with respect to its risk, safety, reliability and availability is of great practical aid to system designers and operators. Indeed, the identification of which components mostly contribute to the system failure behavior allows one to trace system design bottlenecks and provides guidelines for effective operation and maintenance actions for system performance improvement.

In this regard, risk Importance Measures (IMs) are a fundamental outcome of the risk analysis of any complex technological system, as they are intended to quantify the contribution of the system components or their basic failure events to the system unreliability, unavailability or risk. For example, in the Probabilistic Safety Assessment (PSA) of nuclear power plants, IMs evaluate the importance of the components (or their basic failure events) with respect to their impact on the relevant risk measure, e.g., the Core Damage Frequency (CDF) or the Large Early Release Frequency (LERF). In other system engineering applications, such as aerospace and transportation, the impact of the components and their failure events is considered on the system unreliability or, for renewal systems such as the manufacturing production and power generation ones, on the system unavailability.

IMs were first introduced by Birnbaum [7]. The Birnbaum importance measure gives the contributions to the system reliability due to the reliability of the various system components. Components for which a variation in reliability results in the

E. Zio (✉)
Politecnico di Milano, Milan, Italy
e-mail: enrico.zio@supelec.fr; zio@ipmce7.cesnef.polimi.it

Ecole Centrale Paris et Supelec, Paris, France
e-mail: enrico.zio@ecp.fr

H. Pham (ed.), *Safety and Risk Modeling and Its Applications*,
Springer Series in Reliability Engineering, DOI: 10.1007/978-0-85729-470-8_6,

largest variation of the system reliability have the highest importance. Fussell and Vesely later proposed a measure based on the cut sets importance [12]. According to the Fussell–Vesely measure, the importance of a component depends on the number and on the order of the cutsets, in which it appears.

Other concepts of importance measures have been proposed and used, based on different views of the influence of the components on the system performance. Structural importance measures account for the importance of the logic position of the components in the system [21, 22]. Criticality importance measures consider the conditional probability of failure of a component, given that the system has failed [11, 16]. Joint importance measures account for the interactions of components in their contribution to system performance [2, 15].

Lately, IMs are being widely used in risk-informed applications of the nuclear industry to characterize the importance of basic events, i.e., component failures, human errors, common cause failures, etc., with respect to the risk associated to the system [9, 10, 24, 25]. In this framework, two other measures are frequently used: the risk reduction worth and the risk achievement worth [10]. The former one is a measure of the 'worth' of the basic event in achieving the present level of system risk and, when applied to components, it highlights the importance of maintaining the current level of reliability with respect to the basic failure events associated to such components. The latter one, the risk reduction worth, is associated to the maximum decrease in risk consequent to an improvement of the component associated with the basic event considered.

The use of importance measures in risk-informed applications relates to the ranking, or categorization of components or, more generally, basic events, with respect to their risk-significance or safety-significance. A distinction is made between ranking and categorization [10]: the purpose of ranking is generally to arrange items in order of increasing or decreasing importance; the purpose of categorization is to allocate items into groups, according to some pre-set guidelines or criteria.

Distinction is made also between risk-significance and safety-significance [10, 29]. Depending on the application, it may be appropriate to rank or categorize components or basic events, with respect to risk-significance or with respect to safety-significance. Risk-significance and safety-significance are regarded as complementary ways of identifying the role of components or basic events, in determining the risk from operation of the system. On one side, an individual component (basic event) can be identified as being risk-significant if its failure or unavailability (occurrence) contributes significantly to the measures of system risk, e.g., the core damage frequency, the large early release frequency, the unreliability or the unavailability. Safety-significance, instead, is related to the role that the component or the basic event plays in the prevention of the occurrence of the undesired system end state; in other words, safety-significance refers to the significance of a contribution to the probability of system success.

In general, there are two types of applications in risk-informed regulation. The first focuses on the high significance group, with the aim of gaining a reduction in the risk associated to system operation. A categorization according to risk-significance is more appropriate in this case since it focuses on the components that contribute most to the chosen measure of risk.

The second type of application aims at rendering more effective and less costly the requirements and Operation and Maintenance (O&M) activities by focusing them on what is risk-important, while relaxing those on the low significance group [10], provided that at most only a small risk increase results, well within the limits.

In this Chapter, some typical IMs used to characterize the importance of binary components, i.e., components which can be either functioning or faulty, are presented. The definitions of the IMs will be given with respect to the system reliability $R(t)$ and failure probability $F(t) = 1 - R(t)$, which will be taken as the measures of the risk associated to the system. The interpretations and uses of the different importance measures are critically discussed and their advantages and limitations highlighted. Topical issues are addressed, such as the importance of groups of components or basic events and the influence of uncertainty on the importance rankings. The extension of the definitions of the IMs to multi-state components and systems is also addressed.

A number of numerical and application examples are provided to substantiate the underlying definitions, concepts and interpretations.

The material presented draws from the abundant literature on the subject, organizing it in what is hoped to be a systematic presentation of the matter. Part of the base material has been organized by the Author in a similar format for a Chapter of the book Computational Methods for Reliability and Risk Analysis, World Scientific, 2009.

Importance measures are used in various fields to evaluate the relative importance of components in a system with respect to some criteria. This Chapter provides an overview on various types of importance measures relative to reliability and risk criteria. The different importance measures are critically discussed and their advantages and limitations highlighted with reference to the specific application context. The importance measures of groups of components are introduced and discussed. The influence of uncertainty on the importance ranking is considered. The extension to IMs for multi-state components and systems is given.

2 Birnbaum's Measure

Consider a system of n components. Let $\underline{r}(t) = (r_1(t), r_2(t), \ldots, r_n(t))$ be the vector of the reliabilities of the individual components at time t and let $R(\underline{r}(t))$ be the system reliability, dependent on the individual component reliabilities and on the system configuration. Birnbaum's measure of importance of the jth component is defined as:

$$I_j^B(t) = \frac{\partial R(\underline{r}(t))}{\partial r_j(t)} \tag{1}$$

Table 1 Birnbaum IMs of Example 1

Configuration no.	System configuration (system components)	R	I_1^B	I_2^B	I_3^B
I	Series (1–2)	0.9408	$r_2 = 0.96$	$r_1 = 0.98$	/
II	Parallel (1–2)	0.9992	$1 - r_2 = 0.04$	$1 - r_1 = 0.02$	/
III	2-out-of-3 (1–2–3)	0.9957	0.0952	0.0776	0.0584

If $I_j^B(t)$ is large, a small change in the reliability $r_j(t)$ of component j will lead to a large change in the system reliability R at time t.

Properties of Birnbaum's measure are:

i. $0 \le I_j^B(t) \le 1$
ii. When $R(\underline{r}(t))$ is a linear function of $\underline{r}(t)$ and if all components j are independent, then $I_j^B(t)$ does not depend on $r_j(t)$, $j = 1,2,\ldots,n$.

2.1 Example 1

Consider three system configurations composed of two or three binary components with the following reliabilities: $r_1 = 0.98$, $r_2 = 0.96$, $r_3 = 0.94$. Table 1 reports the Birnbaum importances of the three components in the different configurations [16].

In the series system configuration I, the less reliable component 2 is ranked highest. This result is general: in a series system the Birnbaum measure prioritizes components according to increasing values of reliability. The result is also reasonable from the logic structure of the system: in a series system the reliability is driven by the least reliable component, which constitutes the system bottleneck.

Dually, in the parallel configuration II, the importance measure I_j^B ranks highest component 1, the most reliable one. Again, this is reasonable from the point of view of the system logic structure: in a parallel system, the system behavior is driven by the most reliable components and I_j^B prioritizes the components accordingly. This is confirmed also by the ranking produced by I_j^B for the components of the system in configuration III, where the three components 1, 2 and 3 are in a 2-out-of-3 logic of operation. Again, Birnbaum's importance decreases with decreasing reliability of the components.

2.2 Analysis of Birnbaum's Measure in Terms of the System Structure Function

The system state can be represented by an indicator variable X_T which assumes the value of 1 when the system is functioning and 0 when it is faulty. X_T is a Boolean function of the Boolean variables $X_1, X_2,\ldots, X_n$ describing the states of the n components of the system ($X_j = 1$ if the component is functioning, 0 otherwise) [30]:

$$X_T = \Phi(X_1, X_2, \ldots, X_n) = \Phi(\underline{X}) \quad (2)$$

Such function is called a structure function and incorporates all the causal relations among the components states which lead to the system failure event. It maps an n-dimensional vector $\underline{X} = (X_1, X_2, \ldots, X_n)$ of Boolean variables (equal to 0's and 1's) onto a Boolean variable X_T (equal to 0 or 1).

The structure function can be written expliciting the indicator variable X_j of component j [30]:

$$\begin{aligned}\Phi[\underline{X}(t)] &= X_j(t)\Phi[\underline{X}(t), X_j = 1] + (1 - X_j(t))\Phi[\underline{X}(t), X_j = 0] \\ &= X_j(t)\{\Phi[\underline{X}(t), X_j = 1] - \Phi[\underline{X}(t), X_j = 0]\} + \Phi[\underline{X}(t), X_j = 0] \quad (3)\end{aligned}$$

By applying the expectation operator $E[\cdot]$ to $\Phi[\underline{X}(t)]$ and assuming that the components are independent, the system reliability is computed as [30]:

$$\begin{aligned}R(\underline{r}(t)) &= r_j(t) \cdot \{E[\Phi[\underline{X}(t), X_j = 1]] - E[\Phi[\underline{X}(t), X_j = 0]]\} + E[\Phi[\underline{X}(t), X_j = 0]] \\ &= r_j(t) \cdot \{R(r_j = 1, \underline{r}(t)) - R(r_j = 0, \underline{r}(t))\} + R(r_j = 0, \underline{r}(t)) \\ &= r_j(t) \cdot \left\{R_j^+(t) - R_j^-(t)\right\} + R_j^-(t) \quad (4)\end{aligned}$$

where,

- $R_j^+(t) = R(r_j = 1, \underline{r}(t)) = E[\Phi[\underline{X}(t), X_j = 1]]$ is the system reliability when component j is always in its functioning state, i.e., $X_j = 1$ and $r_j = 1$, throughout the time interval [0, t]. It represents the maximum reliability achievement if component j is considered perfect, i.e., always in the functioning state.
- $R_j^-(t) = R(r_j = 0, \underline{r}(t)) = E[\Phi[\underline{X}(t), X_j = 0]]$ is the system reliability when component j is always in the faulty state, i.e., $X_j = 0$ and $r_j = 0$, throughout the time interval [0, t]. It represents the maximum reduction in reliability if component j is considered failed with certainty and permanently, or, which is equivalent, removed from the system.

Hence, we can write:

$$I_j^B(t) = \frac{\partial R(\underline{r}(t))}{\partial r_j(t)} = R[r_j = 1, \underline{r}(t)] - R[r_j = 0, \underline{r}(t)] = R_j^+(t) - R_j^-(t) \quad (5)$$

Since $\Phi[\underline{X}(t), X_j = 1] - \Phi[\underline{X}(t), X_j = 0]$ can assume a value equal to 1 or 0 only, then

$$\begin{aligned}I_j^B(t) &= E\{[\Phi[\underline{X}(t), X_j = 1] - \Phi[\underline{X}(t), X_j = 0]]\} \\ &= P\{\Phi[\underline{X}(t), X_j = 1] - \Phi[\underline{X}(t), X_j = 0] = 1\} \\ &= P\{\text{the system state } (\underline{X}(t), X_j = 1) \text{ is a ``critical''; path vector (i.e., probability} \\ &\quad \text{that the system functions only if } X_j = 1 \text{, i.e., when component } j \text{ functions})\end{aligned}$$

Note that the fact that component j is critical tells nothing about the state of component j. The statement concerns only the states of the other system components, $\underline{X}(t)$, i.e., the system must be in such a state that component j being failed ($X_j = 0$) leads to the failure of the system ($\Phi(\underline{X}(t)) = 0$) and component j being functioning ($X_j = 1$) leads to the success of the system ($\Phi(\underline{X}(t)) = 1$).

For example, in the previous series system configuration I, component 1 is critical only if component 2 is functioning, regardless of the state of component 1 and thus $I_1^B = P[X_2 = 1] = r_2$, the reliability of component 2. Vice versa, for the parallel system configuration II, component 1 is critical only if component 2 is failed and thus $I_2^B = P[X_2 = 0] = 1 - r_2$.

Note also that, denoting by $q_j(t) = 1 - r_j(t)$ the probability that component j fails before t, the Birnbaum importance measure can be defined dually with respect to the system failure probability:

$$
\begin{aligned}
I_j^B(t) &= \frac{\partial R(\underline{r}(t))}{\partial r_j(t)} = R[r_j = 1, \underline{r}(t)] - R[r_j = 0, \underline{r}(t)] \\
&= \frac{\partial F\left(\underline{q}(t)\right)}{\partial q_j(t)} = F\left[q_j = 1, \underline{q}(t)\right] - F\left[q_j = 0, \underline{q}(t)\right] = F_j^+(t) - F_j^-(t)
\end{aligned}
\tag{6}
$$

where,

- $\underline{q}(t) = (q_1(t), q_2(t), \ldots, q_n(t))$ is the vector of the unreliabilities at time t of the individual components;
- $F\left[\underline{q}(t)\right] = 1 - R[\underline{r}(t)]$ is the system failure probability or unreliability (or, more generally, risk) at time t;
- $F_j^+(t) = F\left[q_j = 1, \underline{q}(t)\right] = P\{\Phi[\underline{X}(t), X_j = 0] = 0\}$ is the system failure probability or unreliability when component j is in its faulty state ($X_j = 0$) throughout the time interval [0, t]. It represents the maximum risk achievement if component j is considered failed with certainty and permanently, or, which is equivalent, removed from the system;
- $F_j^-(t) = F\left[q_j = 0, \underline{q}(t)\right] = P\{\Phi[\underline{X}(t), X_j = 1] = 0\}$ is the system failure probability or unreliability when component j remains in the functioning state ($X_j = 1$) throughout the time interval [0, t]. It represents the maximum reduction in risk if component j is considered perfect, i.e., always in the functioning state.

3 Criticality Importance

Birnbaum's importance for the component j at time t is independent of the reliability of component j itself, i.e., $I_j^B(t)$ is not a function of $r_j(t)$.

Let $C[\underline{X}(t), X_j = 1]$ be the event that the system is in a state such that j is critical. Such event is independent of the state of j. Then,

$$P\{C[\underline{X}(t), X_j = 1]\} = I_j^B(t) \tag{7}$$

The probability that j is critical and ($\cap$) failed at time t is

$$P\{C[\underline{X}(t), X_j = 1] \cap [X_j(t) = 0]\} = I_j^B(t) \cdot [1 - r_j(t)] \tag{8}$$

The "criticality importance" $I_j^{cr}(t)$ of component j at time t is defined as the probability that component j is critical for the system and failed at time t. given that the system has failed at time t:

$$\begin{aligned} I_j^{cr}(t) &= P\{C[\underline{X}(t), X_j = 1] \cap [X_j(t) = 0] \,|\, \Phi[\underline{X}(t)] = 0\} \\ &= \frac{P\{C[\underline{X}(t), X_j = 1] \cap [X_j(t) = 0]\}}{P\{\Phi[\underline{X}(t)] = 0\}} = \frac{I_j^B(t) \cdot [1 - r_j(t)]}{1 - R(\underline{r}(t))} = \frac{I_j^B(t) \cdot q_j(t)}{1 - R(\underline{r}(t))} \end{aligned} \tag{9}$$

In other words, $I_j^{cr}(t)$ is the probability that component j has caused the system failure, given that the system is failed at time t. When component j is repaired, the system will start functioning again.

3.1 Example 2

Let us consider the same three system configurations as in Example 1. Table 2 reports the criticality importance measures I_j^{cr} of the three components $j = 1, 2, 3$ [16].

A number of considerations follows:

- The numerator of I_j^{cr} is the probability that the system failure has been caused by component j. Indeed, for instance in case of the series system configuration I, the numerator is $r_2(1 - r_1) = P(\text{component 2 working}) \cdot P(\text{component 1 failed})$, i.e., the probability that the system failure has been caused by component 1. Similarly to I^B, in a series system the most important component according to I^{cr} is the least reliable one, for it will most probably be the cause for the system failure.
- In the parallel system configuration II, $I_1^{cr} = I_2^{cr}$ as it should be since if the system is failed, it will start functioning again irrespective of which of the components is repaired. Obviously, this result is general and applies to all simple parallel systems.

Table 2 Criticality IMs of Example 2

Configuration no.	System configuration (system components)	I_1^{cr}	I_2^{cr}	I_3^{cr}
I	Series (1–2)	$\frac{I_1^B(1-r_1)}{1-r_1r_2} = 0.3243$	$\frac{I_2^B(1-r_2)}{1-r_1r_2} = 0.662$	–
II	Parallel (1–2)	$\frac{I_1^B(1-r_1)}{1-r_1-r_2+r_1r_2} = 1$	1	–
III	2-out-of-3 (1–2–3)	0.4428	0.7219	0.8149

- In the 2-out-of-3 system configuration III, the importance is now increasing with decreasing component reliability, opposite to the Birnbaum's measure. Physically, the more unreliable the component is the more its contribution to the system failure.

4 Fussell–Vesely Importance Measure

The Fussell–Vesely importance measure of component j at time t, $I_j^{\mathrm{FV}}(t)$, takes into account the fact that a component may contribute to the system failure without being critical. A cut set is a set of events (component failures) whose representing vector of indicator variables $\underline{X}$ is such that the structure function $X_T = \Phi(\underline{X}) = 0$. A minimal cut set (mcs) is a cut set that does not have another cut set as a subset. Physically, a minimal cut set is an irreducible cut set: repairing one element (component) of the set, repairs the system [30] (i.e., when the indicator variable X of the element returns to 1 upon repair also the system indicator variable X_T returns to 1). A component contributes to system failure when a minimal cut set (mcs) containing its failure event occurs.

We then define the Fussel–Vesely importance measure as follows:

$I_j^{\mathrm{FV}}(t)$ = probability that at least one mcs containing j is verified at time t, given that the system is failed at t.

Let:

m_j number of mcs containing component j, $j = 1, 2,\ldots, n$,

$M_{jh} =$ hth mcs among those containing component j, verified at time t,

$D_j(t) =$ event that at least one mcs that contains component j is verified at time t

$$= M_{j1}(t) \cup M_{j2}(t) \cup \cdots \cup M_{jm_j}(t), (\cup = \text{logic OR operator})$$

Then,

$$I_j^{\mathrm{FV}}(t) = P\{D_j(t)|\Phi[\underline{X}(t)] = 0\} = \frac{P\{D_j(t) \cap \Phi[\underline{X}(t)] = 0\}}{P\{\Phi[\underline{X}(t)] = 0\}} = \frac{P\{D_j(t)\}}{P\{\Phi[\underline{X}(t)] = 0\}} \tag{10}$$

Assuming independent components,

$$\begin{aligned} P\{\Phi[\underline{X}(t)] = 0\} &= 1 - R(\underline{r}(t)) \\ P\{M_{jh}(t)\} &= \prod_{l \in M_{jh}} (1 - r_l(t)) \end{aligned} \tag{11}$$

However, since component j may belong to more than one minimal cut set, the mcs-events M_{jh}, $j = 1, 2,\ldots, n$, $h = 1, 2,\ldots, m_j$ may not be disjoint and independent, even if all the components are independent. Taking for simplicity the M_{jh}-independence assumption as valid [30],

$$P\{D_j(t)\} \cong 1 - \prod_{h=1}^{m_j} [1 - P\{M_{jh}(t)\}] \tag{12}$$

and the Fussell–Vesely importance measure may be written as:

$$I_j^{\mathrm{FV}}(t) \cong \frac{1 - \prod_{h=1}^{m_j} [1 - P\{M_{jh}(t)\}]}{1 - R(\underline{r}(t))} \tag{13}$$

Adopting the rare-event approximation, neglecting the situation of two or more mcs containing j verified at the same time [30]:

$$I_j^{\mathrm{FV}}(t) \cong \frac{\sum_{h=1}^{m_j} P\{M_{jh}(t)\}}{1 - R(\underline{r}(t))} = \frac{\sum_{h=1}^{m_j} P\{M_{jh}(t)\}}{F(t)} \tag{14}$$

Note from Eq. 14 that the numerator of $I_j^{\mathrm{FV}}(t)$ can be interpreted as the sum of the terms in the risk equation containing component j, i.e., the fraction of the risk pertaining to j. Then, $I_j^{\mathrm{FV}}(t)$ can be alternatively computed as:

$$I_j^{\mathrm{FV}}(t) \cong \frac{F(t) - F_j^-(t)}{F(t)} \tag{15}$$

where the numerator actually yields the part of F containing the term q_j.

4.1 Example 3

Let us consider the same three system configurations as in Example 1. Table 3 reports the Fussel–Vesely importance measures I_j^{FV} of the three components $j = 1, 2, 3$ [16].

A number of considerations follows:

- The values of $I_j^{\mathrm{FV}}, j = 1, 2,\ldots, n$, for the three considered system configurations are very close, if not equal, to the corresponding values of $I_j^{\mathrm{cr}}, j = 1, 2,\ldots, n$. This should not surprise since both I_j^{FV} and I_j^{cr} aim at quantifying the contribution of a component to the system failure probability, though from slightly different

Table 3 Fussel-Vesely IMs of Example 3

Configuration no.	System configuration (system components)	I_1^{FV}	I_2^{FV}	I_3^{FV}
I	Series (1–2)	$\frac{1-r_1}{1-r_1 r_2} = 0.3378$	$\frac{1-r_2}{1-r_1 r_2} = 0.6757$	–
II	Parallel (1–2)	1	1	–
III	2-out-of-3 (1–2–3)	0.4651	0.7442	0.8372

perspectives. Compare, for instance, the numerators of the two measures I_1^{FV} and I_1^{cr}, i.e., $(1 - r_1)$ and $r_2(1 - r_1)$, respectively: according to the Fussell–Vesely measure, component 1 contributes to the system failure when it fails (i.e., when its single-component cut set is verified), independently on whether component 2 is functioning or not; instead, the criticality measure considers the contribution to the system failure of component 1 only when it is this component responsible for the system failure, i.e., when 1 is failed, but 2 is working.

- In the parallel system configuration II, the system itself constitutes a minimal cut set, that is $D_1(t) = D_2(t) = \{\Phi[\underline{X}(t)] = 0\}$; it then follows that $I_1^{FV} = I_2^{FV} = 1$.

5 Risk Achievement Worth

The mathematical definition of the Risk Achievement Worth (RAW) of component j at time t is:

$$\mathrm{RAW}_j(t) = \frac{F\left[q_j = 1, \underline{q}(t)\right]}{F(t)} = \frac{F_j^+(t)}{F(t)} \tag{16}$$

In words, the risk achievement worth is the ratio of the risk when component j is considered always failed in $(0, t)$ $(q_j = 1, X_j = 0)$ to the actual value of the risk. It is a measure of the 'worth' of the basic event in achieving the present level of system risk and when applied to the system components, it highlights the importance of maintaining the current level of reliability with respect to the basic failure event associated to such components. The RAW is a very discriminating measure and it has to be interpreted very carefully. While it can be an appropriate measure for assessing the effect of a temporary change in the component operative condition in which it is made unavailable, if it is used in the context of assessing permanent changes, it is an extreme bounding measure since it considers only complete unavailability as a change.

5.1 Example 4

Let us consider the same three system configurations as in Example 1. Table 4 reports the RAW values of the three components $j = 1, 2, 3$, respectively.

A number of considerations follows:

- Components in series logic have the same value of RAW, e.g., components 1 and 2 in system configuration I. Indeed, the system is failed if any of the series components is failed, so that $F_j^+ = 1, j = 1, 2, \ldots, n$.

Table 4 Risk Achievement Worth IMs of Example 4

Configuration no.	System configuration (system components)	RAW_1	RAW_2	RAW_3
I	Series (1–2)	$\frac{1}{q_1+q_2-q_1q_2}=16.89$	$\frac{1}{q_1+q_2-q_1q_2}=16.89$	–
II	Parallel (1–2)	$\frac{q_2}{q_1q_2}=\frac{1}{q_1}=50$	$\frac{1}{q_2}=25$	–
III	2-out-of-3 (1–2–3)	22.67	18.31	13.75

- Components in parallel logic are ranked by RAW opposite to their failure probability. In other words, the more reliable components are ranked first, as suggested by the Birnbaum measure. Reasonably, for parallel components the achievement in risk is highest if the most reliable component is taken out of service.

6 Risk Reduction Worth

The mathematical definition of the Risk Reduction Worth (RRW) of component j at time t is:

$$\mathrm{RRW}_j(t)=\frac{F(t)}{F\left[q_j(t)=0,\underline{q}(t)\right]}=\frac{F(t)}{F_j^-(t)} \tag{17}$$

In words, the risk reduction worth is the ratio of the nominal value of the risk to the risk when component j is always available ($q_j=0$, $X_j=1$). It measures the potential of component j in reducing the risk, by considering the maximum decrease in risk achievable when the component j is always perfectly operating. This measure is useful for identifying improvements which can most reduce risk.

6.1 Example 5

Let us consider the same three system configurations as in Example 1. Table 5 reports the RRW values of the three components $j = 1, 2, 3$, respectively.

A number of considerations follows:

- Components in series logic are ranked by RRW in increasing order of failure probability. Reasonably, for series components the reduction in risk achievable by improving the component to perfection is highest for the components which contribute most to the system failure, i.e., the least reliable, bottleneck components.
- Components in parallel logic have the same RRW values. Indeed, a simple parallel system cannot fail if any of its constituting components cannot fail so that $F_j^-=0$ and $\mathrm{RRW}_j=\infty, j=1,2,\ldots,n$.

Table 5 Risk Reduction Worth IMs of Example 5

Configuration no.	System configuration (system components)	RRW_1	RRW_2	RRW_3
I	Series (1–2)	$\frac{q_1+q_2-q_1q_2}{q_2} = 1.48$	$\frac{q_1+q_2-q_1q_2}{q_1} = 2.96$	-
II	Parallel (1–2)	$\frac{q_1q_2}{0} = \infty$	$\frac{q_1q_2}{0} = \infty$	–
III	2-out-of-3 (1–2–3)	1.79	3.58	5.38

7 Different Importance Measures for Different Uses

Different importance measures may lead to different rankings of the components. The analyst should be aware of such differences for a proper use of the informative content provided by the measures.

A most important area of application of importance measures is in support to the establishment of test and maintenance programs [24], which greatly influence the unavailability of components. In this respect, the ranking produced by the Birnbaum importance measure I^B seems to be the most appropriate one. On the other hand, the question on 'what will be the result in terms of risk, when a certain component is taken out of service' seems to be best addressed by RAW.

The other traditional area of application of IMs is in the design of systems and plants. Significant components are selected with the aid of IMs and improvement in the design is introduced for decreasing the unavailability of the selected components and improving the defense-in-depth against their failures. Two IMs are often used for these purposes: $I^{FV}$and I^B. The I^{FV} importance is used for the selection of components candidate for improvement because it contributes to risk the most; then, the information from the Birnbaum importance measure I^B allows identifying for which components the improvements are most effective.

The I^{FV} importance measure is also the most appropriate one for identifying the components that most probably are the cause of system failure and, therefore, it can be used to set up a repair priority list.

Another important issue concerns IMs which are most appropriate to rank or categorize components with respect to risk-significance or with respect to safety-significance (as defined in Sect. 1). Let us represent the risk metric F by the following linear equation [29], which can be derived proceeding as for the dual Eq. 4 for the system reliability:

$$F = \alpha_j \cdot q_j + \beta_j \tag{18}$$

where q_j is the unavailability of the generic component j, $\alpha_j = F_j^- - F_j^+$ is the coefficient with which q_j appears in the risk equation and $\beta_j = F_j^-$represents the collection of all the other terms of F that do not contain q_j. Eq. 18 holds when component j is independent from the other components.

Using Eqs. 18 and 15 one can write the expression for the Fussel–Vesely IM as [29]:

$$I_j^{\text{FV}} = \frac{\alpha_j q_j + \beta_j - \beta_j}{\alpha_j q_j + \beta_j} = \frac{\alpha_j q_j}{\alpha_j q_j + \beta_j} \approx \frac{\alpha_j q_j}{\beta_j}, \quad \text{when } \alpha_j q_j \ll \beta_j \tag{19}$$

The Fussell–Vesely importance measure is often used as a measure of risk-significance. The assumption $\alpha_j q_j \ll \beta_j$ is verified in high-risk installations, such as the nuclear ones, which are designed with significant redundancy according to the defense-in-depth principle, so that it is very unlikely that a single component alone contributes much to the risk. I_j^{FV}, then, turns out to be proportional to the unavailability of component j and represents the contribution of component j to the risk metric F: in this sense, I_j^{FV} represents the risk-significance of component j. Note that $\alpha_j q_j$ is the probability of the union of all the minimal cut sets containing component j, so that I_j^{FV} can be alternatively interpreted as the relative contribution to risk of all the minimal cut sets containing component j.

Similarly, the expression (16) of RAW can be written as:

$$\text{RAW}_j = \frac{\alpha_j + \beta_j}{\alpha_j q_j + \beta_j} \approx \frac{\alpha_j}{\beta_j} + 1, \quad \text{when } \alpha_j q_j \ll \beta_j \tag{20}$$

Thus, when $\alpha_j q_j \ll \beta_j$, RAW_j is independent on q_j and represents the degree of defense against failure provided by the rest of the installation. The risk achievement worth RAW is thus typically used to characterize components according to their safety-significance, as it is a measure of the impact of setting to one the unavailability of the particular component, i.e., of removing it. A high value of RAW_j means that component j is highly safety-significant since the increase in risk due to the unavailability of the component is high. Note that RAW_j represents a somewhat extreme measure of the amplification of the system risk due to component j, since it assumes its complete unavailability. Hence, its use to rank components and define changes in the technical specifications (surveillance and/or test frequencies, etc.) must be very careful: there are very few components, if any, for which the impact of a proposed change is to render them totally ineffective. For this reason, the use of RAW as a safety-significance measure is still controversial and a debate is ongoing among the practitioners on whether other measures could be more suitable.

Finally, note that in general all measures are time-dependent: at different times one may get different rankings of the importance of the components depending on the current state of the system and its components.

8 Open Issues on Importance Measures

The following concerns have been raised on the IMs [10]:

1. IMs produce risk rankings that are not necessarily related to the risk change that results from credible changes to the contributor probabilities q_j. Indeed, IMs

deal with changes in reliability or risk only at the extremes (0,1) of the defined range of probability.

2. IMs rank only individual components or basic events whereas they are not directly applicable to combinations or groups of components or basic events. Indeed, there is no simple relationship between the importance measures evaluated at the single component or basic event level and those evaluated at the level of a group of components or basic events. In practice, different basic events may represent different modes of failure or unavailability of a single component and in order to determine the importance of such component one has to consider all the related basic events as a group. Furthermore, many risk-informed applications deal with evaluating the risk change associated to changes in the plant technical specifications (surveillance and/or test frequencies, etc.) which impact a group of components.
3. IMs do not typically consider the credible uncertainty range of components' unavailabilities or basic event probabilities and this raises a doubt on the robustness of the conclusions drawn from importance analyses.
4. IMs have been mainly applied to systems made up of binary components (i.e., components that can be in two states: functioning or faulty). This kind of systems have many practical applications; yet the hypothesis of dichotomizing the components and system states is often over-simplified and insufficient for describing the real functioning of many systems, whose performance can settle on different levels (e.g., 100, 80, 50% of the nominal capacity) depending on the operative conditions of the constitutive multi-state components.

Research efforts are being performed to address the above concerns. For example, to address the first issue, i.e., the fact that the importance measures deal with changes in the probabilities of the basic events only at the extremes 0 or 1 of their ranges, a generalized risk importance measure has been proposed [10] which depends also on the actual value of a proposed change in the probability of the basic failure event (see Sect. 9).

Furthermore, a Differential Importance Measure, DIM, has been introduced to partially overcome the second issue [9]. The DIM is a first-order sensitivity measure that ranks the parameters of the risk model according to the fraction of the total change in risk due to a small change in the parameters' values, taken one at a time. The DIM bears the important property of additivity: the DIM of a group of components or basic events is the sum of the DIMs of the single components or basic events of the group. The concepts underpinning the DIM definition will be illustrated in details in the later Sect. 12.

The need for importance measures capable of considering combinations of components arises also when planning a budget-constrained improvement in the reliability of a system design, for example by replacing one of its components with a better-performing one or by inspecting and maintaining it more frequently. Due to the budget constraints, the improvement may need to be accompanied by the sacrifice of the performance of another, less important component.

The interactions of these coupled changes to system design must be accounted for when assessing the importance of the system components. To this aim, second-order sensitivity measures such as the Joint Reliability Importance (JRI) and Joint Failure Importance (JFI) measures have been introduced [2, 15].

A second-order extension of the DIM, named $\mathrm{DIM}^{\mathrm{II}}$, has been proposed for accounting the interactions of pairs of components when evaluating the change in system performance due to changes of the reliability parameters of the components. The extension aims at supplementing the first-order information provided by DIM with the second-order information provided by JRI and JFI for use in risk-informed decision-making [32].

As for the third issue, in general the importance measures are random variables. For example, the definition of the risk reduction worth RRW_j is given by Eq. 17 as the ratio of the nominal value of the risk F to the risk when component j is always available, F_j. F and F_j^- are to be considered as random variables characterized by given probability distributions; then, RRW_j is a random variable for which specific statistics can be calculated, e.g., mean, median, etc. The probability distribution for RRW_j describes the random variability due to the intrinsic randomness of the system performance states. Uncertainties in the reliability parameters of the system components (epistemic uncertainties) have also to be included to give the total probability distribution. When the importance measures are treated as random variables, the non-trivial inter-comparison of their distribution for component ranking must be carried out. Methods have been proposed in the literature, but their application in practice is rare or non-existent, mainly due to the computational difficulty and burden [6, 23]. Some of these methods will be illustrated in Sect. 13.

Finally, as for the last issue, some research results on the generalization of IMs for application to multi-state systems made up of multi-state components will be illustrated in details in Sect. 14.

9 Generalized Risk Importance Measure

To overcome the fact that the importance measures deal with changes in the probabilities of the basic events only at the extremes 0 or 1 of their ranges, a generalized risk importance measure has been defined by considering the following relative change in risk due to a change in the probability of the basic failure event j from the value q_j to the value q_j^n [10]:

$$\frac{\Delta F_j}{F} = \frac{F_j^n - F}{F} = \left\{F\left[q_j = 1, \underline{q}(t)\right] - F\left[q_j = 0, \underline{q}(t)\right]\right\}\left(\frac{q_j^n - q_j}{F}\right) \quad (21)$$

where:

q_j^n = the considered new value for the probability of the basic failure event j,

F_j^n = the system risk measure with the new value for the probability of the basic failure event j.

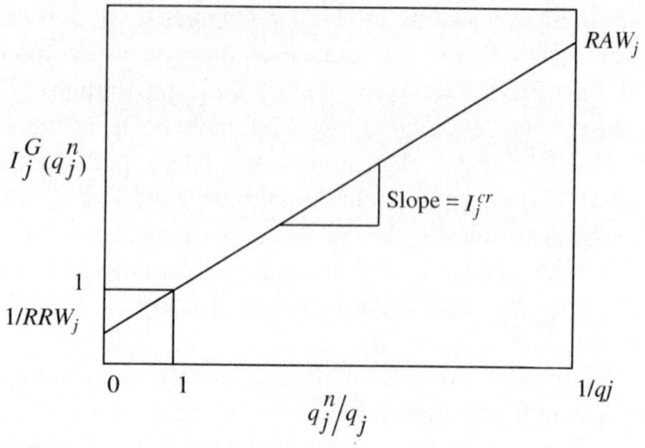

Fig. 1 Risk impact curve [10]

Rearranging the equation slightly, yields the generalized importance measure, $I_j^G(q_j^n)$:

$$I_j^G\left(q_j^n\right) = \frac{F_j^n}{F} = \left\{F\left[q_j = 1, \underline{q}(t)\right] - F\left[q_j = 0, \underline{q}(t)\right]\right\}\left(\frac{q_j^n - q_j}{F}\right) + 1 \quad (22)$$

Further rearrangement and the use of the importance measures definitions previously introduced yields:

$$I_j^G\left(q_j^n\right) = \frac{F_j^n}{F} = I_j^{cr}\left(\frac{q_j^n}{q_j}\right) + \frac{1}{\mathrm{RRW}_j} \quad (23)$$

The defined importance measure is considered general since q_j^n can take any value and is not restricted to a value of 0 or 1 as is the case in the definitions of RRW, RAW, I^{FV}. The continuous relationship between q_j^n and the generalized importance measure $I_j^G(\cdot)$ gives rise to the so-called 'risk impact curve'. The relationship is linear with a slope equal to the criticality importance and the y-axis intercept equal to the inverse of the risk reduction worth. Furthermore, when q_j^n equals unity, the generalized importance measure attains the value of the risk achievement worth. Figure 1 illustrates this relationship [10].

10 Importance Measures for Multiple Basic Events

An issue of concern is that importance measures rank only individual components or basic events whereas they are not directly applicable to combinations or groups of components or basic events. In practice different basic events may, for example, represent different modes of failure or unavailability of a single component and in order to determine the importance of such component one has to consider all the related basic events as a group. Furthermore, many risk-informed applications deal with evaluating the risk change associated to changes in the plant technical specifications (surveillance and/or test frequencies, etc.); such changes may indeed impact a group of components.

In general there is no simple relation between the importance measures of the individual components and the group taken as a whole.

Suppose that the expression of the indicator variable T of the failure event of a system, in terms of the indicator variables of its basic Boolean events $A, B, C_1, C_2, C_3, C_4, D, E, F, G, H$ is of the form [10]:

$$T = AB(C_1 + C_3) + DE(C_2 + C_4) + F(C_1 + C_3)(C_2 + C_4) + GH \tag{24}$$

where as usual the product of two indicator variables denotes the intersection of the two associated events and the sum denotes the union [30]. Let us calculate some importance measures for the components of group C, constituted by two pairs of components in series, $C_1 - C_3$ and $C_2 - C_4$.

10.1 Risk Achievement Worth

For a single component, the calculation of the risk achievement worth entails calculating the risk of the system when the failure event of the considered component is verified with certainty. For example, the expression for RAW(C_1) in terms of basic events is [10]:

$$\mathrm{RAW}(C_1) = \frac{AB(1 + C_3) + DE(C_2 + C_4) + F(1 + C_3)(C_2 + C_4) + GH}{AB(C_1 + C_3) + DE(C_2 + C_4) + F(C_1 + C_3)(C_2 + C_4) + GH} \tag{25}$$

where the indicator variable C_1 has been replaced by the value which indicates occurrence of the associated event of failure of the component C_1. Then, the numerical value of RAW(C_1) can be calculated by replacing the indicator variables with the probabilities of the corresponding basic events.

Analogously, for the components of group C, the calculation of the group RAW entails calculating the system risk when the failure event of all of the components belonging to the group is verified with certainty. An approach for calculating the group RAW could be the simple setting at the value 1 of all the indicator variables related to the basic events belonging to group C. This would result in [10]:

$$\mathrm{RAW}(C) = \frac{2AB + 2DE + 4F + GH}{AB(C_1 + C_3) + DE(C_2 + C_4) + F(C_1 + C_3)(C_2 + C_4) + GH} \tag{26}$$

Note that this way of proceeding leads to the fact that each of the two pairs of components in series $C_1 - C_3$ and $C_2 - C_4$ contributes with a term 2 in the system structure function. Generalizing, the effect of this simple substitution is that for n components in series, the unavailability of the series group is n times the unavailability of the single component, which has been set to 1 and thus it leads to an unavailability equal to n. Indeed, this is not the correct approach to calculate the system risk when all of the failure events of the group are verified: the effect of this simple substitution is the generation of non-minimal cut sets.

Another approach could be to evaluate the RAW for each of the basic events and then add all the resulting RAWs. This approach, however, has problems similar to the above formulation since the numerator of this sum would contain a contribution of 4*GH*, since each of the four single-component RAWs would contribute a term *GH*.

The correct way of proceeding for calculating the group RAW is to evaluate the structure function as a probability equation with the appropriate Boolean reduction [30]. The step-by-step procedure to be followed is:

1. Take the structure function of the system failure event corresponding to the measure of interest (CDF, LERF, unreliability, etc.).
2. Rename the basic events in the group under investigation so that they all have the same identifier.
3. Boolean-reduce the thereby obtained structure function.
4. Calculate the risk index for the new basic event with its value set at unity.

This gives an importance measure for the group. In this case, the group is totally correlated in its effect, since the probabilities of the individual members of the group are set to one. The difference between this approach and the previous one setting individually to one of the group event probabilities in the not-reduced expression for T can be seen in the following.

The substitution of C_1, C_2, C_3, C_4 by C and the re-reducing of the Eq. 24 would give [10]:

$$T = ABC + DEC + FC + GH \tag{27}$$

and

$$\mathrm{RAW}(C) = \frac{AB + DE + F + GH}{AB(C_1 + C_3) + DE(C_2 + C_4) + F(C_1 + C_3)(C_2 + C_4) + GH} \tag{28}$$

It can be easily seen that following this approach, the group RAW cannot be expressed simply in terms of combinations of the RAW measures for the individual members of the group. A consequence of this fact is that the risk equation must be re-reduced and evaluated each time a group RAW is to be evaluated.

10.2 Birnbaum Importance Measure

The Birnbaum importance measure for an individual basic event is evaluated by

$$I_j^B = F\left[q_j = 1, \underline{q}(t)\right] - F\left[q_j = 0, \underline{q}(t)\right] \tag{29}$$

For the single jth basic event, the sensitivity of the risk measure to the probability of that event, q_j, can be parameterized as

$$F(q_j) = I_j^B q_j + F\left[q_j = 0, \underline{q}(t)\right] \tag{30}$$

The meaning of the Birnbaum importance measure therefore is that it represents the sensitivity coefficient of the risk measure to the probability of that basic event and provides one way of looking at the defense-in-depth issue in a probabilistic sense.

One may think of applying the procedures discussed above for the RAW to produce meaningful Birnbaum importance measures of groups of events. For example, using the not-reduced structure function (24) if a substitution of 1 were made for each member C_k of the group C, the resulting 'group Birnbaum measure' would be [10]:

$$I_C^B = 2AB + 2DE + 4F \tag{31}$$

On the contrary, following the same substitution and re-reduction procedure, which leads to (27) would result in a 'group Birnbaum measure' of $AB + DE + F$. That neither of these Birnbaum importance measures is an appropriate sensitivity measure will be shortly shown in the following Sect. 11.

10.3 Fussell–Vesely Importance

The Fussell–Vesely measure of importance for a single basic event represents the fraction of the risk measure to which the basic event contributes, i.e., it is the sum of the cut sets involving such basic event divided by the sum of all the cut sets. The Fussell–Vesely measure obtained by including all cut sets that contain one or more basic events of the group C is given by [10]:

$$I_C^{FV}(C) = \frac{AB(C_1 + C_3) + DE(C_2 + C_4) + F(C_1 + C_3)(C_2 + C_4)}{AB(C_1 + C_3) + DE(C_2 + C_4) + F(C_1 + C_3)(C_2 + C_4) + GH} \tag{32}$$

This is a measure that assesses the contribution of the group C in such a way that any cut set that has a contribution from any one member C_k of the group is included, $k = 1, 2, 3, 4$. Note, however, that this is not the same result that would be obtained by adding the individual Fussell–Vesely measures $I_{C_k}^{FV}$, $k = 1, 2, 3, 4$. Since this measure does not involve assessing changes, but is a simple ratio of contributors, this is an appropriate measure of group importance.

10.4 Risk Reduction Worth

The risk reduction worth importance of a single basic event is the ratio of the risk value to that calculated with the probability of such basic event set to 0. Substituting 0 for each member of the group to calculate the RRW of the group is an appropriate way since in this case there is no problem with non-minimal cut sets.

11 Relationship of Importance Measures to System Risk Changes

Paradoxically, importance measures are for the most part not directly related to the risk changes associated with the change in the system which is considered. That this is true for those importance measures which are based on taking parameter values or basic event probabilities to their extremes should be obvious. Thus, there is concern in identifying sensitivity measures related to importance measures that can fill the role of characterizing directly the change in risk, particularly when a group of components is affected by the change. As a simple example, take a cut set equation in terms of indicator variables and treat it as an algebraic equation, replacing each of the events in the group C of interest by a common indicator variable, and without performing a Boolean reduction, differentiate the equation with respect to that variable. For the Eq. 24, replacing C_k with C, $k = 1, 2, 3, 4$, the algebraic equation would become [10]:

$$T = 2ABC + 2DEC + 4FC^2 + GH \tag{33}$$

and differentiating:

$$\frac{\partial T}{\partial C} = 2AB + 2DE + 8FC \tag{34}$$

This is a sensitivity parameter that is valid when the changes in the value for C are small, and the approximation

$$\Delta E[T] = \Delta F = \frac{\partial F}{\partial C}\Delta C \tag{35}$$

is appropriate if the impact of the change on each member of the group is the same. This sensitivity parameter is, however, different from any of the importance measures presented in the previous sub-chapters. As the magnitude of the changes in C increases, higher order derivatives are needed to assess the change in F. Thus, it can be concluded that the sensitivity of risk to a multi-component change cannot easily be related to single-component importance measures.

12 The Differential Importance measure (DIM)

As highlighted in the previous Sections, a limitation of the above mentioned importance measures is that they rank only individual components or basic events whereas they are not directly applicable to combinations or groups of components or basic events [10]. To partially overcome this limitation, the Differential Importance Measure, DIM, has been introduced for use in risk-informed decision making [8, 9]. The DIM is a first-order sensitivity measure that ranks the parameters of the risk model according to the fraction of the total change in the risk that is due to a small change in the parameters' values, taken one at a time. The DIM bears an important property of additivity: the DIM of a group of components or basic events is the sum of the DIMs of the single components or basic events of the group.

In what follows, we briefly recall the concepts underlying the definition of the DIM introduced in [9].

Consider the generic risk metric F. In general, the risk metric of interest can be expressed as a function $F(p_1, p_2, \ldots, p_{N_p})$ of the parameters p_i, $i = 1, 2, \ldots, N_p$ of the underlying stochastic model (components failure rates, repair rates, ageing rates, maintenance intervals, human error probabilities, etc.). The total variation of the function of interest due to small variations in its parameters, one at a time, is given by the differential

$$dF = \frac{\partial F}{\partial p_1} \cdot dp_1 + \frac{\partial F}{\partial p_2} \cdot dp_2 + \cdots + \frac{\partial F}{\partial p_{N_P}} \cdot dp_{N_P} \tag{36}$$

The DIM of parameter p_i is then defined as the fraction of total change in F that is due to a change in the parameter value

$$\mathrm{DIM}(p_i) = \frac{dF_{p_i}}{dF} = \frac{\frac{\partial F}{\partial p_i} \cdot dp_i}{\frac{\partial F}{\partial p_1} \cdot dp_1 + \frac{\partial F}{\partial p_2} \cdot dp_2 + \cdots + \frac{\partial F}{\partial p_{N_P}} \cdot dp_{N_P}} \tag{37}$$

Because of its definition, once all the individual sensitivities $\partial F/\partial p_i$, $i = 1, 2, \ldots, N_p$ have been computed, the DIM enjoys the additivity property, i.e., the DIM of a subset of parameters, p_i, $p_j, \ldots, p_k$ is the sum of the DIMs of the individual parameters:

$$\begin{aligned} \mathrm{DIM}(p_i, p_j, \ldots, p_k) &= \frac{\frac{\partial F}{\partial p_i} \cdot dp_i + \frac{\partial F}{\partial p_j} \cdot dp_j + \cdots + \frac{\partial F}{\partial p_k} \cdot dp_k}{dF} \\ &= \mathrm{DIM}(p_i) + \mathrm{DIM}(p_j) + \cdots + \mathrm{DIM}(p_k) \end{aligned} \tag{38}$$

Viewing the definition of DIM in Eq. 37 in terms of a limit for the parameter variation going to zero, allows defining the operational steps for its computation. Two different hypotheses can be considered:

1. all the parameters change by the same small value (uniform changes);
2. the parameters are changed by the same percentage (uniform percentage changes).

Under hypothesis (1), DIM(p_i) measures the importance of parameter p_i with respect to a small equal change in all parameters; under hypothesis (2), DIM(p_i) measures the importance of parameter p_i when all the parameters are changed by the same fraction of their nominal values.

Clearly, the two assumptions address different situations and should lead to different importance values. The conditions under which to apply one hypothesis or the other depend on the problem and risk metric model at hand. In particular, when investigating the effects of changes at the parameter level, hypothesis (1)

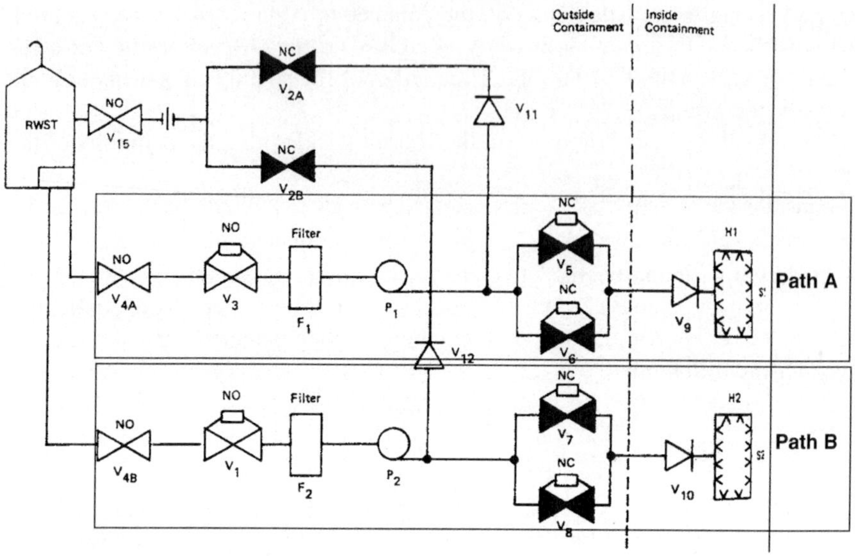

Fig. 2 CSIS simplified flow diagram [27]

cannot be used since the parameters may have different dimensions (e.g., failure rates have inverse-time units, maintenance intervals have time units and human error probabilities are dimensionless numbers).

12.1 Example 6

An application of DIM is considered with reference to the Containment Spray Injection System (CSIS) of a nuclear power plant. The function of the Containment Spray Injection System (CSIS) is to deliver cold water containing boron through spray heads from the Refuelling Water Storage Tank (RWST) to the containment volume during the first half hour after a large Loss Of Coolant Accident (LOCA). Refer to [27] for a comprehensive description of the system [19].

The principal objective of CSIS is to reduce the pressure in the containment. The CSIS also provides the preferred path for delivery of sodium hydroxide to the containment for initial fission product removal. Figure 2 shows a simplified flow diagram of the system. The CSIS consists of two redundant spray subsystems from the RWST to the containment. The valves colored in black in Fig. 2 are normally closed during plant operation. In order to operate both subsystems of the CSIS, valves V5 or V6 and V7 or V8 must be opened and pumps P1 and P2 must be started. In the event of a large LOCA this would normally be done by a signal from the Consequence Limiting Control System (CLCS). It should be noted that valves V1 and V3 also receive a CLCS signal to prevent those valves from being closed

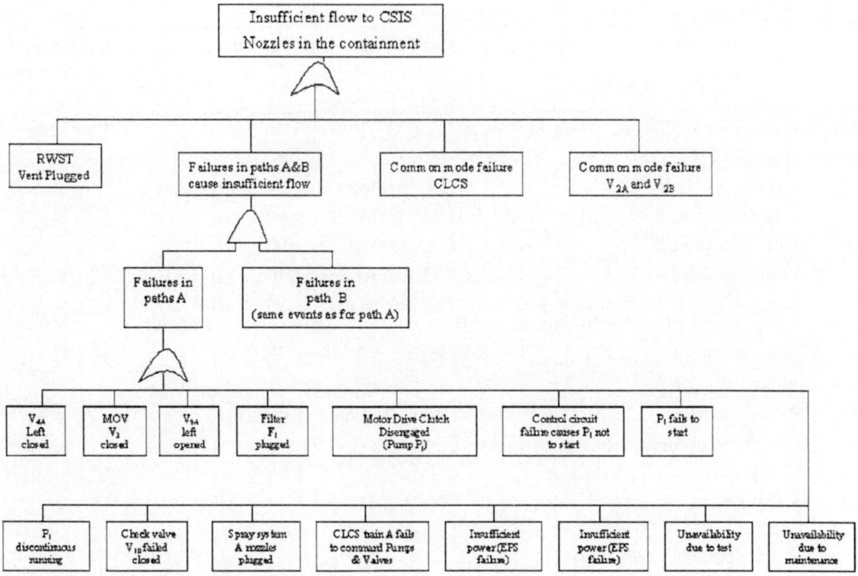

Fig. 3 Fault tree for the CSIS, adapted from [27]

during the CSIS operation or to open them should they have been inadvertently closed.

The CSIS is designed on the following basis:

a. Either spray subsystem S_1 or S_2 will provide sufficient spray to the containment atmosphere.
b. The CSIS is required to function only until the water supply in the RWST is exhausted.

The CSIS is considered to fail in its function when it is incapable of delivering spray fluid from the RWST to the containment atmosphere at a rate at least equivalent to the full delivery from one of the two containment spray pumps. The fault tree for the CSIS failure event is reported in Fig. 3. The unavailability data of the basic events are reported in Table 6, with the original reference coding from [27]. The minimal cut sets can be readily found by inspection of the fault tree. There are three first-order cut sets, one consisting of a failure related to the RWST (Event 1 in Table 6), which is the only water supply for the CSIS, whereas the others are common mode failures (Events 32 and 33 in Table 6). The first common mode failure refers to the CLCS and accounts for the miscalibration of several sensors that prevent the proper CLCS signal from reaching the CSIS in the event of a LOCA. The second common mode failure refers to the possibility that both CSIS flow recirculation valves V_{2A} and V_{2B} were left open after the monthly pump test due to an operator error. Several second-order cut sets also contribute to the CSIS unavailability, arising from the combination of all the failure events related to one of the two injection paths with all those of the other.

Table 6 Unavailability data of the failure events of the CSIS [27]

	Event/component	Code [27]	Unavailability, q_i	Occurrence rate, λ_i (day^{-1})
1	RWST Vent Plugged	CVT0001P	4.40×10^{-7}	1.21×10^{-9}
Failures on path A				
2	V_{4A} left closed	CXVA004X	1.00×10^{-3}	2.74×10^{-6}
3	MOV V_3 closed	CMV100AC	1.00×10^{-4}	2.74×10^{-7}
4	V_{2A} left opened	CXVA002X	1.00×10^{-2}	2.75×10^{-5}
5	Filter F_1 plugged	CFLA01AP	1.10×10^{-4}	3.01×10^{-7}
6	Motor Drive Clutch disengaged (Pump P_1)	CCL1A01G	3.00×10^{-7}	8.22×10^{-7}
7	Control circuit failure causes P_1 not to start	CST1A01F	1.00×10^{-3}	2.74×10^{-6}
8	P_1 fails to start	CPMA01AA	1.00×10^{-3}	2.74×10^{-6}
9	P_1 discontinuous running	CPMA01AF	1.50×10^{-5}	4.11×10^{-8}
10	Check valve V_{10} failed closed	CCVA001C	1.00×10^{-4}	2.74×10^{-7}
11	Spray system A nozzles plugged	CNZA001P	1.30×10^{-4}	3.56×10^{-7}
12	CLCS train A fails to command pumps and valves	GCL01	4.60×10^{-3}	1.26×10^{-5}
13	Insufficient power (EPS failure)	JD00	4.10×10^{-5}	1.12×10^{-7}
14	Insufficient power (EPS failure)	JK00	1.10×10^{-6}	3.01×10^{-7}
15	Unavailability due to test	No code	1.94×10^{-3}	5.32×10^{-6}
16	Unavailability due to maintenance	No code	2.20×10^{-3}	6.06×10^{-6}
Failures on path B				
17	V_{4B} left closed	CXVB004X	1.00×10^{-3}	2.74×10^{-6}
18	MOV V_1 closed	CMV100BC	1.00×10^{-4}	2.74×10^{-7}
19	V_{2B} left opened	CXVB002X	1.00×10^{-2}	2.75×10^{-5}
20	Filter F_2 plugged	CFLB01AP	1.10×10^{-4}	3.01×10^{-7}
21	Motor Drive Clutch disengaged (Pump P_2)	CCL1B01G	3.00×10^{-4}	8.22×10^{-7}
22	Control circuit failure causes P_2 not to start	CST1B01F	1.00×10^{-3}	2.74×10^{-6}
23	P_2 fails to start	CPMB01BA	1.00×10^{-3}	2.74×10^{-6}
24	P_2 discontinuous running	CPMB01BF	1.50×10^{-5}	4.11×10^{-8}
25	Check valve V_9 failed closed	CCVB001C	1.00×10^{-4}	2.74×10^{-7}
26	Spray system B nozzles plugged	CNZb001P	1.30×10^{-4}	3.56×10^{-7}
27	CLCS train B fails to command pumps and valves	GCL02	4.60×10^{-3}	1.26×10^{-5}
28	Insufficient power (EPS failure)	JC00	4.10×10^{-5}	1.12×10^{-7}
29	Insufficient power (EPS failure)	JJ00	1.10×10^{-6}	3.01×10^{-9}
30	Unavailability due to test	No code	1.94×10^{-3}	5.32×10^{-6}
31	Unavailability due to maintenance	No code	2.20×10^{-3}	6.06×10^{-6}
Common mode failures				
32	CLCS signal fail to reach CSIS	No code	1.00×10^{-3}	2.74×10^{-6}
33	Both V_{2A} and V_{2B} left open after test	No code	9.00×10^{-4}	2.47×10^{-6}

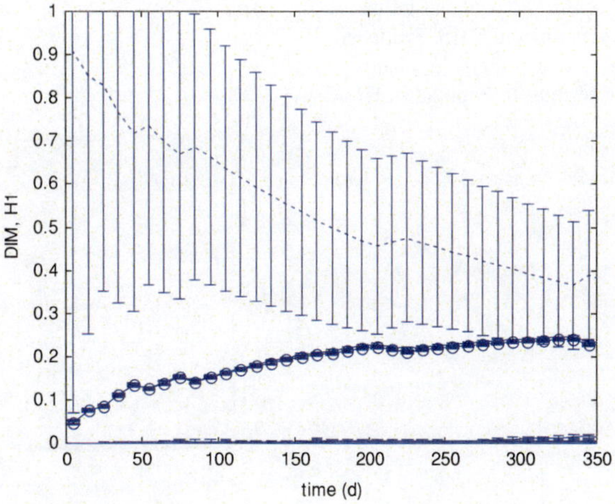

Fig. 4 Time-dependent DIMs of the failure events of the CSIS. Parameter changed according to hypothesis H1. The MC *error bars* are also reported. (······: event 1, —: event 32, ○: event 33; ——: other events)

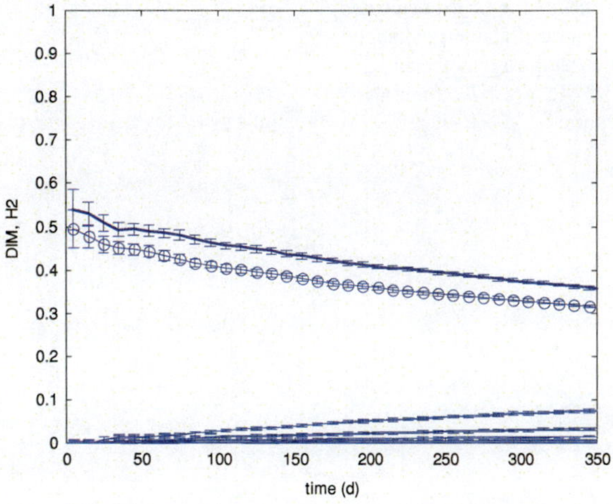

Fig. 5 Time-dependent DIMs of the failure events of the CSIS. Parameter changed according to hypothesis H2. The MC *error bars* are also reported. (······: event 1, —: event 32, ○: event 33; ——: other events)

In Figs. 4 and 5 we report the time-dependent behavior of the DIM computed by MC simulation (10^7 trials), with parameters changed under hypothesis H1 and H2, respectively. Then, for sake of clarity, Figs. 6 and 7 report only the values of the DIM at 1 year. The number of MC trials is 10^7 in both cases. The CPU time required for the simulation was of about 2 min on an ATHLON 1400 MHz processor. The additional burden in the simulation due to the computation of the 33 (number of parameters) $\times$ 36 (time points) first-order sensitivities was of a factor 1.5.

The ranking produced by the DIM at 1 year under hypothesis H1 (Fig. 6) assigns the highest importance to the three events constituting the three first-order cut sets: the RWST vent plugged and the two common mode failures of operator errors on CLCS calibration and after test of valves V_{2A} and V_{2B}. Then, the other

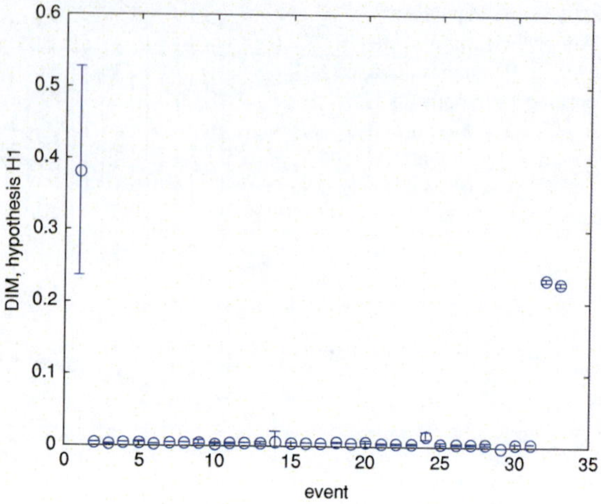

Fig. 6 DIMs of the failure events of the CSIS, evaluated at 1 year. Parameter changed according to hypothesis H1

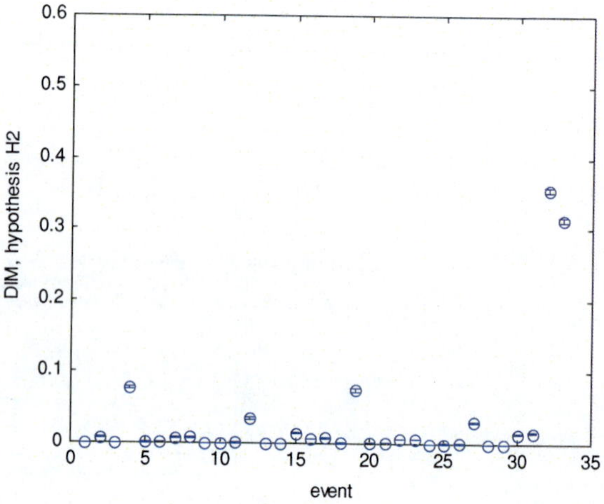

Fig. 7 DIMs of the failure events of the CSIS, evaluated at 1 year. Parameter changed according to hypothesis H2. The MC *error bars* are also reported

events contributing to the second-order cut sets are assigned low importance, without much difference among themselves. The reason for this ranking stands in that, under hypothesis H1, the DIM gives indications on the relevance of an event with respect to its logical role in the system, regardless of the probability that the event actually occurs. This viewpoint is similar to that of the Birnbaum IM. As a consequence, event 1 in Table 6 is ranked as the other two common mode failures even though its occurrence rate is four orders of lower magnitude. The very low likelihood of such event is also the reason for the great uncertainty in the estimate of its DIM.

In Fig. 7, the ranking according to the DIM under the hypothesis H2 at 1 year is reported. Compared to that produced under hypothesis H1, such ranking reflects that under hypothesis H2 the logical importance of the events is weighed by their actual occurrence probability. Thus, under such hypothesis, only common mode

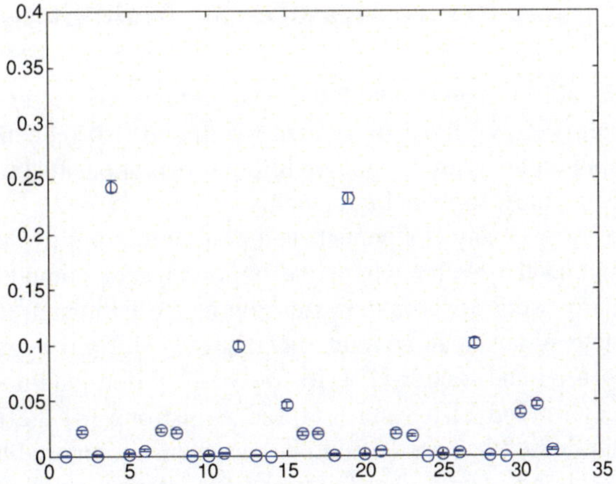

Fig. 8 DIMs of the failure events of the CSIS after the mitigating actions, evaluated at 1 year. Parameter changed according to hypothesis H2. The MC *error bars* are also reported

failures are assigned high importance, while the plugging of the RWST vent drops to the group of events of lowest criticality. Furthermore, when adopting hypothesis H2, the DIM allows discriminating the relative importance among the events contributing to the second-order cut sets. Indeed, events 4 and 19, i.e., valves V_{2A} and V_{2B} respectively left open, stand out from the low importance group, due to their relatively high occurrence probability (Table 6). Likewise, events 12 and 27, related to failures of the CLSC to command the proper tripping of the CSIS pumps and valves, are ranked in accordance to the fact that they have the second highest occurrence probability among the events referred to each injection path.

Importance measures are useful to identify possible weaknesses in the system design and operation, and to suggest improvement actions, e.g., the introduction of a redundancy or of a more reliable component, aiming at reducing the criticality degree of the identified situation.

With reference to the example of the CSIS, one can suggest the introduction of corrective actions to limit the criticality of the high-importance events. In order to prioritize the corrective actions, reference is made to the ranking produced under the hypothesis H2, which seems more informative for the present purposes. Both common mode events are related to human errors; the effects on the system of two possible corrective actions are analyzed. The first one refers to the case that the CLSC miscalibration can be detected immediately upon occurrence and a mitigating repair action is undertaken. For simplicity, a constant repair rate of arbitrary value $\mu = 0.5$ day^{-1} is assumed. The second one considers the possibility of resorting to two independent testing teams for valves V_{2A} and V_{2B} so that the contribution to the failure of V_{2A} and V_{2B} due to a common mode event vanishes and only the two independent failure events 4 and 19 remain. Figure 8 shows the effects on the values of the DIMs after the mitigation action has been done: as expected, the importance of the event of CLSC miscalibration is significantly reduced and the independent failures of valves V_{2A} and V_{2B} assume top relevance. The modification results in a decrease of a factor of 5 in the system unavailability.

13 Importance Measures for Multi-State Systems

As above mentioned, classical importance measures, such as the Birnbaum, Fussell–Vesely, risk achievement worth and risk reduction worth, have been mainly applied to systems made up of binary components (i.e., components that can be in two states: functioning or faulty).

Since many components and systems are multi-state, efforts have been made to extend the evaluation of the importance of components to multi-state systems. Early work towards the extension of the Birnbaum measure to the case of multi-state systems can be found in Griffith [13], for the case of finitely many states and in Kim and Baxter [17], for the case of continuum structure functions. Later, in Levitin and Lisnianski [18] and Armstrong [3] the Birnbaum measure has been applied to the case of multi-state systems composed by binary components and to components with dual failure-modes, respectively.

Importance measures related to the occupancy of a given state by a component have been proposed in Griffith [13] and Wu and Chan [28]: these measures characterize the importance of a given component being in a certain state or degrading to the neighboring state with respect to the expected system performance. The IM of a given component is, therefore, represented by a vector of values, one for each state of the component. Such representation may be of difficult interpretation to the reliability analyst in practice.

Other measures have been defined in order to prioritize multi-state components with respect to the MSS availability. In [5], two measures are proposed to identify the components with the highest potential of improvement in the system availability and the components responsible for the unutilized capacity of the system. Furthermore, in [20] measures are proposed in order to characterize how much a performance level of a component is responsible for the achievement or non-achievement of a given system performance.

A generalization of some commonly used importance measures has been proposed for application to multi-state systems constituted by multi-state components [20]. Physically, these measures characterize the importance for a multi-state component of achieving a given level of performance and their definitions entail evaluating the system output performance measure when the functioning of the component of interest is restricted in performance. In [20], an analysis of the generalized measures is presented when the performance of the components is restricted according to different models and when different system output performance measures are considered.

13.1 Multi-State Systems

Consider a system made up of n components. Let $X_j(t)$ be a random variable representing the performance level of component j at time t, $j = 1, 2, \ldots, n$. $X_j(t)$ can assume one of $m_j + 1$ values:

$$x_{j0}, x_{j1}, \ldots, x_{jm_j} \left(0 = x_{j0} \leq x_{j1} \leq \ldots \leq x_{jm_j}\right) \tag{39}$$

The value x_{jk} is the level of performance of component j when in state k ranging from complete failure ($x_{j0} = 0$) to the maximum performance value (x_{jm_j}).

Let $W(t)$ be a non-negative random variable representing the performance level of the system at time t. The system performance $W(t)$ is determined on the basis of the individual components' performances, $X_j(t)$, $j = 1, 2,\ldots, n$. and depends on the system logic of operation of the considered system. $W(t)$ can assume one of $m + 1$ values, ranging from complete system failure (state $i = 0$) to perfect functioning (state $i = m$):

$$w_0, w_1, \ldots, w_m (0 = w_0 \leq w_1 \leq \ldots \leq w_m) \tag{40}$$

In practice, MSS may be requested to work at different performance levels at different times. For example, the production of electrical and thermal power plants varies according to the daily and seasonal load demands. Assume that at time t a given level of system performance $D(t)$ is required.

The behavior of a MSS is generally judged in terms of a measure of output system performance O [4]. For example, the system steady-state availability or the system steady-state performance are frequently used. A detailed description of the various measures of MSS output performance can be found in [4]. In the following illustration of the MSS importance measures, we shall often refer to the availability at time t, which for a MSS is the probability that at that time the system has performance $W(t) \geq D$.

13.2 Importance Measures for MSS

One of the first notions of IMs for multi-state components in MSS has been introduced in the early eighties by Griffith [13]. Consider a system made up of n components having $m + 1$ possible levels of performance w_i such that $0 \leq w_0 \leq w_1 \leq \cdots \leq w_m$. Each component j, $j = 1, 2, \ldots, n$, has m_j possible states. The performance of component j when in state k is x_{jk} $(0 \leq x_{j0} \leq x_{j1} \leq \cdots \leq x_{jm_j})$.

Physically, the measure of the Griffith's importance of component j being in state k, $I_k^G(j)$, represents the variation in the expected system performance due to a degradation of component j from the performance state k to the performance state $k - 1$. Note that, when applied to binary systems, the Griffith's IM reduces to the Birnbaum's importance measure.

In this sense, the Griffith's importance measure allows one to identify those performance states of the components for which a single-step decrement in performance has major effects on the system.

The Griffith's importance measure of component j is the vector $\underline{I}^G(j) = \left(I_1^G(j), \ldots, I_{m_j}^G(j)\right)$. In Griffith [13], it is shown that $\underline{I}^G(j)$ can be interpreted as the rate of improvement of the MSS performance following an improvement in the performance of its multi-state components.

Later, the performance utility importance function $I_k^U(j)$ has been introduced in order to identify which levels of components performance contribute the most to the system performance [28]. Such information is indeed not easily retrievable from the Griffith's IM.

The performance utility importance function, $I_k^U(j)$, of state k of component j is the expected value of the system performance when component j resides in state k times the probability that the component j actually resides in that state k. In this sense, $I_k^U(j)$ can be interpreted as the contribution of state k to the overall system performance.

The utility importance of component j is defined by the vector $I_k^U(j)$ of the importances of its individual states, i.e., $\underline{I}^U(j) = \left(I_0^U(j), I_1^U(j), \ldots, I_{m_j}^U(j)\right)$. The performance utility importance function is useful to determine which state of a component contributes the most to the overall system performance, compared to the other states of that component. If a state k of component j has a high value of $I_k^U(j)$, it significantly contributes to the system performance. Note that, by definition, a state k of component j is important according to the measure I^U either if the system has high performances when component j is in state k (high value of $E[W_{jk}]$) or if the probability p_{jk} of component j being in state k is high.

Other importance measures characterize the role of multi-state components with respect to the MSS availability. In particular, in [5] the system availability improvement potential, I_1 and expected unutilized capacity of component j, I_2 are introduced. The system availability improvement potential of component j, $I_1(j)$, indicates which components should receive attention in order to increase the system availability the most. This measure is useful in the identification of system bottlenecks. Physically, $I_1(j)$ equals the variation in the system availability obtained by fixing the performance of component j to its highest achievable one, e.g., 100%. It can be verified that $I_1(j)$ equals the probability that component j acts as a system bottleneck.

The expected unutilized capacity of component j, $I_2(j)$, expresses how much the performance of the component can be reduced without effects on the system availability. This measure is useful in the system design phase to identify components having too much or too little extra performance with respect to that actually required by the system.

In [20] two other importance measures are introduced, characterizing the contribution to the system availability of state k of component j. The first measure, $I^{M1}(j, k, t)$, is the probability that the performance $W(t)$ of the system at time t is smaller than the required performance $D(t)$ when component j is in its lower performance state 0 and greater than $D(t)$ just when component j is in state k. Such measure characterizes how much state k of component j is responsible for the system providing at least the required performance level $D(t)$. The second measure, $I^{M2}(j, k, t)$, is the probability that the performance $W(t)$ of the system at time t is smaller than the required performance $D(t)$ when component j is in the state $k-1$ with performance just lower than that of state k and that $W(t) \geq D$ when component j is in its best performing state $k = m_j$. Such measure characterizes how much state k of component j is responsible for the missed delivery by the system of the required performance $D(t)$.

13.3 Importance Measures of Multi-State Components Performance

The Birnbaum, Fussell–Vesely, risk achievement worth and risk reduction worth IMs have been generalized considering the contribution to the MSS output performance W given by a component j that achieves a pre-defined level of performance α [32].

Let us denote by $k_{j\alpha}$ the state in the ordered set of states of component j whose performance $x_{jk_{j\alpha}}$ is equal to or immediately below α, i.e., $x_{jk_{j\alpha}} \leq \alpha < x_{jk_{j\alpha}+1}$. Then, the following quantities can be defined:

- $W_j^{\leq \alpha} = W(\underline{X}|X_j \leq \alpha \text{ in } [0,\ \tau])$: system output performance when the performance X_j of the j-th component is restricted to be below or equal to α (i.e., component j is restricted in states $k \leq k_{j\alpha}$) in $t \in [0, \tau]$.
- $W_j^{> \alpha} = W(\underline{X}|X_j > \alpha \text{ in } [0,\ \tau])$: system output performance when the performance X_j of the j-th component is restricted to be above α (i.e., component j is restricted in states $k > k_{j\alpha}$) in $[0, \tau]$.

By doing so, the complete ordered set of states of the generic jth component is divided into two ordered subsets, thus re-introducing a collectively binary logic for the states *functioning* above performance level α and *faulty* below level α, respectively. In this framework, the following IMs can be defined:

Birnbaum measure of α-level

$$bW_j^{\alpha} = W_j^{> \alpha} - W_j^{\leq \alpha} \tag{41}$$

The bW_j^{α} is the maximum change in system output performance W when the performance of component j is changed from always above the α-level ($X_j > \alpha$, i.e., states $k > k_{j\alpha}$) to always below or equal to the α-level of performance ($X_j \leq \alpha$, i.e., states $k \leq k_{j\alpha}$).

Fussell–Vesely measure of α-level

$$fW_j^{\alpha} = \frac{W - W_j^{\leq \alpha}}{W} \tag{42}$$

The fW_j^{α} is the ratio of the decrement in the system output performance W due to the component j operating with a level of performance below or equal to ($X_j \leq \alpha$, i.e., states $k_j \leq k_{j\alpha}$) in $[0, \tau]$ to the nominal value of W.

Achievement Worth of α-level

$$aW_j^\alpha = \frac{W_j^{>\alpha}}{W} \tag{43}$$

The aW_j^α depends on the system output performance achieved by the system when component j is obliged to operate with a performance above α ($X_j > \alpha$, i.e., states $k > k_{j\alpha}$) in $[0, \tau]$.

Reduction worth of α-level

$$rW_j^\alpha = \frac{W}{W_j^{\leq\alpha}} \tag{44}$$

The rW_j^α represents the reduction in W which can be achieved when the output performance of component j is maintained below or equal to level ($X_j \leq \alpha$, i.e., states $k \leq k_{j\alpha}$). Also in the case of MSS, yrW^α and fW^αproduce the same ranking of component importance.

13.4 Remarks on MSS Importance Measures

Let us compare the IMs characterizing how components contribute to the expected value of the MSS performance, i.e., those proposed in Griffith [13] and [28], and those just illustrated in the Sect. 13.3 based on the limitation of the performance of the multi-state components, in which the availability is taken as output performance measure W [31].

For brevity's sake, we do not give the details of the analytical relationships that can be shown to hold among the IMs. Yet, it is worth mentioning that all the measures can be derived from the knowledge of the performance utility importance function and of the probabilities of the components being in their states [33]. This entails a significant reduction in the computation time needed to perform the importance analyses, since the computation burden is only in the estimation of the $I_k^U(j)$ and p_{jk}. Such feature is particularly profitable when dealing with complex systems whose modeling often requires the use of time-consuming simulation codes. In these cases, there is no need to repeat the system simulation each time a different IM is considered, since the value of the IM can be derived from $I_k^U(j)$ and p_{jk}.

From the physical point of view, the IMs considered are related to the occupancy of a given state by a component. However, the various IMs refer under different perspectives to the event that a component occupies a given state or a subset of states. In particular, a distinction can be made between measures referring to 'occurring events' and to 'existing conditions' as pointed out in [26] with reference to IMs for binary states. An example of an occurring event is the failure of a component, whereas an example of an existing condition is a component being faulty. The importance of an occurring event depends both on the effect of the event on the system as well as on the probability of that event actually occurring. On the other hand, the importance of an existing condition is related to the fallbacks of that condition on the system, regardless of its occurrence probability. The $I_k^U(j)$ measure considers the event of component j residing in state k as an 'occurring event'. Indeed, by definition $I_k^U(j)$ is the expected system performance $E[W_{jk}]$ when component j resides in state k times the probability p_{jk} of component j of actually residing in that state. Instead, the other measures refer to existing conditions: the one-step degradation addressed by I^G or the confinement of the components into states with performance always below or above are given conditions and their effect on the system are analyzed regardless of their actual occurrence probability.

IMs have different applications for decision-making depending on whether they refer to occurring events or existing conditions. Consider a case in which the goal is to prioritize actions for system improvement, such as increasing inspection/maintenance frequencies or allocating redundancies. Then, it seems reasonable to consider IMs adopting the existing condition perspective: indeed, the analyst has to judge the performance of the system after the improvement has been done. For example, if the aim is that of achieving the maximum improvement in system performance, the analyst's decision should be driven by the measures aW_j^α; on the contrary, if the aim is that of reducing the likelihood of low- or zero-performance system configurations then one should follow the prioritization suggested by the performance reduction measures rW_j^α, fW_j^α.

On the contrary, from the perspective of the occurring event, the $I_k^U(j)$ measures identify which components in which state contribute or not to the performance of the system without any change. In this view, the system performance can be increased by acting on component j with respect to either $E[W_{jk}]$ or p_{jk}, depending on which factor is responsible for the low value of $I_k^U(j)$.

A more detailed discussion of the analytical and physical relationships holding among the measures is contained in [33].

13.5 Example 7

Let us consider a system made up of a series of $\eta = 2$ macro-components (nodes), each one performing a given function (Fig. 9). Node 1 is constituted by $n_1 = 2$ components in parallel logic, whereas node 2 is constituted by a single component ($n_2 = 1$) so that the overall number of components in the system is $n = \sum_{l=1}^{\eta} n_l = 3$. The mission time τ is 1,000 h [5].

For each component $i = 1, 2, 3$ there are $m_i = 5$ possible states, each one corresponding to a different hypothetical level of performance $x_{i,j}$, $j = 1, 2, \ldots, 5$. Each component is assumed to move stochastically from one state j to another state k, according to exponential time distributions with rate $\lambda_{i,j \to k}$ (h^{-1}).

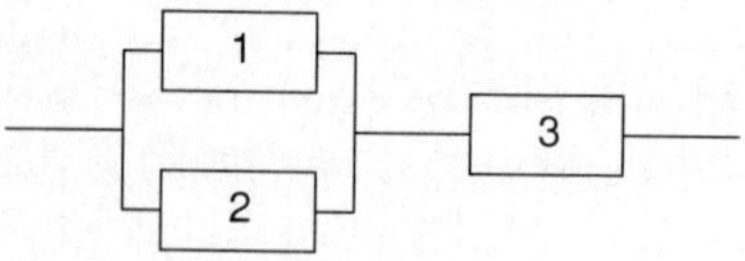

Fig. 9 Reliability block diagram of the system

To each component $i = 1, 2, 3$, there corresponds a transition matrix Λ^i of the values of the transition rates:

$$\Lambda^1 = \begin{bmatrix} - & 5\times10^{-3} & 0 & 0 & 5\times10^{-4} \\ 5\times10^{-3} & - & 5\times10^{-3} & 0 & 6\times10^{-3} \\ 0 & 5\times10^{-3} & - & 5\times10^{-3} & 8\times10^{-3} \\ 0 & 0 & 5\times10^{-3} & - & 8\times10^{-3} \\ 1\times10^{-2} & 5\times10^{-3} & 5\times10^{-3} & 5\times10^{-3} & - \end{bmatrix} \quad (45)$$

$$\Lambda^2 = \begin{bmatrix} - & 5\times10^{-3} & 0 & 0 & 1.5\times10^{-3} \\ 5\times10^{-3} & - & 5\times10^{-3} & 0 & 2\times10^{-3} \\ 0 & 5\times10^{-3} & - & 5\times10^{-3} & 3\times10^{-3} \\ 0 & 0 & 5\times10^{-3} & - & 4\times10^{-2} \\ 1\times10^{-2} & 5\times10^{-3} & 5\times10^{-3} & 5\times10^{-3} & - \end{bmatrix} \quad (46)$$

$$\Lambda^3 = \begin{bmatrix} - & 5\times10^{-4} & 0 & 0 & 5\times10^{-5} \\ 5\times10^{-3} & - & 5\times10^{-4} & 0 & 6\times10^{-5} \\ 0 & 5\times10^{-3} & - & 5\times10^{-4} & 7\times10^{-5} \\ 0 & 0 & 5\times10^{-3} & - & 8\times10^{-5} \\ 1\times10^{-1} & 5\times10^{-2} & 5\times10^{-2} & 5\times10^{-2} & - \end{bmatrix} \quad (47)$$

Table 7 gives the values of the performances $x_{i,j}$ (in arbitrary unit) of the three components in correspondence of all the possible states $j = 1, 2,\ldots, 5$. Note that state 5 corresponds to zero-performance, i.e., component failure.

The output performance $W_{\underline{j}}$ associated to the system state $\underline{j} = (j_1, j_2, \ldots, j_n)$ is obtained on the basis of the performances $x_{i,j}$ of the components $i = 1, 2,\ldots, n$ constituting the system. As in [18], it is assumed that the performance of node 1 constituted by two components in parallel is the sum of the performances of the two individual components and that the performance of the two-nodes series system is that of the node with the lowest performance. For example, when the system is in configuration $\underline{j}^* = (1, 3, 2)$, the first node is characterized by a performance equal to $x_{1,1} + x_{2,3} = 120$ and the second node by a performance $x_{3,2} = 75$: hence, the value of the system performance is $W_{\underline{j}^*} = 75$.

Let Γ_i^α and $\overline{\Gamma}_i^\alpha$ be the sets of states of component i characterized by a performance level below or equal to and above α, respectively. Then, $\overline{W}_i^{\le\alpha} = \overline{W}(j_i \in \Gamma_i^\alpha \text{ in } [0, \tau])$ is the system mean performance over the period τ when the performance of the ith component is below or equal to α (i.e., $j_i \in \Gamma_i^\alpha$) in $[0, \tau]$ and $\overline{W}_i^{>\alpha} = \overline{W}(j_i \in \overline{\Gamma}_i^\alpha \text{in } [0, \tau])$ is the system mean performance over τ when the performance of the ith component is above α (i.e.,$j_i \in \overline{\Gamma}_i^\alpha$) in $[0, \tau]$.

Table 7 Performance values of the system components (in arbitrary units)

	Performance ($x_{i,j}$)				
Component (i)	$j = 1$	$j = 2$	$j = 3$	$j = 4$	$j = 5$
1	80	60	40	20	0
2	80	60	40	20	0
3	100	75	50	25	0

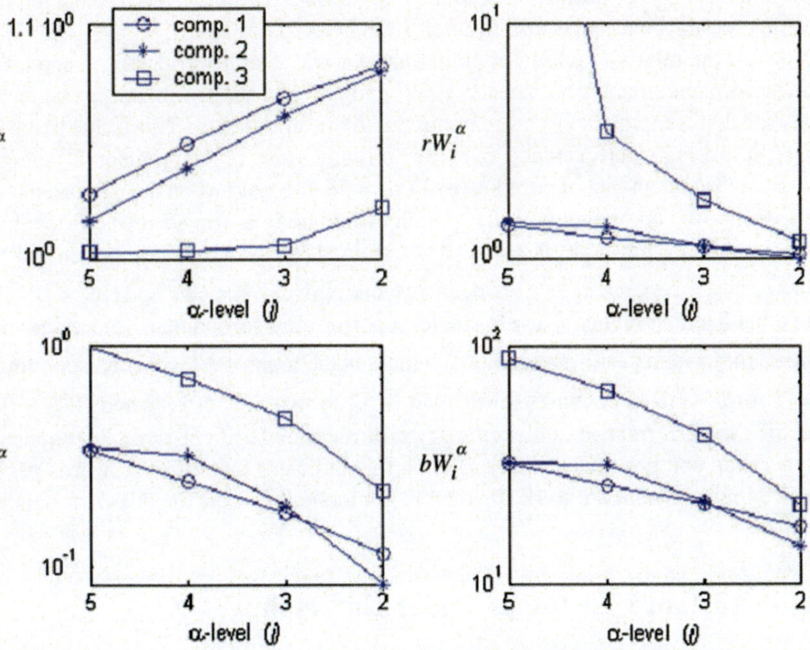

Fig. 10 MSS component performance importance measures as a function of the reference α-level (state j)

Figure 10 reports the results pertaining to the calculation of the MSS performance importance measures for each of the three components and all possible α-levels. The system mean performances $\overline{W}$, $\overline{W}_i^{\leq\alpha}$, $\overline{W}_i^{>\alpha}$, $i = 1, 2,\ldots, n$, for the evaluation of the MSS importance measures have been estimated by Monte Carlo simulation. Component 3 is the most important according to the measures rW_j^α, fW_j^α and bW_j^α and the least important according to aW_j^α at any α-level. The measures rW_3^α, fW_3^α are indicators of the reduction in mean system performance $\overline{W}_3^{\leq\alpha}$ due to component 3 providing at most a level of performance. Indeed, the performance $x_{3,j}$ of component 3 in its state $j = 1, 2,.., 5$, determines the system performance which cannot exceed the value $\overline{W}_3^{\leq\alpha} = \alpha = x_{3,j}$. The large values of $bW_3^\alpha = \bar{W}_3^{>\alpha} - \bar{W}_3^{\leq\alpha}$ also follows from the above considerations. Thus, as expected, to achieve a satisfactory system performance, it is very important to assure a sufficient α-level for component 3. Note that rW_3^0 equals infinity, given that the performance reduction $\overline{W}_3^{\leq 0}$ of component 3 at 0-level (i.e., with component 3 always in the zero performance state $j = 5$, $x_{3,5} = 0$), equals zero (correspondingly, $fW_3^0 = 1$). On the contrary, the small values of the performance achievement worth measure obtained at any α-level indicate that actions for improving the performance of components above are more effective if devoted to component 1 or 2. Indeed, component 3 is already characterized by an high average performance over the mission time ($\bar{x}_3 = 97.531$).

As for the relative ranking of components 1 and 2, at lower performance levels, the performance reduction measures rW^α and fW^α and the Birnbaum measure bW^α indicate that component 2 is more important than component 1. This is due to the fact that the average performance over the mission time of component 2 is higher ($\bar{x}_2 = 59.085$) than that of component 1 ($\bar{x}_1 = 58.116$) and when the performance of one of the two components in the parallel logic node is forced to the 0-level, the performance of the node is entirely determined by the other component so that $\overline{W}_1^{\leq 0} > \overline{W}_2^{\leq 0}$ because $\bar{x}_2 > \bar{x}_1$ and, correspondingly, $rW_1^0 < rW_2^0$ and $fW_1^0 < fW_2^0$. On the other hand, at higher α-levels, the least performing component 1 more and more affects the system performance $\overline{W}_1^{\leq \alpha}$, which, at the highest α-level corresponding to $j = 1$ for $i = 1, 2, 3$ becomes lower than $\overline{W}_2^{\leq \alpha}$ so that $rW_1^0 > rW_2^0$ and $fW_1^0 > fW_2^0$. As for the performance achievement worth measure, aW^α, it ranks component 1 higher than component 2 for any α-level: this indicates that efforts towards performance improvement are to be devoted to component 1, with lower performance.

14 Importance Measures Under Uncertainty

Uncertainties inevitably enter the modeling of the behavior of a system and it is customary in practice to categorize them into two types: aleatory and epistemic [1]. The former type (also referred to as irreducible, stochastic or random uncertainty) describes the inherent variation associated with the physical system or the environment (e.g., variation in atmospheric conditions, in fatigue life of compressor and turbine blades, etc.); the latter (also referred to as subjective or reducible uncertainty) is, instead, due to lack of knowledge of quantities or processes of the system or the environment (e.g., lack of experimental data to characterize new materials and processes, poor understanding of coupled physics phenomena, poor understanding of accident initiating events, etc.). In practice, IMs are typically calculated without due account of the uncertainties.

On the contrary, this sub-chapter shows that uncertainties can influence IMs and proposes a procedure to account for them in the resulting ranking of the basic events or components. The uncertainties considered are of epistemic type on the values of the parameters of the system model and are represented by probability density functions.

Table 8 Range of the IMs uniform distributions

	Uniform distribution range
I_A	[0.0141, 0.0155]
I_B	[0.0020, 0.0178]

For simplicity, the illustration is limited to the case of uniformly distributed uncertainty affecting directly the IMs of two components A and B of a hypothetical system (denoted as I_A and I_B, respectively). Table 8 reports the ranges of the IMs uniform distributions (Fig. 11a, b). It can be observed that the IM of component B (I_B) is significantly more uncertain than that of component A (I_A), but the expected value of I_A, $E[I_A]$ is larger than that of B, $E[I_B]$. On the other hand, there is a range in which the I_B quantiles are larger than the I_A ones. For example, if one were to perform the ranking based on the IMs 95th quantile values the conclusion would be that component B is more important than A contrarily to what would happen if the ranking were based on the expected values.

The drawback of comparing the expected values or specific quantiles lies in the loss of information about the distribution. For example with reference to Fig. 11b, the fact that the 95th quantile of I_A (0.015) is lower than that of I_B (0.017) only means that the point value which I_A is lower than with probability of 0.95 is lower than the analogous point value for I_B; the full information on the actual difference between the distributions of I_A and I_B does not play any role.

A natural way to give full account of the difference between the distributions of I_A and I_B is to consider the random variable (rv) $I_A - I_B$ whose pdf and cdf are shown in Fig. 11c and d, respectively. In order to establish if component A is more important than B, one can consider the exceedance probability that I_A is larger than I_B, $r_{AB} = P(I_A > I_B) = 1 - F_{AB}(0)$; for example, in the present case $r_{AB} = 0.81$, which means that with high probability component A is more important than B.

To decide on the relative importance of the two components A and B, it is necessary to fix a threshold $T \in [0.5, 1]$ on the r_{AB} value such that if r_{AB} is larger than T then A is more important than B otherwise no conclusion can be drawn. Obviously, the lower the threshold, the higher the risk associated with the decision.

On the other hand, the choice of a simple-valued threshold has some limitations when considering multiple components. For example, if the IMs of three components, A, B and C are such that their differences all fall very close to T it could happen that $I_A > I_B$, $I_B > I_C$ and $I_C > I_A$. Moreover, r_{AB} could fall very close to T in which case no robust conclusion can be drawn on the components importance given the inevitable approximations and uncertainties related to the estimation of the IMs distributions.

These limitations can partially be overcome by referring the comparison to a threshold range $[T_l, T_u]$ in such a way that for the two components A and B:

- If $r_{AB} > T_u$, then A is more important than B
- If $r_{AB} < T_l$, then B is more important than A.

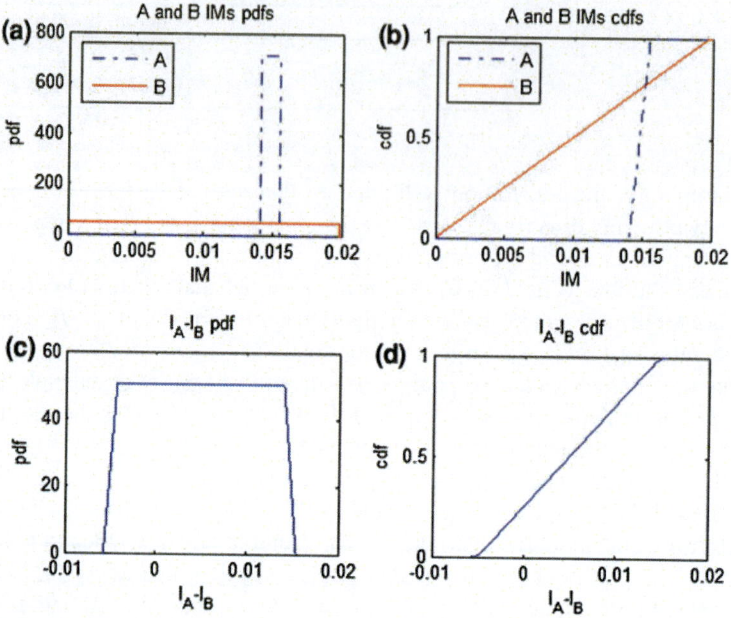

Fig. 11 Probability density functions (pdfs) and cumulative distribution functions (cdfs) of the uniform random variables I_A, I_B (**a** and **b**) and of $I_A - I_B$ (**c** and **d**)

- If $T_l < r_{AB} < T_u$, then A is equally important to B. In this case, different kinds of additional constraints/targets can guide the ranking order (costs, times, impacts on public opinion, etc.).

Let us examine the behavior of the *probabilistic exceedance measure* $r_{AB} = P(I_A > I_B)$ in relation to the values of the standard deviations of the IMs distributions, σ_{I_A} and σ_{I_B}. Figure 12 shows the variation of r_{AB} for increasing values of the standard deviation σ_{I_B}, keeping fixed the mean values of I_A and I_B and the ratio $k = \sigma_{I_A}/\sigma_{I_B}$ for different values of k. In the extreme case of no uncertainties on the knowledge of I_A and I_B ($\sigma_{I_A} = 0$ and $\sigma_{I_B} = 0$), component A is more important than B (because $E[I_A] > E[I_B]$) and thus $r_{AB} = 1$. Increasing the standard deviation σ_{I_B} (and thus also σ_{I_A}, for keeping the ratio k constant), as expected $r_{AB} = 1$ holds as long as the pdfs of I_A and I_B do not overlap, i.e., I_A and I_B are uncertain quantities but it is not uncertain that $I_A > I_B$. The higher the ratio k, the lower the set of points for which $r_{AB} = 1$. Finally, r_{AB} decreases as the overlapping between pdfs increases.

When considering systems with large numbers of components, a procedure for successive ranking must be introduced to avoid the combinatorial explosion of pairwise comparisons. A possible method consists in applying the Quicksort algorithm [14, 23], which is a divide-and-conquer sorting algorithm, that relies on a partition of the elements based on a quantitative indicator of their 'size'. One of the elements in the set to be sorted is selected as pivot, i.e., as reference for moving

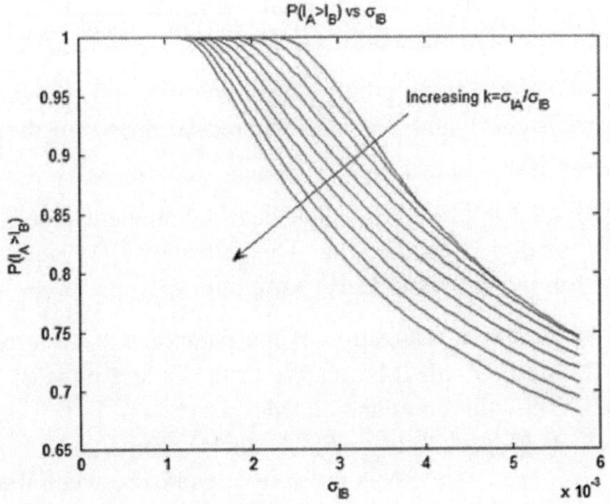

Fig. 12 r_{AB} versus σ_{I_B}, keeping $k = \sigma_{I_A}/\sigma_{I_B}$, $E[I_A]$ and $E[I_B]$ constant

all elements of size smaller before the pivot and all elements of size larger after it; the sublists of smaller and larger elements are then recursively sorted.

In the component importance ranking of interest here, the pivot element p is chosen as the component in the middle position of the components importance rank list obtained looking only at their mean values of reliability/availability. The exceedance probability value of $r_{pj} = P(I_p > I_j)$ is then calculated for each component $j \neq p$ in the list and the pre-defined threshold range $[T_l, T_u]$ defines the rank order between p and j. The basic steps of the procedure are as follows:

1. Rank the components according to their IMs computed by considering the mean values of their reliability/availability parameters, i.e., without considering uncertainties.
2. Define the range $[T_l, T_u]$ of values of the probabilistic exceedance measure r_{pj}; for values r_{pj} in this range, it is not possible to decide if $I_p > I_j$ or $I_p < I_j$ and this leads to consider components p and j as equally important, unless additional constraints/targets (e.g., costs, times, impacts on public opinion, etc.) allow defining an importance rank between the two.
3. Apply the Quicksort algorithm based on $r_{pj} = P(I_p > I_j)$:

3.1. List the components in the rank order found in step 1.
3.2. Choose the middle element of the list (sublist) as pivot element, p.
3.3. For each j in the sublist compute the cdf, F_{pj}, of $I_p - I_j$ and evaluate $r_{pj} = 1 - F_{pj}(0)$:

- If $r_{pj} > T_u$, then put j in the sublist of elements less important than p.
- If $r_{pj} < T_l$, then put j in the sublist of elements more important than p.
- If r_{pj} falls in $[T_l, T_u]$, then p is equally important to j.

3.4. Append the sublist of less important elements to the right of p and the sublist of more important elements to the left of p.

3.5. Recursively apply to each sublist steps 3.2–3.4 until no sublist with more than one element exists.

An alternative ranking procedure has been proposed in [23] which follows the same steps 1 and 2 above, whereas it differs in the steps 3 and 4, which are as follows:

3. Find the probability that each component $i = 1, 2,\ldots, N$ occupies a specific position in the ranking. This is achieved by repeating for $\upsilon = 1, 2,\ldots, M,$ the following Monte Carlo sampling:

3.1. Sample a realization of the components' failure rates $\lambda_1^\upsilon, \lambda_2^\upsilon, \ldots, \lambda_N^\upsilon$.
3.2. Find the υth IMs relative to the failure rates of 3.1.
3.3. Rank the components IMs.
3.4. The probability $P(R_i)$ that component i is in the rank position $R_i = 1, 2,\ldots, N$ is given by the ratio between the number of simulations with component i resulting in position R_i and the number of samples M.

4. To rank the component:

4.1. List the components in the rank order found in step 1.
4.2. Choose the most important component as pivot P i.e., the component with largest probability of being the most important.
4.3. Compute the measure of exceedance r^*_{pj} between the components p and j with $j = p + 1, p + 2$:

$$r^*_{pj} = P(R_p \geq R_j) = \sum_{R_p=1}^{N} p(R_i) \sum_{R_j=1}^{R_p} p(R_j) \tag{48}$$

where R_p = rank of p and R_j = rank of j.
4.4. If $r^*_{pj} > T_u$, then leave component p in the actual position; else, if $T_l < r^*_{pj} < T_u$ then put the component j in position R_p; otherwise, if $r^*_{pj} < T_l$ swap the rank orders of components p and j.
4.5. $p = p + 1$, repeat steps 4.1–4.3 until $p = N$.

As an example, consider the system in Fig. 13. Table 9 reports the components reliabilities and the values of the Birnbaum, Fussel–Vasely, Criticality, RAW and RRW IMs.

Let us now assume that the components are exponential, i.e., with constant failure rates λ_i, $i = A, B, C$ and that epistemic uncertainties affect their failure rates, as described by lognormal distributions.

$$f_{\lambda_i}(\lambda_i) = \frac{e^{-\frac{[\ln(\lambda_i - \mu_i)]^2}{z\sigma_i^2}}}{\lambda_i \cdot \sigma_i \cdot \sqrt{2\pi}} \tag{49}$$

At each time instant t the reliability of component i is:

Fig. 13 Reliability block diagram of the system

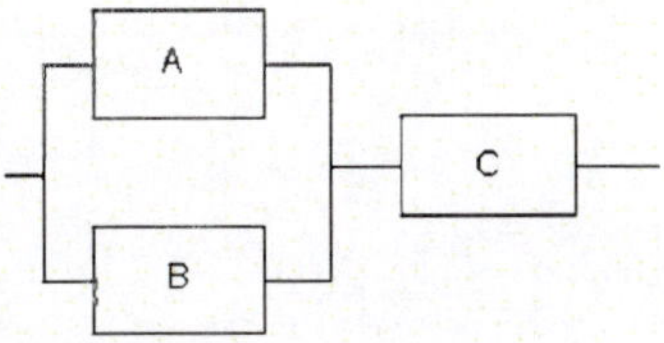

Table 9 Components reliability and importance measures

	Reliability	Birnbaum	Fussel–Vasely	Criticality	RAW	RRW
A	0.985	0.009	0.002	0.001	1.094	1.001
B	0.990	0.014	0.002	0.001	1.141	1.001
C	0.905	0.9999	0.999	0.998	10.5	634

Table 10 Parameters of the lognormal distributions of the components failure rates

Component i	$E[\lambda_i]$	$\mathrm{Var}[\lambda_i]$
A	1.00e-007	5.00E-08
B	1.50e-007	5.00E-08
C	1.00e-006	5.00E-07

$$r_i(t, \lambda_i) = e^{-\lambda_i t} \tag{50}$$

with pdf (for $0 < \lambda_i < 1$):

$$f_{\lambda_i}(\lambda_i) = \frac{e^{-\frac{\left[\ln\left(\frac{\ln(\lambda_i)}{t} - \mu_i\right)\right]^2}{2\sigma_i^2}}}{\lambda_i \cdot \ln(\lambda_i) \cdot \sigma_i \cdot \sqrt{2\pi}} \tag{51}$$

The parameters of the distributions of the failure rates are reported in Table 10; the values have been chosen such that the means of the reliabilities at time $t = 10^5$ (in arbitrary units of time) are equal to the values in column 2 of Table 9. As an example, Fig. 14 reports the lognormal distribution of the failure rate of component A (left) and the pdfs of its reliability at various times (right). In Fig. 16a and b, the pdfs of the failure rates and reliabilities at time $t = 10^5$ (in arbitrary units of time) are reported for all three components.

For brevity's sake, let us limit the discussion of the uncertainties to the Birnbaum measure, the reasoning remaining the same for the other IMs. Note that in spite of the simplicity of the considered system, finding the Birnbaum IM distributions by an analytical approach is impracticable. To overcome this difficulty, Monte Carlo sampling has been applied. The resulting distributions at the fixed time instant $t = 10^5$ are plotted in Figure 16c and d.

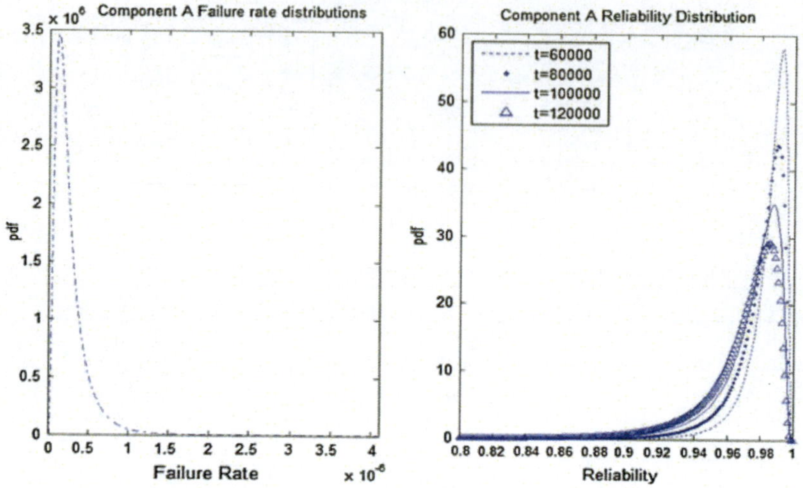

Fig. 14 Lognormal distributions of the failure rate of component A (*left*) and corresponding pdfs of the reliability at different time instants (*right*)

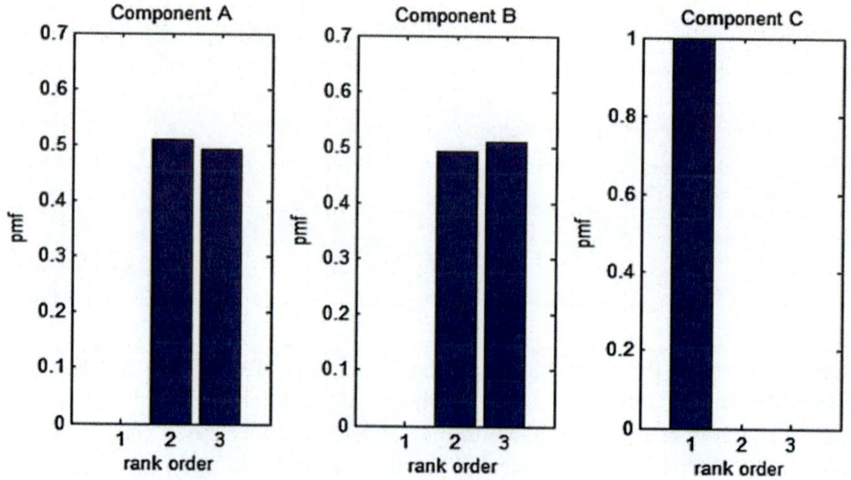

Fig. 15 Probability mass functions of the three components rank orders obtained by the ranking method of [23]

It can be noted that the distribution of the IM of component C is displaced to larger values than that of components A and B which leaves no doubt that the most important component is C, as expected from the structure of the system and the components reliability values. As for the ranking of A and B one must compute the exceedance reliability r_{AB}. The result obtained by Monte Carlo sampling is $r_{AB} = 0.52$, which with respect to $T_l = 0.3$ and $T_u = 0.7$ leads to conclude that

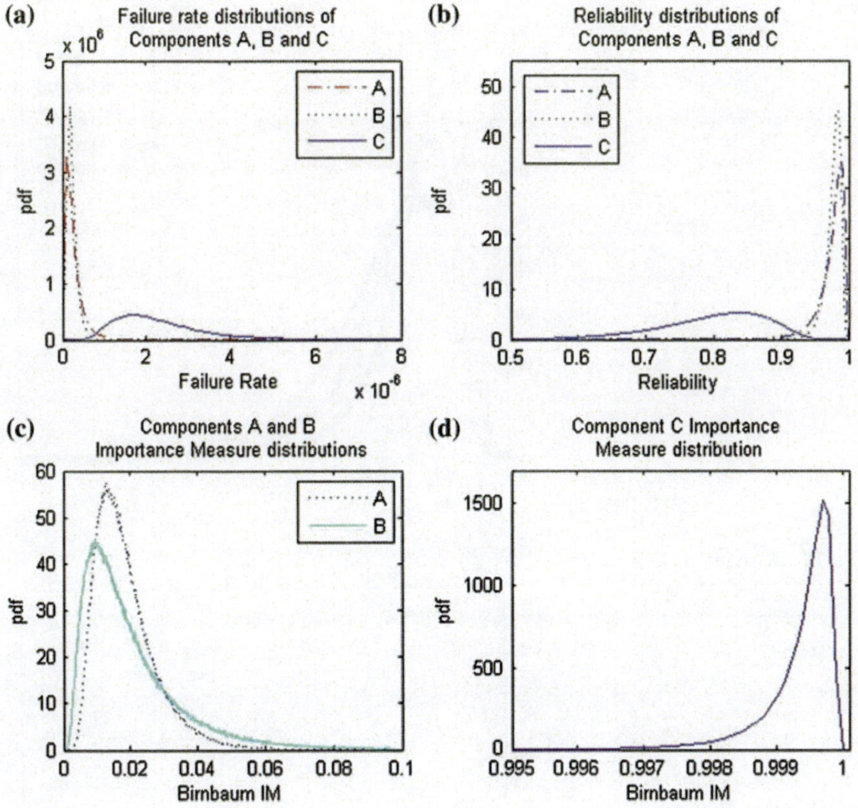

Fig. 16 Pdfs of the failure rates (**a**) reliabilities (**b**) and Birnbaum IMs (**c** and **d**) of the three components

$I_A = I_B$. Hence, the final components ranking provided by the procedure here proposed sees C as the most important element, followed by A and B equally important.

As a term of comparison, the procedure in [23] has been applied. The probability mass functions (pmfs) of the ranks of the three components obtained by Monte Carlo sampling of their uncertain failure rates are reported in Fig. 15. Notice that $r^*_{CA} = r^*_{CB} = 1 > r^*_{AB}$, which implies that component C is more important than both A and B also for this method. On the other side, A results more important than B giving that the exceedance measure $r^*_{AB} = 0.52 \cdot (0.48 + 0.52) + 0.48 \cdot 0.52 = 0.77$. Notice, however, that if one considers $r^*_{AB} = 0.48 \cdot (0.52 + 0.48) + 0.52 \cdot 0.48 = 0.73$, B results more important than A; this shows that, in general, the exceedance measure $r^*_{ij} \neq 1 - r^*_{ji}$ is dependent on the choice of the pivot, and so is the final rank.

It is interesting to investigate the relation between the two exceedance measures calculated, r^*_{ij} and r_{ij}. Given that component C results the most important in all the

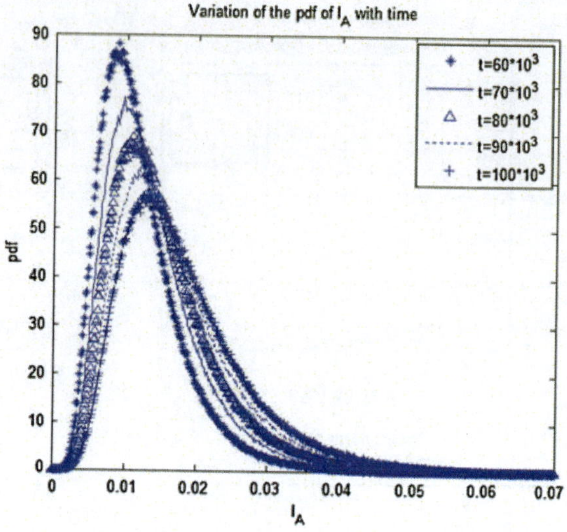

Fig. 17 Variation of the pdf of I_A with time

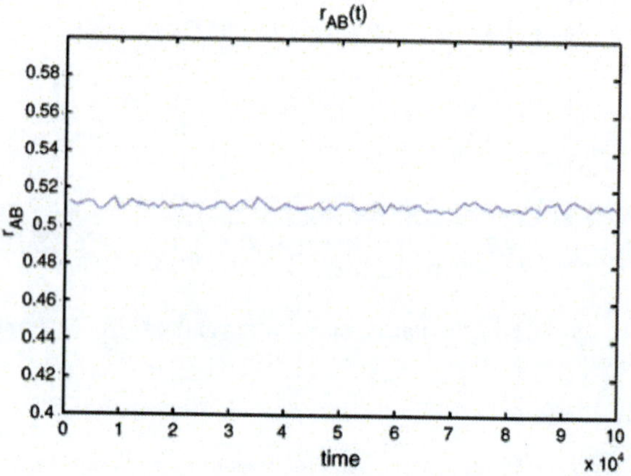

Fig. 18 Behaviour in time of the exceedance measure r_{AB}

M Monte Carlo samples ($P(R_C = 1) = 1$), the probability that A occupies a specific rank order between 2 and 3 is equivalent to the probability that B gains the only other rank order available, i.e., the probability mass value of rank order 2 of component A is the exceedance measure r_{AB} and, vice versa, the probability mass value of rank order 2 of component B is the exceedance measure r_{AB}.

On the contrary, the Birnbaum IM values in Table 9, obtained neglecting uncertainties, would lead to the conclusion that A is more important than B.

Equation 51 and Fig. 14 clearly demonstrate that the components reliabilities depend on both the time t and the failure rate λ_i, the latter being affected by epistemic uncertainty. Since the components IMs are functions of their reliabilities, they also are dependent on t and λ_i, and are affected by epistemic uncertainty. Figure 17 shows how the pdf of I_A varies with time; in particular, at $t = 10^5$ one obtains the curve in Fig. 16c. The dependence of the IM on time implies that, in general, the exceedance measure r_{ij} too is time-dependent ($r_{ij} = r_{ij}(t)$). In Fig. 18, the behavior in time of r_{AB} is reported. In the particular case of the system of Fig. 13, the exceedance measure remains constant and so does the components' ranking; in general this is not true and due account has to be given to the dependence on time but more sophisticated condensation operators might be needed for that scope. The simplest way is to take the average or median value of the exceedance measure over the mission time.

The same reasoning can be repeated for the exceedance measure r^*_{ij} of [23].

Acknowledgments The Author would like to thank Dr. Luca Podofillini of the Paul Scherrer Institute, Switzerland, for the preparation of the material and in particular for contributing to the development of the work related to Examples 6 and 7 and Mr. Michele Compare for contributing to the development of Chap. 14. Finally, many thanks go to Mrs. Lucia Golea for assisting in the preparation of the manuscript.

References

1. Apostolakis GE (1990) The concept of probability in safety assessments of technological systems. Science 250:1359–1364
2. Armstrong MJ (1995) Joint reliability-importance of elements. IEEE Trans Reliab 44(3):408–412
3. Armstrong MJ (1997) Reliability-importance and dual failure-mode elements. IEEE Trans Reliab 46(2):212–221
4. Aven T (1993) On performance measures for multistate monotone systems. Reliab Eng Syst Saf 41:259–266
5. Aven T, Østebø R (1986) Two new importance measures for a flow network system. Reliab Eng 14:75–80
6. Baraldi P, Zio E, Compare M (2008) Importance measures in presence of uncertainties. In: Proceedings of SSARS 2008, Gdańsk/Sopot, Poland
7. Birnbaum LW (1969) On the importance of different elements in a multi-element system. Multivariate analysis, vol 2. Academic Press, New York
8. Borgonovo E (2006) Measuring uncertainty importance: investigation and comparison of alternative approaches. Risk Anal 26(5):1349–1361
9. Borgonovo E, Apostolakis GE (2001) A new importance measure for risk-informed decision making. Reliab Eng Syst Saf 72:193–212
10. Cheok MC, Parry GW, Sherry RR (1998) Use of importance measures in risk informed applications. Reliab Eng Syst Saf 60:213–226
11. Elsayed EA (1996) Reliability engineering. Addison Wesley Longman, England
12. Fussell JB (1975) How to calculate system reliability and safety characteristics. IEEE Trans Reliab R-24(3):169–174
13. Griffith WS (1980) Multistate reliability models. J Appl Prob 17:735–744
14. Hoare CA (1962) Quicksort. Comput J 5:10–15
15. Hong JS, Lie CH (1993) Joint reliability-importance of two edges in an undirected network. IEEE Trans Reliab 42(1):17–23
16. Høyland A, Rausand M (1994) System reliability theory: models and statistical methods. Wiley, NJ
17. Kim C, Baxter LA (1987) Reliability importance for continuum structure functions. J Appl Prob 24:779–785

18. Levitin G, Lisnianski A (1999) Importance and sensitivity analysis of multi-state systems using the universal generating function method. Reliab Eng Syst Saf 65:271–282
19. Marseguerra M, Zio E, Podofillini L (2005) First-order sensitivity analysis of a nuclear safety system by Monte Carlo simulation. Reliab Eng Syst Saf 90:162–168
20. Meng FC (1993) Element-relevancy and characterization results in multi-state systems. IEEE Trans Reliab 42(3):478–483
21. Meng FC (1995) Some further results on ranking the importance of system elements. Reliab Eng Syst Saf 47:97–101
22. Meng FC (1996) Comparing the importance of system elements by some structural characteristics. IEEE Trans Reliab 45(1):59–65
23. Modarres M (2006) Risk analysis in engineering: probabilistic techniques, tools and trends. CRC Press, USA
24. van der Borst M, Shoonakker H (2001) An overview of PSA importance measures. Reliab Eng Syst Saf 72(3):241–245
25. Vasseur D, Llory M (1999) International survey on PSA figures of merit. Reliab Eng Syst Saf 66:261–274
26. Vesely WE (1998) Supplemental viewpoints on the use of importance measures in risk-informed regulatory applications. Reliab Eng Syst Saf 60:257–259
27. Wash-1400 (NUREG 75/014) (1975) Reactor safety study: an assessment of accident risks in US. Commercial Nuclear Power Plants, Appendix 2, Fault Trees
28. Wu S, Chan L (2003) Performance utility-analysis of multi-state systems. IEEE Trans Reliab 52(1):14–20
29. Youngblood RW (2001) Risk significance and safety significance. Reliab Eng Syst Saf 73:121–136
30. Zio E (2007) An introduction to the basics of reliability and risk analysis, Series in quality, reliability and engineering statistics, vol. 13, World Scientific, Singapore
31. Zio E, Podofillini L (2003) Importance measures of multi-state components in multi-state systems. Int J Reliab Qual Safety Eng 10(3):289–310
32. Zio E, Podofillini L (2004) A second-order differential importance measure for reliability and risk applications, SAMO, Sensitivity Analysis of Model Output, March 8–11, 2004, Santa Fe. Available on CD-rom
33. Zio E, Marella M, Podofillini L (2004) A comparison of different importance measures for multistate systems, MMR, mathematical methods in reliability, June 21–25, Santa Fe, New Mexico, USA

Dependent Competing-Risk Degradation Systems

Yaping Wang and Hoang Pham

1 Introduction

The failure of many units or systems, such as components, parts, machines, can be generally classified into two kinds of failure modes: one is catastrophic failure in which units break down by some sudden external shocks; the other is degradation failure in which units fail to function due to the physical deterioration. There are a great number of such cases for this kind of competing failure modes in our real life.

- A battery supplies electric power by chemical reaction. It is gradually weakened by usage and finally turns out to be useless when the substances in the battery are exhaustive. On the other hand, overheating or overvoltage can also cause the damage of the battery.
- An electronic component may have two kinds of failure modes. One failure mode is due to the overloading stress to the system caused by random voltage spikes. The other is due to wear-out process, which usually happens when it has run for many cycles.

In the catastrophic failure case the units break down as soon as cumulative shock damage or some extreme shock exceeds some predetermined failure threshold, while in the degradation case the unit fails when the total degradation amount drops below a critical failure level. Therefore, it would be necessary to formulate probabilistic or stochastic models of such competing risk systems combining degradation and random shock that outline the features of the two-process complex phenomenon. The reliability analysis for this model is a key step

Y. Wang · H. Pham (✉)
Department of Industrial & Systems Engineering, Rutgers University,
96 Frelinghuysen Road, Piscataway, NJ 08854-8018, USA
e-mail: hopham@rci.rutgers.edu

H. Pham (ed.), *Safety and Risk Modeling and Its Applications*,
Springer Series in Reliability Engineering, DOI: 10.1007/978-0-85729-470-8_7,

for the manufacturing engineers to make warranty plans and maintenance policies for the products featured by the two competing failure modes. Numerous mathematical formulations and probabilistic models are proposed to combine these two competing risk failure modes.

2 Literature Review

2.1 Random Shock Model

Shock Models, one of the most important subjects in reliability modeling, are usually defined by the time interval between two consecutive shocks, the damage size caused by individual random shock, and the system failure function. Usually, in shock models the system is subject to shocks of random magnitude occurring at random times. The basic setup in the shock models is pairs of the i.i.d two-dimension random variables $\{(A_n, B_n)\}_{n=0}^{\infty}$, where A_n denotes the magnitude of the nth shocks and B_n the time interval between the $(n - 1)$st and nth shocks, or alternatively, the time between the nth and the $(n + 1)$st shocks, called model I and model II, respectively. However, model II differs significantly from model I in the following two aspects: (1) in model II the magnitude A_n impacts the time interval B_n until the $(n + 1)$st shock; (2) there exists a first shock at time zero in model II while in model I both A_0 and B_0 equal to zero. Shock models have been studied by a number of literatures [1–25] for the purpose of providing the mathematical formulations for modeling the system reliability in the random environments. Traditionally, three classic random shock models are widely used, listed as follows:

- Cumulative Shock Model.
- Extreme Shock Model.
- δ-Shock Model.

Furthermore, some extensions including mixed shock model and run shock model are developed in recent years.

2.1.1 Cumulative Shock Model

Cumulative shock model means the system breaks down when the cumulative shock magnitude exceeds some given threshold. Let $T \geq 0$ denote the lifetime of the system, $\{N(t), t \geq 0\}$ the counting process generated by the renewal sequences $\{B_n\}_{n=0}^{\infty}$, and $Z \subset R$ some prefixed critical region. A suitable representation of the cumulative shock model is indicated in the form of

$$\{T \leq t\} \Leftrightarrow \left\{ \sum_{i=0}^{N(t)} A_i \in Z \right\},$$

where A_i denotes the magnitude of the nth shocks.

Thus, $F(t) = \Pr\{T \leq t\}$ and $\overline{F}(t) = 1 - F(t)$ denote the lifetime distribution and the survival function for the system, respectively.

In terms of the cumulative model, the properties of the lifetime distribution are studied in some early literatures. Esary and Marshall [1] justify the properties of the lifetime distribution for a device subject to accumulative shocks governed by a Poisson process with a probability $\bar{P}_k$ of surviving the first k shocks. A-Hameed and Proschan [2] extend the results from [1] to a non-homogeneous Poisson process in series system consisting of finite components and obtained the bound for the mean life of the device. Agrafiotis and Tsoukalas [3] discuss the first passage time and asymptotic properties of a correlated cumulative shock model with the excess level increments for individual shock. In the model the shock interval and shock magnitude are correlated for given n. Gut and Husler [4] put forward a generalized cumulative shock model in which only the summation of a suitable portion of the most recent shocks will contribute to the system stopping time, that is

$$\{T \leq t\} \Leftrightarrow \min\left\{n : S_{k_n,n} = \sum_{j=n-k_n+1}^{n} X_j > t\right\}, t > 0,$$

where X_j denotes the magnitude of the jth shock; $S_{k_n,n}$ the summation of the magnitude from the k_nth shocks to the nth shock.

In addition, some results of the maintenance policies for a system subjected to cumulative shocks have been studied by several researchers. Nakagawa and Kijima [5] propose a periodic replacement policy with minimal repair at failure for the cumulative shock model in order to obtain the optimal solution for the time T^*, shock N^*, and damage Z^* at which time point the replacement is done. Qian et al. [6] focus on the analysis of maintenance policies for an extended cumulative shock model with shocks occurring at a non-homogeneous Poisson process. The system will be maintained when cumulative shock does not exceed the failure threshold, repaired when cumulative shock exceeds the threshold, and replaced at failure N or time T. Wortman et al. [7] examine the maintenance strategy with the inspection time modulated by a renewal process for a non-self-announcing failure system subject to deterioration governed by random shocks. Chelbi and Ait-Kadi [8] develop the expression of the time-stationary availability for a hidden failure system subject to the transient shocks with a predetermined inspection time in order to generate an optimal solution for the target availability level with limit resources.

2.1.2 Extreme Shock Model

Extreme shock model means that the system fails as soon as the magnitude of an individual shock goes into some given critical region $Z \subset R$, shown as:

$$\{T \leq t\} \Leftrightarrow \{\min\{i : A_i \in Z\}, \leq N(t)\}$$

where T denotes the lifetime of the system, $N(t)$ the counting process of random shock occurring, and A_i the magnitude of the nth shocks.

In the early work on this model, Esary and Marshall [1] study the properties of the maximum shock threshold underlying the assumption of homogeneous Poisson process, while Ross [9] extends the analysis of Esary and Marshall [1] to the non-homogeneous condition. One important setting of the extreme shock model is "general shock model" developed by Shanthikumar and Sumita [10]. The model extends the extreme shock model by considering a sequence of random shocks governed by a correlated pair of renewal sequences (X_n, Y_n), where X_n denotes the magnitude of the nth shock and Y_n the time interval between two consecutive shocks. Chen and Li [11] analyze a deteriorating system subject to extreme shock with the deterioration process governed by both the external shocks and internal loading under the assumptions such that: (1) the magnitude of the random shock the system can bear will be decreasing with the numbers of repairs; (2) the repair time will be increasing upon each repair. Finally, an optimal replacement policy N^*, at which failure number the system will be repaired, is determined by minimizing the long-run average cost.

2.1.3 δ-Shock Model

The δ-shock model in Li [12] is defined as the system fails when the time lag between the two successive shocks falls into some critical region determined by a prefixed parameter δ. Therefore, a suitable representation for the δ-shock model can be given as:

$$T \leq t \Leftrightarrow \{\min\{i : B_i \in C(\delta), i \in N\}, \leq N(t)\},$$

where $C(\delta)$ is the critical region for the system and B_i the time interval between the $(n-1)$st and nth shocks.

In the work of Li and Kong [13], some useful results for the δ-shock model underlying homogeneous Poisson process are given, such as survival function, class properties, and asymptotic behavior of the lifetime distribution. Moreover, the analytical survival function for the non-homogeneous Poisson process is also discussed.

Furthermore, Li and Zhao [14] derive some useful properties for the reliability estimation of the coherent structure of series system, parallel system, and k-out-of n system, such as bounds for system mean lifetime and limiting probability. Lam and Zhang [15] study a replacement policy for two systems embedded in the δ-shock model: one is deteriorating system with a non-decreasing threshold after repair times and geometrically increasing repair times; the other is improving system with decreasing threshold after repair and geometrically decreasing repair times. Rangan and Tansu [16] extend the δ-shock model of [15] by considering the random threshold failure in the context of renewal process.

2.1.4 Some Extensions and Variations

Ross [9] defines a general shock model with a damage function D_t in which $D_t(x_1,\ldots, x_n,\ldots, 0)$ represents the damage at time t if exactly n shocks occur with the magnitude of $x_1,\ldots, x_n$. By assuming $D_t(x_1,\ldots, x_n, 0) = \max(x_1,\ldots, x_n, 0)$ or $D_t(x_1,\ldots,x_n,0) = \sum_{i=1}^{n} x_i$, the cumulative shock and extreme shock model are obtained respectively. Another realistic variation of shock model is provided by Fan et al. [17] for the purpose of studying the lifetime distribution of the multi-component system that suffers from the interplay of aging and random shocks. In the model the probability that a shock with magnitude x arriving at time u is fatal to the system is given by $1 - \exp[-\delta(a + u) - x]$, where a is the initial age of the system at time zero and δ is the system aging rate.

Igaki et al. [18] generalize the cumulative and extreme shock model by considering a trivariate stochastic process including the magnitude of the shocks, the shock intervals, and the random system state. In the model, the magnitude of the shocks and the shock intervals are correlated with each other with a joint distribution affected by the transition of system state modulated by a Markov renewal process. Gut [19] proposes a mixed shock model to combine the extreme model and cumulative shock model. In the model the system is supposed to break down either due to a cumulative effect of many small shocks or one single large shock, whichever comes first. Further realistic analysis of the model is stated by Gut and Husler [4].

A new case, called the run shock model, is introduced by Mallor and Omey [20], which models a system operating normally until k consecutive shocks with critical magnitude, which is expressed as

$$T \leq t \Leftrightarrow \min\{n|A_{n-j} \in Z, j = 0, 1, \ldots, k - 1\}, \leq N(t),$$

where A_{n-j} represents the magnitude of the $(n - j)$th shock and k is some given positive integer.

One extension of the run shock model is proposed by Mallor and Santos [21], in which the system breaks down when the cumulative damage due to the shock in a critical k run, where all of shocks are critical and not contained in any of $k + 1$ consecutive sequence, exceeds a fixed threshold z. This model is governed by three correlated variables, including the magnitude of shock, shock occurring interval, and cumulative damage, respectively. Further analysis of the lifetime distribution and mean time to failure for the extended shock model is provided by Mallor and Santos [22]. Moreover, Mallor et al. [23] generalize the asymptotic behavior for the lifetime of the mixed shock model that is the combination of the accumulative shock model and run shock model.

In the work of Finkelstein and Zarudnij [24], two types of non-cumulative shock processes are considered. The first model divides the shocks into three levels according to their magnitude, shocks with small level that are harmless to the system, shocks with intermediate level that lead to the system failure with some probability, and shocks with large level that result in the immediately system failure. The second model assumes that the system will not fail if two successive small shocks do not occur in a short time period. The system failure function and its exponential approximation are derived analytically. Bai et al. [25] derive the asymptotic lifetime distribution for a new shock model based on a marked point process with cluster mark and then illustrate its application to an example with the insurance background.

2.2 Degradation Model

Nowadays the products are developed to be more reliable with a longer lifetime and higher quality, so it is really a significant challenge to obtain the sufficient and accurate time-to-failure data in a cost-effective manner prior to the product release. The design engineers may be unable to obtain the time-to-failure data of the new products by testing them under the normal operating conditions, either because their lifetimes are too long, or the time between design and product release is too short.

Two approaches are introduced to overcome these obstacles to obtain the information of the lifetime regarding the warranty periods and the reliability specifications of the products. One is Accelerated Life Testing (ALT); the other is degradation analysis. Also, by combining the ALT and degradation analysis, a new approach, called Accelerated Degradation Testing (ADT), is created.

Degradation analysis involves the measurement of the degradation of a product at various time points and this information is then used to estimate the eventual failure lifetime for the product. One of the major advantages of performing reliability analysis based on degradation data is that it relates the reliability analysis directly to the physics of failure mechanism. Cutting tools, hydraulic structures, brake linings, airplane engine compressor blades, corroding pipelines, and rotating equipment, all of these structural systems suffer from increasing wear with usage and age. Various physical deterioration processes can be observed, such as cumulative wear, crack growth, erosion, corrosion, fatigue, consumption, etc. The deterioration of these systems might incur high cost due to production losses, delays, and safety hazards if the resistance of a deterioration structure drops below the applied failure threshold. Therefore, many papers have been focused on the research and applications for the reliability degradation model [26–51].

2.2.1 Methodology

Based upon the methodology, the studies of the degradation analysis can be divided into three main categories.

One of the most widely used methods to model the degradation data is called "General Path Model". Lu and Meeker [26] introduce a two-stage method of estimating the parameters of the mixed-effect path model to describe the system degradation performance. The general path model can be represented as

$$x_{ij} = f(t; \phi, \theta_i) + \varepsilon_{ij},$$

where x_{ij} denotes the observation of degradation measure from the jth measurement on the ith unit; θ_i random effect parameter for the ith unit follows normal distribution, that is, $\theta_i \sim N(\mu_\theta, \Sigma_\theta)$; ϕ the fixed-effect parameter.

The two steps for estimating the parameters in the general path model are given as:

- The 1st stage: for each sample unit, the degradation model is individually fitted to the sample path to obtain n estimates of the model parameters;
- The 2nd stage: the estimates of ϕ, μ_θ, and Σ_θ are computed using the n estimates obtained by the 1st stage.

Yuan and Pandey [27] elaborate the limitation of the linear regression for degradation analysis and then propose a general nonlinear mixed-effect (NLME) model to analyze and predict the degradation process. An alternative approach of two-stage maximum likelihood estimation for the general path model is proposed by Robinson and Crowder [28], where a Bayesian estimation model is used to estimate the distribution function of failure time in the general path model. Wu and Shao [29] develop the statistical inference using the ordinary least square and weighted least squares to estimate the parameters of degradation path in a nonlinear mixed-effect model.

The second approach to perform the degradation analysis is "Stochastic Process Model", such as Markov process and Brownian motion. Van Noortwijk et al. [30] model a Bayesian failure model of degradation analysis, in which the average amounts of deterioration are l_1-isotropic, and then explicitly obtain the probability of preventive repairs per unit time, failure with and without inspection conditional on the average amount of degradation. Van Noortwijk and Pandey [31] generalize a stochastic gamma process model for a stochastically deteriorating system. In the model, the system may fail when its deteriorating resistance drops below certain critical stress s. The degradation path is modulated by gamma distribution, and thus a suitable representation of the lifetime distribution is in the form of

$$F(t) = \Pr(T \le t) = \Pr(X(t) \ge r_0 - s\} = \int_{x=r_0-s}^{\infty} f_{X(t)}(x)dx = \frac{\Gamma(v(t), [r_0 - s]u)}{\Gamma(v(t))},$$

where T is the lifetime for the system; r_0 is the initial resistance; s is the deterministic critical stress; and Γ is the incomplete gamma function.

The corresponding maintenance policies for the gamma deterioration model are provided by [47, 48]. Nicolai et al. [32] compare three stochastic processes for degradation analysis, including Brownian motion with nonlinear drift, gamma process with nonlinear shape function and two-stage hit-and-grow (TSHG) process.

After that, the parameter estimation of these three processes for the inspection data and expert data is provided. In Hsieh et al. [33] a non-homogeneous compound Poisson process is utilized to model the discrete degradation. Then the first passage time distribution, the likelihood estimation of the model parameters, and confident interval approximation are derived. Xue and Yang [34] extend the two-state reliability measure to the multi-state reliability by combing the Markov process and *s*-coherent system structure in order to derive the dynamic reliability for the multi-state system with complex structure, including the series systems and parallel systems.

Kharoufeh [35] analytically derives the failure time distribution and the moments of the lifetime for a single-unit system with cumulative wear damage that is affected by the external environment by using the Markov additive process. Saassouh et al. [36] propose a two-mode stochastically deteriorating model with a sudden change point in the degradation path, where the increments of deterioration follow a gamma law when the system is in the first mode, and the mean deterioration rate increases when it flips into the second mode. Based on the definition of the model, the decision rules for an online maintenance policy are determined to optimize the system performance from the angle of asymptotic unavailability.

Kharoufeh and Cox [37] develop a degradation-based model for assessing the distribution and the moments of the lifetime function using the hybrid approach including two models: Model I describes a degradation process with the rate of degradation affected by the state of random environment governed by a homogeneous Markov process; Model II establishes a new path function by estimating the degradation rates using differential equation where the number of degradation status is approximated by *K*-nearest clustering method. Ebrahimi [38] proposes a general stochastic model to estimate the reliability of the system in terms of a deterioration process with covariate. The survival and hazard functions are analytically derived for two semi-parametric models using the differential equation by taking the exponential form of the survival function.

The third method is "Statistical Method", including both parametric and nonparametric estimation. Huang and Dietrich [39] present an extended graphical approach for degradation analysis by considering the ordering of degradation data. In the model, the degradation path is modulated by a truncated Weibull distribution and the analytical log likelihood expression is discussed. Bae et al. [40] examine the relationship between the degradation path function and the lifetime distribution where the additive and multiplicative function are used to model the degradation path. The results verify that the degradation path will have a great influence on the form of the lifetime distribution in terms of the class properties of lifetime distribution and failure rate.

Zue et al. [41] extend the results of Yang and Xue [34] from the *s*-normal process to the random process with a general distribution by using three approaches for degradation analysis. The first is the random process to fit the degradation process with a data free-distribution; the second is the traditional general path function; the third method is to use a multiple linear regression. Finally, a mixture model is introduced to model both hard failure and soft failure for the continuous state device. Bae and Kvam [42] present a nonlinear random coefficients model to fit the degradation path for the non-monotonically deteriorating light displays, and then consider four different methods to approximate the log-likelihood estimation, including a first-order method, Lindstrom-bates algorithm, adaptive importance sampling, and adaptive Gaussian quadrature method.

2.2.2 Maintenance Policy

Also increasing interest has been put upon the probabilistic models and mathematical formulations of the maintenance and replacement policies for the degradation or multi-state systems [43–51]. Grall et al. [43] propose a condition-based maintenance model including both the inspection and replacement policies based on a multi-level control-limit rule of a stochastically deteriorating system for the purpose of obtaining the optimal replacement threshold and inspection scheduling to minimize the long-run expected cost. Wang et al. [44] consider a novel maintenance model combing the condition-based replacement, periodical inspections, and (S, s) type provisioning policy, noted as (T, S, s, L_p) policy, where T denotes the inspection interval, S the maximum inventory level, s the reorder point, and L_p the replacement threshold. Furthermore, a simulation model is established to modulate the uncertain deterioration process and finally the maintenance scheduling is optimized to minimize the cost rate using genetic algorithm. Kiessler et al. [45] examine the limiting average availability of a hidden-failure deterioration system with the periodic inspections where deterioration rate is governed by a Markov model.

Yang and Klutke [46] characterize the properties of the lifetime distribution for the Levy degradation process and then illustrate the implement of the results to inspection scheduling for the maintained system with non- self-announcing failures. Zhao [47] presents a preventive maintenance policy for a deteriorating system with a critical reliability level to satisfy the preference of field managers with the imperfect PM effect modulated by a parameter of degradation ratio. Pandey et al. [48] consider an age replacement policy for the gamma deterioration model where the component is replaced when the system fails or reaches a specific age, whichever occurs first.

Van Noortwijk and Frangopol [49] describe two maintenance models of condition-based maintenance and reliability-based maintenance for the deteriorating civil infrastructures for the purpose of minimizing the life-cycle cost under the constraint of adequate reliability level. Delia and Rafael [50] analyze the maintenance policies with two types of repair modes, including preventive and corrective repairs, and phase-type distributed repair times for a cold standby system subject to multi-stage degradation. A later work of Dellia and Rafael [51] studies a maintenance model with failure and inspection following arrival processes and two types of repair modes, minimal and perfect, distributed as different phase-type distributions for a deteriorating system suffering from both internal and external failures.

2.3 Multiple Competing Risks of Degradation and Random Shocks

In real world, system typically deteriorates as a result of both the graceful degradation and discrete random shocks. However, most of the papers regarding the multiple competing risks of degradation and random shocks are under the assumption that these two processes are independent with each other. The incorrect independent assumption may underestimate the reliability and lifetime behaviors of the system. Therefore, on dealing with the relationship of competing risk of degradation and random shocks, the dynamic behaviors of the dependent structure between them may have a non-trivial impact on the reliability function estimation or on the maintenance and warranty policies for the deteriorating systems.

The dependence of these two processes can be exhibited in two aspects: (1) degradation process will make the system more vulnerable to the environment factors, such as temperature, pressure and random shocks; (2) random shocks will accelerate the degradation process with two modes, sudden jump or minor acceleration of the degradation rate. Furthermore, although in many studies of reliability model, the multiple degradation processes are assumed to have independent lifetimes, it may be more realistic to assume some sort of dependence among different degradation processes. For example, a system may have multiple components with its own degradation process or even one component may be subject to multiple degradations, in which the lower status of one degradation process could result in an increasing load on some of the other degradation processes. As a sequence, a more systematic probability model for the dynamic dependent structure underlying these two processes should be called for.

2.3.1 Independent Model for Degradation and Shocks

Sim and Endrenyi [52] consider a Markov process to model the maintenance policy with periodically minimal maintenance and major maintenance after a number of minimal maintenances for a continuously operating system subject to degradation and Poisson failures. The optimal solution to minimize either cost rate or unavailability is derived. A probabilistic model of the reliability analysis for the deteriorating structural systems subject to Poisson shocks is introduced by Ciampoli [53], where a stochastic differential equation is employed to model the degradation process in order to obtain the total damage of the system. A generalized Petri Net is proposed by Hosseini et al. [54] to formulate a new condition-based maintenance model for a system subject to deterioration failures and Poisson failures. In order to maximize the system throughout, an optimal inspection policy based on minimal maintenance, major maintenance and major repairs is obtained.

Zhu et al. [55] examine the maintenance model for a competing risk of degradation and sudden failure, where the unit is renewed when it reaches a predetermined degradation level or comes to a sudden failure within the limit of certain degradation threshold. Also a preventive maintenance (PM) is done at the scheduled time and the consecutive repair times are modulated as an increasing quasi-renewal process due to system aging. The maintenance scheduling variables of degradation threshold and scheduled time to preventive maintenance are determined by maximizing the system availability with the constraint of repair cost. Weibe et al. [56] derive the reliability estimation and the optimal solution for calculating the discounted cost based on both condition-based and age-based policy for a maintained system that deteriorates due to both transient shocks and cumulative degradation process governed by a stochastic point process.

In practice, there is not only the system operating cost which will increase with the system aging, but also the cost of time of the inspection, repair and replacement. Chiang and Yuan [57] present a state-dependent maintenance policy $R_{i,j}(T, N, \alpha)$ for a continuously deteriorating system subject to degradation and fatal shocks using a continuous-time Markov process, where T is the system inspection interval, N is the system boundary for replacement, and α is the probability that repair will restore the system to a better state. Delia and Rafael [58] examine the replacement policy for a Markovian degraded system submitted to internal or external failures with holding time on various system levels, external repair time and internal repair time, all of which follow the phase-type distribution. In the work by Kharoufeh et al. [59], the lifetime distribution as well as the limiting availability for a periodically inspected single-unit system with hidden failure is explicitly derived by utilizing the Laplace-Stieltjes transform. The system is submitted to two failure mechanisms, the degradation wear that is governed by its random environment characterized as a continuous Markov chain and random shocks modulated as a homogeneous Poisson process.

2.3.2 Dependent Model for Degradation and Shocks

Tang and Lam [60] study a δ-shock maintenance model for a deteriorating system with shocks occurring according to a renewal process, where the interarrival time of shocks follows a Weibull or gamma distribution. Because the system is deteriorating, the deadlock threshold for the δ-shock is geometrically non-decreasing after each repair, and the repair time is modulated by an increasing geometric process. Frostig and Kenzin [61] derive the limiting average availability in a maintenance model for a hidden-failure system that suffers from the wear- out and cumulative shock damage with a Poisson process. Two models are discussed: model I assumes the wear out process and shock which will not receive any impact from the external environment; in model II, the shock magnitude, the shock rate and wear- out process, all of them are dependent on the external environment modulated by a Markov process. Based on the approach proposed by Mori and Ellenwood [62], Van Noortwijk et al. [63] put forward a novel approach to combine two stochastic processes of deteriorating resistance and fluctuating load for the reliability analysis of a structural component. In the model, the deteriorating process is modulated as a gamma process and the random loading exceed is given by a generalized Pareto distribution with loading arriving according to a Poisson process.

Chiodo and Mazzanti [64] deal with the problem of the reliability function assessment for power system devices due to repeated shocks. The systems may survive under the condition that the individual stress load is less than the remaining degradation resistance. Lehman [65] surveys two classes of degradation-threshold-shock models (DTS), including general DTS and DTS with covariates, where the system failure may be due to the competing risk of degradation and trauma. In the general DTS model, the traumatic failure is assumed to be modeled as a stochastic Poisson process with intensity factor that is governed by system aging level. In the DTS with covariates, a dynamic environment random variable is included in the model. Fan et al. [66] consider a single-component system that suffers from the non-homogeneous Poisson process shocks. In the system, aging will increase the magnitude of shock sizes, thereby resulting in a larger fatal probability. After that, an extension of that shock model to multi-component system is examined.

Cha and Finkelstein [67] extend the Brown-Proschan model by assuming that the random shocks will result in an immediate system failure with a probability $p(t)$, but accelerate the system aging process by certain random increment with probability $q(t)$. Finkelstein [68] introduces a generalized Strehler-Mildvan model to estimate the first passage time of the survival function for the system subject to cumulative damage due to biological aging and sudden killing event. The asymptotic aging properties for the repairable system are discussed. Satow et al. [69] focus on the replacement policy for one single unit that suffers from cumulative damage due to aging process and shocks in order to obtain the optimal replacement level k^* which minimizes the expected cost rate.

Deloux et al. [70] propose a maintenance policy that combines the statistical process control (SPC) and condition-based maintenance (CBM) for a continuously deteriorating system with two kinds of failure mechanisms, deteriorating and random shocks. The system failure is governed by deteriorating process as a function of the deterioration level and the system time but an associated failure acceleration factor due to stress is taken into account when the stress intensity exceeds some critical level λ. Klutke and Yang [71] present a maintenance policy for the periodically inspected systems with non- self-announcing failure, submitted to cumulative damage due to both graceful degradation and random shocks for the purpose of optimizing the system performance from the limiting average availability point of view.

2.3.3 Multiple Degradation Model

Li and Pham [72] focus on the reliability analysis for a generalized multi-state degradation system subject to multiple competing failure processes, consisting of two degradation processes and cumulative random shock. The paper assumes that all of these processes are independent, and any of them causes the system to fail according to the threshold values of each process. No repair or maintenance policies are considered. Based on the definition of multi-state degraded system in [72], a condition-based maintenance model is built by Li and Pham [73], where an average long-run cost rate function is minimized by Nelder-Mead downhill simplex method. The interinspection sequence is generated by a geometric sequence.

Wang and Coit [74] propose a general model of predicting the reliability on the correlated multiple degradation processes and verify that the system reliability might be underestimated because of the incorrect independent assumption by the simulated data. A gamma-based state space model is studied by Zhou et al. [75] to predict the lifetime for a multiple degradation processes with uncertain failure threshold using multivariate normal distribution, where expectation–maximization (EM) algorithm is utilized to estimate parameters of the model and Monte Carlo-based particle smoothing algorithm is used to deal with the expectation estimation of complete likelihood in step E of EM algorithm.

2.4 *Copula Method*

Only a few studies focus on the issue of multiple degradation processes according to the literature review of previous section. These two papers in [72, 73] consider the reliability and maintenance model for two degradation processes and random shocks, but all of them are independent with each other. A traditional way to build correlated multiple degradation model is to utilize the tool of multivariate distribution which will create the restrictions on the same distribution of each marginal degradation path. Recently considerable attention has been paid to the dependence behavior between random variables modeled by copulas, which allow us to link the univariate marginal distributions to obtain a joint probability of the events. Compared with the traditional multivariate distribution, the most attractive advantages of the copula method are listed as follows: (a) the univariate marginal function can be modulated separately from their dependent structure; (b) the marginal probability can be drawn from different kinds of distributions without restriction; (c) the parameter coefficients of the copula model can be time-varying, not constant. Because of these advantages, copula model is a powerful alternative approach of the multivariate distribution to analyze the correlated multiple degradation processes. There are several aspects of copulas method that could be worked with.

2.4.1 Theoretical Model

Embrechts and Puccetti [76] provide analytical procedures to calculate the bounds on the distribution function of the sum of n dependent risks with overlapping margins, that is, the bounds for the sum $S = X_1 + \cdots + X_n$, where $X = (X_1, \ldots, X_n)$ belongs to Frechet class of probability measure. Kojadinovic and

Yan [77] compare the asymptotic properties of three semi-parametric methods of estimating the parameters in copula models, which are maximum pseudo-likelihood estimation method of moment estimator based on Spearman's rho, and method of moment estimator based on Kendall's tau. Monte Carlo simulation is used to examine the performance of the different estimators with finite samples and compute the asymptotic relative efficiency. Rodriguez-Lallena and Ubeda-Flores [78] examine the properties of the conditional distribution of $H_1(X)$ given that the joint distribution of X is H_2, where H_1 and H_2 are the multivariate distribution functions for random vectors $X = (X_1, X_2, \ldots, X_n)$ with common univariate marginal distributions.

Chen and Fan [79] derive the asymptotic properties of estimators for a class of copula-based semi-parametric stationary Markov models characterized by parametric copula functions and nonparametric margins. Hurlimann [80] proposes a modified statistical method of inference functions of margins (IFM) characterized as two-step maximum likelihood estimations of univaraite marginal distributions and copulas, followed by minimizing the chi-square statistic of a bivariate version of the Pearson goodness-of-fit test to determine the dependence parameters for copula fitting in the bivariate cumulative returns. Abegaz and Naik-Nimbalkar [81] introduce an alternative approach based on a copula-based Markov chain to investigate the conditional probability of distributions and utilize one- and two-stage statistical inference method to estimate the parameters. In addition, a parametric pseudo-likelihood ratio test is given to select the copula model for the two-stage estimation. Zezula [82] illustrates how to use the special variance structure of Gaussian copulas to facilitate the parameter estimations under the condition that the data dimension is large.

2.4.2 Application

Goorbergh et al. [83] apply dynamic copula model to better-of-two-markets and worse-of-two-markets options on the S&P500 and NASDAQ to examine the dependent behavior of bivariate option pricing with association between the assets. The aim of the study in Zhang and Singh [84] is to derive the bivariate joint distribution of rainfall frequency using four Archimedean copulas in order to determine the return periods. Based on the data from US stock, Fernandez [85] verifies that using tail-dependency tests to select the copula model may be misleading especially when the data are featured with conditional volatility and series correlation. With the help of Monte Carlo simulation Al-Harthy et al. [86] illustrate how the copula method can be suitable to model the dependencies in oil and gas evaluations and come to the conclusion that compared with some more commonly used approaches to model dependence the copula method can accurately detect the tail dependence structure of the variable distribution. Dalla Valle [87] suggests a new methodology for studying the dependent relationships of operational risk management by combining the copula and Bayesian models computed using simulation methods, especially Markov Chain Monte Carlo. Dakovic and

Czado [88] examine the point and interval estimates using joint maximum likelihood and semi-parametric models to estimate the parameters of a bivariate *t*-copula model in financial data.

Applications of copula methods in various fields can also be found in Ning [89] for the dependence structure between the foreign exchange market and equity market, Roch and Alegre [90] for daily equity returns, Ausin and Lopes [91] for multivariate time series using time-varying copulas, and Renard and Lang [92] for design hydrology.

2.4.3 Special Case: Dependent Risk Model

A special case of copula application is dependent risk model, which can be used as a source of reference to construct the dependent competing risk model in reliability using copulas, for instance the studies in [93–98]. Kaishev et al. [93] establish a dependent multiple-decrement model to examine the dependencies among causes of death in order to analyze the impact of complete or partial elimination of causes of death on the survival function from competing risks using copulas. Bedford [94] puts forward two different methods of nonparametric maximum likelihood and bilinear adjustment estimator to perform the quantile tests for copulas in competing risk problems. Lo and Wilke [95] develop a new copula graphic estimator to a model with multiple dependent competing risk and apply the model to the data set of duration of unemployment from Germany.

Cossette et al. [96] derive the discounted penalty function via Laplace transform for a generalized Farlie-Bumbel-Morgenstern copula model in the presence of the associations between the claim sizes and interclaim time in a compound Poisson risk model. Another study of Cossete et al. [97] presents two approaches of a class factor method and copula method to construct the dependent risk models for the insurance portfolio. Embrechts et al. [98] provide the properties and computational procedures of distributional bounds for the dependent risks functions using copulas.

3 Some Recent Studies

3.1 Cumulative Competing Risks of Degradation and Random Shocks

The detailed description of the problem could be obtained in Wang and Pham [99]. The paper includes three parts to elaborate the problem from a deteriorating system suffering from random shocks:

Part I: A two-process cumulative combination model of competing failures between degradation and random shocks is introduced with additive and multiplicative degradation path.

Part II: Two numerical examples for additive and multiplicative degradation path are used to illustrate the combination model and also the sensitivity analysis is discussed.

Part III: Based on the definition of the model, an imperfect preventive maintenance (PM) policy is put forward to obtain the optimum pair (N^*, T^*) using differential evolution algorithm in order to minimize the expected total cost rate.

Different from the traditional competing model, a two-process cumulative model may be more suitable to describe the problem. It is because both of these damage works directly on the unit of system, and a cumulative damage can clearly reflect the system's status. Also, the perfect preventive maintenance is not so practical in real world. In the paper they model the imperfect preventive maintenance by an improvement factor to reduce the system cumulative damage proportionally after each imperfect PM.

3.2 *Dependent Competing Risks of Degradation and Random Shocks*

Wang and Pham [100] contribute to the knowledge of the multi-state cumulative dependent modeling of degradation and random shocks by adding time-scaled covariate factors from random shocks into the degradation path function, and also modulating the system's resistance as a Quasi-renewal process in order to capture the aging influence. Concretely, two kinds of random shocks are considered: (1) fatal shock if an individual shock magnitude exceeds critical threshold, which results in the system's immediate failure; (2) non-fatal shock, which leads to both a sudden increment jump and a minor acceleration of the degradation rate. Therefore, the system survives only under the condition that non-fatal shock happens and also the cumulative degradation amount does not drop below certain critical failure threshold. Furthermore, the system resistance will decrease proportionally to the preceding resistance with the change of the system degradation status, modulated by a decreasing Quasi-renewal process. The structure of the paper can be shown as:

Firstly, they introduce the dependent cumulative competing risk model for the degradation wear and random shocks, and then a numerical example with three different forms of degradation path functions will be employed to illustrate the comparison between the dependent and independent models. Secondly, the analytical derivation formulas of the lifetime distribution and the degradation status for the multi-state dependent competing risk model will be explored. Based on the assumption and derivation for the dependent model, a numerical example will be devoted to illustrate the application of the model by analytical approach. Finally, Monte Carlo simulation is used as an alternative method to extend the results from analytical approach.

3.3 Multiple Competing Risks of Degradation and Random Shocks

Wang and Pham [101] propose a dependent model of multiple competing risks of degradation and random shocks by adding time-scaled covariate factors from random shocks into the degradation paths, and also modulating the joint distribution of multiple degradation processes using copula method to link the marginal functions. The traditional way to build multiple degradation models is to utilize the tool of multivariate distribution but it creates some restrictions on the same distribution of each marginal degradation path. Recently considerable attention has been paid to the dependence behavior between random variables modeled by copulas, which allow us to link the univariate marginal distributions to obtain a joint probability of the events.

Although copula is a flexible and powerful technique to build multivariate distribution, widely used in various applications including economics, finance, and actuarial science, the copula method applied to model multiple degradation processes is seldom. Furthermore, few papers consider the two types of dependent structure in one competing risk model, the dependent structure between degradation and shock, among degradation processes. The paper works on the multiple-degradation copula model embedded with random shocks in order to consider two dependent behaviors in one model. In [101], they discuss mathematical models to predict the reliability of the system subject to dependent multiple degradation and random shocks using copula method. Two kinds of random shocks are discussed such as: fatal shock with probability $p(t)$, which results in the system's immediate failure; non-fatal shock, which leads to both a sudden increment jump and a minor acceleration of the degradation rate.

3.4 Multi-objective Optimization of Imperfect Preventive Maintenance Policy for Dependent Competing Risk System with Hidden Failure

In the work of Wang and Pham [102], a multi-objective maintenance optimization embedded within the imperfect preventive maintenance (PM) and replacement for one single-unit system subject to the dependent competing risk of degradation wear and random shocks is studied. The contribution of the paper is listed as follows:

- The competing risks of degradation wear and random shocks are dependent with each other. There are two kinds of random shocks: fatal shock will cause the system failure immediately, but on the other hand non-fatal shock will increase the system virtual age by certain cumulative shock loading.
- A reduction factor is used to simulate the imperfect PM by modulating the system critical threshold with a quasi-renewal process.

- System failure is hidden, that is, the system failure will be only detected at the time of the scheduled maintenance or replacement.
- The imperfect PM cost followed by no failure, degradation failure, and random shock failure can be varied.
- Multi-objective optimization is employed to simultaneously maximizing the system availability and minimizing the system cost rate by the fast elitist non-dominated Sorting Genetic Algorithm (NSGA-II) in Matlab 7.6.0.

The chapter aims to present an overview of competingrisk with respect to dependent system degradation and random shocks. It also discusses briefly maintenance related risk aspect subject to imperfect preventive maintenance with considerations of the degradation critical threshold using quasi-renewal process. Further research has also been discussed regarding the condition based maintenance and the maintenance policies for multi-component system embedded within the framework of dependent competing risk of degradation wear and random shocks.

References

1. Esary JD, Marshall AW (1973) Shock models and wear process. Annu Prob 1(4):627–649
2. Hameed MSA, Proschan F (1973) Nonstationary shock models. Stoch Process Appl 1(10):383–404
3. Agrafiotis GK, Tsoukalas MZ (1987) On excess-time correlated cumulative process. J Oper Res Soc 46:1269–1280
4. Gut A, Husler J (2005) Realistic variation of shock models. Stat Prob Lett 74(2):187–204
5. Nakagawa T, Kijima M (1989) Replacement policies for a cumulative damage model with minimal repair at failure. IEEE Trans Reliab 28:581–584
6. Qian C, Nakamura S, Nakagawa T (2003) Replacement and minimal repair polices for a cumulative damage model with maintenance. Comput Math Appl 46:1111–1118
7. Wortman MA, Klutke G-A, Ayhan H (1994) A maintenance strategy for systems subjected to deterioration governed by random shocks. IEEE Trans Reliab 43(3):439–445
8. Chelbi A, Ait-Kadi D (2000) Generalized inspection strategy for randomly failing systems subjected to random shocks. Int J Prod Econ 64:379–384
9. Ross SM (1981) Generalized poisson models. Annu Prob 9(5):896–898
10. Shanthikumar JG, Sumita U (1983) General shock models associated with correlated renewal sequences. J Appl Prob 20:600–614
11. Chen J, Li Z (2008) An extended extreme shock maintenance model for a deteriorating system. Reliab Eng Syst Safety 93:1123–1129
12. Li Z (1984) Some probability distribution on Poisson shocks and its application in city traffic. J Lanzhou Univ 20:127–136
13. Li Z, Kong X (2007) Life behavior of δ-shock model. Stat Prob Lett 77(6):577–587
14. Li Z, Zhao P (2007) Reliability analysis on the δ-shock model of complex systems. IEEE Trans Reliab 56(2):340–348
15. Lam Y, Zhang YL (2004) A shock model for the maintenance problem of repairable system. Comput Oper Res 31:1807–1820
16. Rangan A, Tansu A (2008) A new shock model for system subject to random threshold failure. Proc World Acad Sci Eng Technol 30:1065–1070

17. Fan J, Ghurke SG, Levine RA (2000) Multicomponent lifetime distribution in the presence of ageing. J Appl Prob 37:521–533
18. Igaki N, Sumita U, Kowada M (1995) Analysis of Markov renewal shock models. J Appl prob 32:821–831
19. Gut A (2001) Mixed shock models. Bernoulli 7(3):541–555
20. Mallor F, Omey E (2001) Shocks, runs and random sums. J Appl Prob 38:438–448
21. Mallor F, Santos J (2003) Classification of shock model in system reliability. Monografias del Semin Matem Garcia de Galdeano 27:405–412
22. Mallor F, Santos J (2003) Reliability of systems subject to shocks with a stochastic dependence for the damages. Test 12(2):427–444
23. Mallor F, Omey E, Santos J (2006) Asymptotic results for a run and cumulative mixed shock model. J Math Sci 138(1):5410–5414
24. Finkelstein MS, Zarudnij VI (2001) A shock process with a non-cumulative damage. Reliab Eng Syst Safety 71:103–107
25. Bai J-M, Li Z-H, Kong X-B (2006) Generalized shock models based on a cluster point process. IEEE Trans Reliab 55(3):542–550
26. Lu CJ, Meeker WQ (1993) Using degradation measures to estimate of time-to-failure distribution. Technometrics 35:161–176
27. Yuan X-X, Pandey MD (2009) A nonlinear mixed-effects model for degradation data obtained from in-service inspections. Reliab Eng Syst Safety 94:509–519
28. Robinson ME, Crowder MJ (2000) Bayesian methods for a growth-curve degradation model with repeated measures. Life Data Anal 6:357–374
29. Wu S-J, Shao J (1999) Reliability analysis using the least squares method in nonlinear mixed-effect degradation models. Stat Sinica 9:855–877
30. van Noortwijk JM, Cooke RM, Kok M (1995) A Bayesian failure model based on isotropic deterioration. Eur J Oper Res 82:270–282
31. van Noortwijk JM, Pandey MD (2003) A stochastic deterioration process for time-dependent reliability analysis. Proceedings of the eleventh IFIP WG 7.5 working conference on reliability and optimization of structural systems, pp 259–265
32. Nicolai RP, Dekker R, van Noortwijk JM (2007) A comparison of models for measurable deterioration: an application to coatings on steel structures. Reliab Eng Syst Safety 92:1635–1650
33. Hsieh M-H, Jeng S-L, Shen P-S (2009) Assessing device reliability based on scheduled discrete degradation measurements. Prob Eng Mech 24:151–158
34. Xue J, Yang K (1995) Dynamic reliability analysis of coherent multistate systems. IEEE Trans Reliab 44(4):683–688
35. Kharoufeh JP (2003) Explicit results for wear processes in a Markovian environment. Oper Res Lett 31:237–244
36. Saassouh B, Dieulle L, Grall A (2007) Online maintenance policy for a depredating system with random change of mode. Reliab Eng Syst Safety 92:1677–1685
37. Kharoufeh JP, Cox SM (2005) Stochastic models for degradation-based reliability. IIE Trans 37(6):533–542
38. Ebrahimi N (2001) A stochastic covariate failure model for assessing system reliability. J Appl Prob 38:761–767
39. Huang W, Dietrich DL (2005) An alternative degradation reliability modeling approach using maximum likelihood estimation. IEEE Trans Reliab 54(2):310–317
40. Bae SJ, Kuo W, Kvam PH (2007) Degradation models and implied lifetime distributions. Reliab Eng Syst Safety 92:601–608
41. Zuo MJ, Jiang R, Yam RCM (1999) Approaches for reliability modeling of continuous-state devices. IEEE Trans Reliab 48(1):9–18
42. Bae SJ, Kvam PH (2005) A nonlinear random coefficients model for degradation testing. Technometrics 46(4):460–469
43. Grall A, Berenguer C, Dieulle L (2002) A condition-based maintenance policy for stochastically deteriorating systems. Reliab Eng Syst Safety 76:167–180

44. Wang L, Chu J, Mao W (2009) A condition-based replacement and spare provisioning policy for deteriorating systems with uncertain deterioration to failure. Eur J Oper Res 194:184–205
45. Kiessler PC, Klutke G-A, Yang Y (2002) Availability of periodically inspected systems subject to Markovian degradation. J Appl Prob 39:700–711
46. Yang Y, Klutke G-A (2000) Lifetime-characteristics and inspection-schemes for levy degradation process. IEEE Trans Reliab 49(4):377–382
47. Zhao YX (2003) On preventive maintenance policy of a critical reliability level for system subject to degradation. Reliab Eng Syst Safety 79:301–308
48. Pandey MD, Yuan XX, van Noortwijk JM (2005) Gamma process model for reliability analysis and replacement of aging structural components, Safety and Reliability of Engineering Systems and Structures. In: Proceedings of the Ninth International Conference on Structural Safety and Reliability (ICOSSAR), Rome, pp 2439–2444
49. van Noortwijk JM, Frangopol DM (2004) Two probabilistic life-cycle maintenance models for deteriorating civil infrastructures. Prob Eng Mech 19:345–359
50. Delia M-C, Rafael P-O (2006) A deteriorating two-system with two repair modes and sojourn times phase-type distributed. Reliab Eng Syst Safety 91:1–9
51. Delia M-C, Rafael P-O (2008) A maintenance model with failures and inspection following Markovian arrival processes and two repair modes. Eur J Oper Res 186:694–707
52. Sim SH, Endrenyi J (1993) A failure-repair model with minimal & major maintenance. IEEE Trans Reliab 42(1):134–140
53. Ciampoli M (1998) Time dependent reliability of structural systems subject to deterioration. Comput Struct 67:29–35
54. Hosseini MM, Kerr RM, Randall RB (2000) An inspection model with minimal and major maintenance for a system with deterioration and poisson failures. IEEE Trans Reliab 49(1):88–987
55. Zhu Y, Elsayed EA, Liao H, Chan LY (2010) Availability optimization of systems subject to competing risk. Eur J Oper Res 202(3):781–788
56. van der Weide JAM, Pandey MD, van Noortwijk JM (2010) Discounted cost model for condition-based maintenance optimization. Reliab Eng Syst Safety 95:236–246
57. Chiang JH, Yuan J (2001) Optimal maintenance policy for a Markovian system under periodic inspection. Reliab Eng Syst Safety 71:165–172
58. Delia M-C, Rafael P-O (2006) Replacement times and costs in a degrading system with several types of failure: the case of phase-type holding times. Eur J Oper Res 175:1193–1209
59. Kharoufeh JP, Finkelstein DE, Mixon DG (2007) Availability of periodically inspected systems with Markovian wear and shocks. J Appl Prob 43:303–317
60. Tang Y-Y, Lam Y (2006) A δ-shock maintenance model for a deteriorating system. Eur J Oper Res 168:541–556
61. Frostig E, Kenzin M (2009) Availability of inspected systems subject to shocks—a matrix algorithmic approach. Eur J Oper Res 193:168–183
62. Mori Y, Ellingwood BR (1994) Maintaining: reliability of concrete structures. I: role of inspection/repair. Struct Eng 120(3):824–845
63. van Noortwijk JM, van der Weide JAM, Kallen MJ, Pandey MD (2007) Gamma processes and peaks-over-threshold distribution for time-dependent reliability. Reliab Eng Syst Safety 92:1651–1658
64. Elio C, Giovanni M (2006) Indirect reliability estimation for electric devices via a dynamic 'stress-strength' model, SPEEDAM international symposium on power electronics, electrical drives, automation and motion
65. Lehmann A (2009) Joint modeling of degradation and failure time data. J Stat Plan Infer 139(5):1693–1706
66. Fan J, Ghurye SG, Levine RA (2000) Multi-component lifetime distributions in the presence of ageing. J Appl Prob 37:521–533

67. Cha JH, Finkelstein M (2009) On a terminating shock process with independent wear increments. J Appl Prob 46:353–362
68. Finkelstein M (2009) On damage accumulation and biological aging. J Stat Plan Infer 139(5):1643–1648
69. Satow T, Teramoto K, Nakagawa T (2000) Optimal replacement policy for a cumulative damage model with time deterioration. Math Comput Model 31:313–319
70. Deloux E, Castanier B, Berenguer C (2009) Predictive maintenance policy for a gradually deteriorating system subject to stress. Reliab Eng Syst Safety 94(2):418–431
71. Klutke G-A, Yang Y (2002) The availability of inspected systems subject to shocks and graceful degradation. IEEE Trans Reliab 51(3):371–374
72. Wenjian L, Pham H (2005) Reliability modeling of multi-state degraded systems with multi-competing failures and random shocks. IEEE Trans Reliab 54(2):297–303
73. Wenjian L, Pham H (2005) An inspection-maintenance model for systems with multiple competing processes. IEEE Trans Reliab 54(2):318–327
74. Peng W, Coit DW (2004) Reliability prediction based on degradation modeling for systems with multiple degradation measures, Reliability and maintainability 2004 annual symposium-RAMS, pp 302–307
75. Yifan Z, Lin M, Rodney C, H-E Kim (2009) Asset life prediction using multiple degradation indicators and lifetime data: a gamma-based state space model approach. The 8th international conference on reliability, maintainability and safety
76. Embrechts P, Puccetti G (2010) Bounds for the sum of dependent risks having overlapping marginals. J Multivar Anal 101:177–190
77. Kojadinovic I, Yan J (2010) Comparison of three semiparametric methods for estimating dependence parameters in copula models. Insur Math Econ 47(1):52–63
78. Rodriguez-Lallena JA (2003) Manuel Ubeda-Flores, distribution functions of multivariate copulas. Stat Prob Lett 64:41–50
79. Chen X, Fan Y (2006) Estimation of copula-based semiparametric time series models. J Econ 130:307–335
80. Hurlimann W (2004) Fitting bivariate cumulative returns with copulas. Comput Stat Data Anal 45:355–372
81. Abegaz F, Naik-Nimbalkar UV (2008) Modeling statistical dependence of Markov chains via copula models. J Stat Plan Infer 138:1131–1146
82. Zezula I (2009) On multivariate Gaussian copulas. J Stat Plan Infer 139:3942–3946
83. Van den Goorbergh RWJ, Genest C, Werker BJM (2005) Bivariate option pricing using dynamic copulae models. Insur Math Econ 37:101–114
84. Zhang L, Singh VP (2007) Bivariate rainfall frequency distributions using Archimedean copulas. J Hydrol 332:93–109
85. Fernandez V (2008) Copula-based measures of dependence structure in assets returns. Physica A 387:3615–3628
86. Al-Harthy M, Begg S, Bratvold RB (2007) Copulas: a new technique to model dependence in petroleum decision making. J Petrol Sci Eng 57:195–208
87. Valle LD (2009) Bayesian copulae distributions, with application to operational risk management. Methodol Comput Appl Probab 11:95–115
88. Dakovic R, Czado C (2009) Comparing point and interval estimates in the bivariate *t*-copula model with application to financial data. Statistical Papers, 1613-9798 (online)
89. Ning C (2010) Dependence structure between the equity market and the foreign exchange market—a copula approach. J Int Money Finance 29(5):743–759
90. Roch O, Alegre A (2006) Testing the bivariate distribution of daily equity returns using copulas. An application to the Spanish stock market. Comput Stat Data Anal 51:1312–1329
91. Concepcion Ausin M, Lopes HF (2010) Time-varying joint distribution through copulas. Comput Stat Data Anal 54(11):2383–2399
92. Renard B, Lang M (2007) Use of a Gaussian copula for multivariate extreme value analysis: some case studies in hydrology. Adv Water Resour 30:897–912

93. Kaishev VK, Dimitrina DS, Haberman S (2007) Modeling the joint distribution of competing risks survival times using copula functions. Insur Math Econ 41(3):339–361
94. Bedford T (2006) Copulas, degenerate distributions and quantile tests in competing risk problems. J Stat Plan Infer 136(5):1572–1587
95. Lo SMS, Wilke RA (2010) A copula model for dependent competing risks. J Roy Stat Soc Ser C (Appl Stat) 59(2):359–376
96. Cossette H, Marceau E, Marri F (2008) On the compound Poisson risk model with dependence based on a generalized Farlie-Gumbel-Morgenstern. Insur Math Econ 43:444–455
97. Cossette H, Gaillardetz P, Marceau E, Rioux J (2002) On two dependent individual risk models. Insur Math Econ 30:153–166
98. Embrechts P, Hoing A, Juri A (2003) Using copulae to bound the value-at-risk for functions of dependent risks. Finance Stoch 7:145–167
99. Wang Y, Pham H (2009) The imperfect preventive maintenance policies for two-process cumulative damage model (submitted to Int J Syst Sci)
100. Wang Y, Pham H (2011) Modeling the dependent competing risks with multiple degradation processes and random shock using time-varying copulas. IEEE Trans Reliab, vol 6(4) to appear
101. Wang Y, Pham H (2010) Dependent competing risk model with multiple-degradation and random shock using time-varying copulas. 16th ISSAT international conference on reliability and quality in design, Washington D.C., pp 100–104
102. Wang Y, Pham H (2011) A multi-objective optimization of imperfect preventive maintenance policy for dependent competing risk system with hidden failure. IEEE Trans Reliab, vol 6(3) to appear

Risk and Design Management Based on Failure Mode

Yuichi Otsuka

1 Introduction

Latent problems in products can cause the accidents in use. Failure modes and effects analysis (FMEA) is normally used to predict these latent problems. However, a process of filling data sheets of FMEA is not obvious in detail and the utility of FMEA sheet in next product plans involves various problems such as difficulties in searching specific data. In this chapter, the revised method of FMEA, called System design review based on failure modes (DRBFM) based on the design concept GD^3, will be introduced. In the System DRBFM method, the deviations from reliable and safe design conditions (Good Design) are firstly specified. This process is a little different from the traditional FMEA method, however it is indispensable. Users of the System DRBFM method can recognize the goal of predicting failure modes with this deviations. Because they can only check whether a current design model should be regarded as the reliable one by taking measures for all failures from the predicted deviations. Without the target, they should judge by themselves that no latent failures are involved in their planning products. Its heavy work. We also present a system management concept in using the System DRBFM method in product quality management fields.

Y. Otsuka (✉)
Nagaoka University of Technology, Room 562, 1603-1 Kamitomioka-cho,
Nagaoka-shi, Niigata 940-2188, Japan
e-mail: otsuka@vos.nagaokaut.ac.jp

H. Pham (ed.), *Safety and Risk Modeling and Its Applications*,
Springer Series in Reliability Engineering, DOI: 10.1007/978-0-85729-470-8_8,

1.1 Problems in Predicting Failure Modes

Failures caused by unconsidered problems in a design stage frequently occur. Some of which are considered in advance, however only insufficient measures have been taken and the failure also occurs. Already-known phenomena [6] normally occupies main causes of the failures. The product that occurs an accident has probably been involving latent problems before the accident happening [3, 19]. These latent problems could not have previously been observed by a designer, then no measures were taken. This oversight or underestimation lastly results in accidents. In order to prevent the accident of products in advance, it is quite important to introduce the concept of "Finding problems" in a design stage and a management process.

Using a knowledge database to support the quality of design review in a design stage is highly recommended [1]. Xijuan et al. [20] proposed one evaluation method for an efficiency of design review using an error indicator. However, these tools are effective only after the designer considers specific errors. Unfortunately they are not very successful in finding the error itself. The causes of errors often exist in the knowledge or experience of a designer. A systematic framework is then necessary to make him notice his lacks in the knowledge or experience that causes the errors in design ideas.

1.2 Design Review Based on Failure Mode

The authors have proposed a DRBFM as the methodology for stimulating creativity of a designer and finding problems in drawing. The previous researches [12, 13] showed the procedure of the System DRBFM and provided case studies for visualizing the effective process of finding problems. These previous reports described the effectiveness of the System DRBFM process in visualizing the latent problems. However, conducting the DRBFM method is not easy work and then the users of DRBFM expect to improve the procedure and supporting policies and tools to help them. To solve these problems, the types of designer's errors in conducting the DRBFM method are classified. Furthermore, the procedure of the DRBFM method is improved in order to find the errors effectively. At first, an organizational learning policy using designer's errors in conducting the DRBFM method is introduced Next, the rearranged procedure of System DRBFM and the case study of conducting the System DRBFM method are introduced.

2 System DRBF

In a hierarchical system, a change in an element belonging to a low class amplifies its effect in affecting a higher component or a sub-system and can finally result in accidents [9]. To visualize this risk, we discuss the process of visualization for the

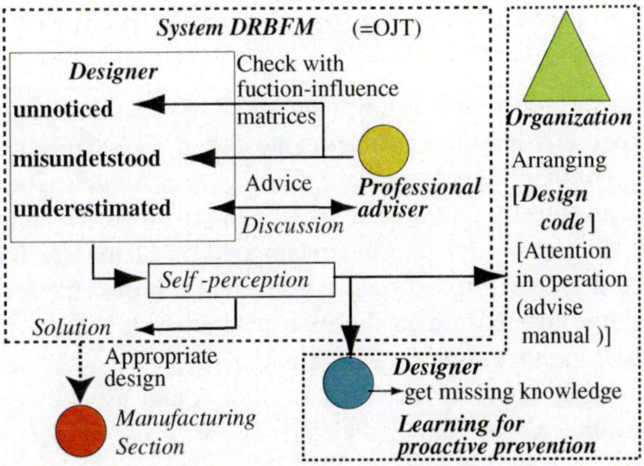

Fig. 1 Concept of timely OJT and precautionary prevention by system DRBFM

effects of changes in an element that affects the functionality of higher components and suffers the reliability or safety of an entire structure. The procedure of discussion for finding latent problems is defined as the System DRBFM. The System DRBFM is the way of discussing the effect flow of latent failures, caused by the changes from past reliable design condition (dimension, environment), to entire fracture in the hierarchical structure of product in order to find them easier by stimulating the participants' focus of thinking.

2.1 Designer's Error as the Target of Finding Problems

Figure 1 shows the management policy of the System DRBFM for organizational learning by introducing the concept of OJT. The designer's error involved in a design idea can be classified into the following three types.

- *Unnoticed*: a problem that the designer did not consider. No measures were taken.
- *Underestimation*: a problem that the designer noticed but took no measures to prevent it, because he thought the problem would not adversely affect the reliability or safety of the product.
- *Misunderstanding*: a problem that the designer noticed and took insufficient measures, because of his lack in knowledge or experience.

In Japan, more than half of the causes in the recall of automobiles has been occupied by problems in design [10]. It is practically necessary to find the latent problems through the process of making design drawings. Most of the technical problems are caused by known phenomena [16]. Previously finding these problems can result in taking preventive actions (because measures are also known by past records). Therefore, the designer's error is classified into the three types according to two viewpoints; (1) noticing problems and (2) taking appropriate measures.

2.2 Methods of Finding Designer's Errors

The designer's errors are found by the System DRBFM according to the following process. Unnoticed problems are found by examining the missing columns in a function–influence matrix [13]. Figure 5, which shows the function of the elements and the possible failure of these functions considered by a designer) by a *professional adviser*. Underestimated problems are found by discussion between the designer, *professional adviser* and a necessary *reviewer* about the validity of taking no measure in the case that possible failure in one element may trigger chain failures in surrounding parts.

Misunderstood problems are also found by collations by the designer using a *design code* or the knowledge of the *professional adviser*. If the *professional adviser* has a rational doubt that a mistake may be involved in the measures, he asks the designer to have him notice the misunderstood problem.

3 Procedures of System DRBFM Method

We discuss the procedure of the System DRBFM that finds latent errors (unnoticed, underestimated, or misunderstood problems) in a design according to the concept shown in Fig. 1. At the same time, the System DRBFM process can visualize some lacks in knowledge of a designer and offer an opportunity of learning to the designer (OJT). The participants are a designer and a corresponding *professional advisor* for the products. The *professional advisor* is expected to possess appropriate expertise regarding the failure of the products. In the case of finding insufficient expertise for a discussion, the *professional advisor* asks a necessary *reviewer* to additionally participate in the System DRBFM.

The procedure of the System DRBFM is the following.

1. Visualization of the hierarchical structure of a product.
2. Presenting the intentional changes in a design and the incidental changes in an environment.
3. Determination of technical causes from the point of concern.
4. Evaluating the effect of the derived points of concern to the function in higher class.
5. Drawing an influence flow diagram and final certification using the worksheet.
6. Evaluating the form of the worksheet and its references.

3.1 Determining the Hierarchical Structure of a Target Product

The hierarchical structure of a product corresponds to the management structure of the design shown in Fig. 2. The minimum unit of the structure is the element involved in the charge by one designer.

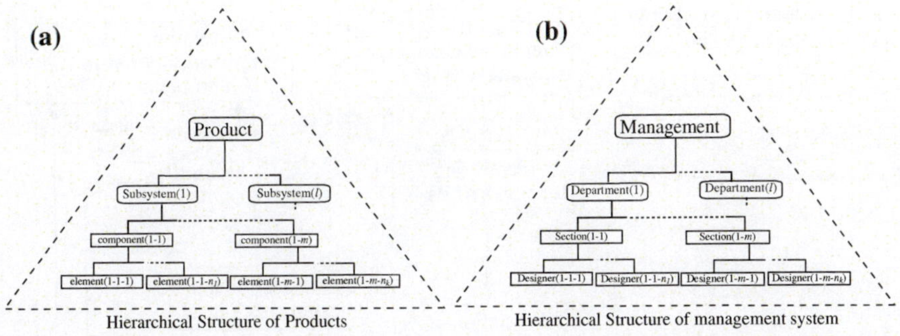

Fig. 2 Definition of a hierarchical structure of a product corresponding with management structure. **a** Hierarchical structure of a product system. **b** Hierarchical structure of a management system

Fig. 3 Definition of positions of intentional changes in product structure by discussion

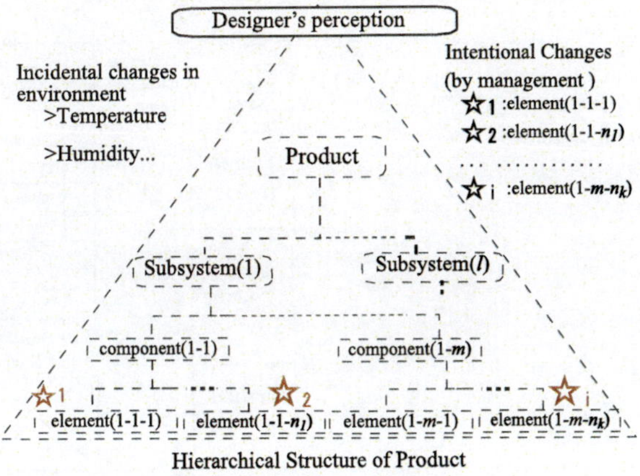

3.2 *Presenting the Intentional Change in a Design and the Incidental Change in the Environment*

A designer in charge of an element presents intentional changes in his design and incidental changes in considering environmental condition as the starting viewpoint of a discussion. We now discuss the case of one intentional change (☆1) in Fig. 3 to simplify this discussion.

3.3 *Determination of Technical Causes from the Point of Concern*

The procedure according to Fig. 4 yields the possible failures by the changes.

1. *Explaining the contents of the changes by the designer*. The designer in charge of element 1–1–1 in Fig. 2a explains the contents of the changes to the participants.

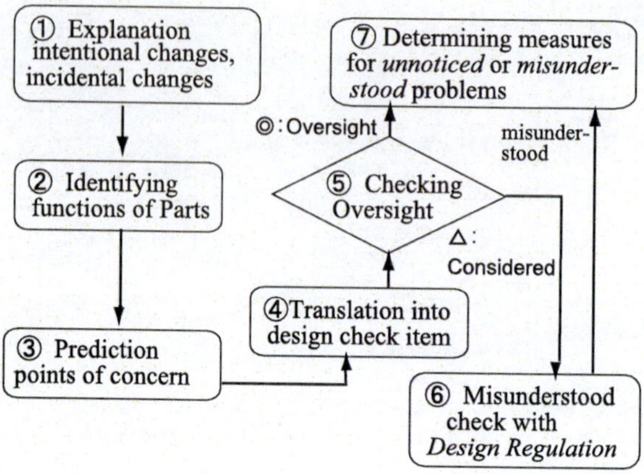

Fig. 4 System DRBFM flow chart for unnoticed and misunderstood problems

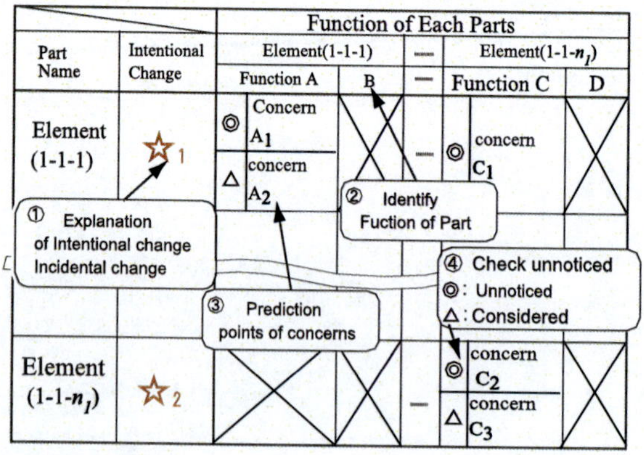

Fig. 5 Part-function influence matrix for discussion of the failures in an element

2. *Determining the function of the target element.* The designer determines a specific function [11] of the target element. The *professional advisor* examines the contents of the determined function.
3. *Deriving the points of concern.* The designer considers the points of concern by the changes, which possibly damages the function. The contents of concern are filled in element-function matrix shown in Fig. 5. If the designer consider no concern due to one change, this should be validated by the discussion among the participants.
4. *Detailed consideration for the technical cause of the points of concern.* If necessary, Fault Tree Analysis [14] is practiced, whose top phenomenon is the point of concern. The branch of the fault tree grows as long as the contents of the cause at the tip node in the tree possesses a sufficient specificity, which are validated by the *professional advisor.*
5. *Inspection of an unnoticed point.* The *professional advisor* determines whether any unnoticed point is lost in the element-function matrix in Fig. 5.

6. *Certification of the content of measures to find misunderstood points*. To find the misunderstood point in the measures considered, the *professional advisor* asks the designer for the following.
   - Whether violations from the design code and attention in operation exist?
   - Does the considered environment in service accurately correspond to a real situation?
   - Are the points of concern and those measures valid for regarding failure modes?
   - Does the *professional advisor* feel a sense of incompatibility in the order of values?
7. *Determining measures for all causes by the points of concern*. The designer checks whether all the causes from the concerns are treated by specific measures. All of the results are filled in the corresponding columns in the DRBFM worksheet. Finally, the *professional advisor* judges whether the contents of the design code or attention in operation cover the contents of the found misunderstanding.

3.4 Evaluating the Effect of the Derived Points of Concern on the Functions of a Higher Component

After the derivation of the points of concern from the changes, the participants also discuss the effect of the concerns on the function of a component in a higher class. In this stage, a designer in charge of the component additionally participates in the process. By participation of the component designer, the consideration process enables keeping the necessary expertise for both the element and the component.

The consideration by participants according to the process shown in Fig. 4 yields the function–effect matrix for the component in a higher class [13]. It should be noted that the function columns in Fig. 6 changed corresponding to those in the component. Continuing the above process up to the top of the products is supposed to obtain function–element matrices such as shown in figures for the all classes. The effect flow diagram in Fig. 7 is the summary of all function–effect matrices that visualize the effects due to the changes.

3.5 Drawing Effect Flow Diagram and Final Certification Using DRBFM Worksheet

In the final stage, all participants, involving the designers in each class, *professional advisors*, necessary reviewers and manager of the DRBFM, certify that solutions for all problems were taken, using Fig. 7 and the DRBFM worksheet.

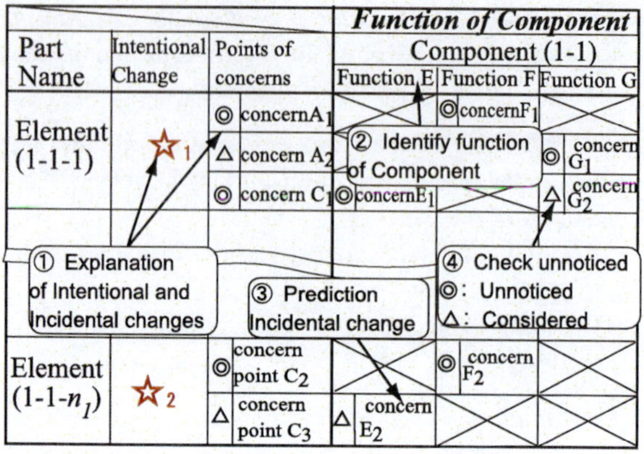

Fig. 6 Component-function matrix with secondary change by initial intentional changes in an element class

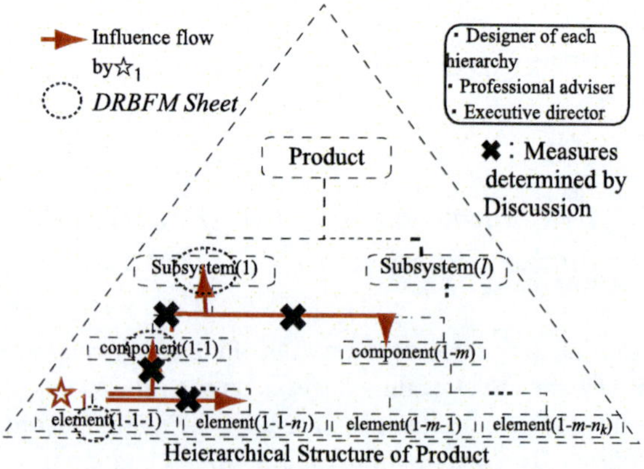

Fig. 7 Influence structure diagram by intentional changes

3.6 Simultaneous OJT Practices

The *professional adviser* point the lack of knowledge or experience in designers that have probably caused the designer's error found through the process. Moreover, he provides the required knowledge or empirical skills to prevent the same errors to the designer at the same time [17]. The designer can understand what he lacked and how to fix it, which helps him to obtain necessary literacy and also encourage his motivation to participate seriously.

3.7 Evaluating the Form of the DRBFM Worksheet

The completed worksheet is evaluated for its sufficiency by a management section based on the following aspects.

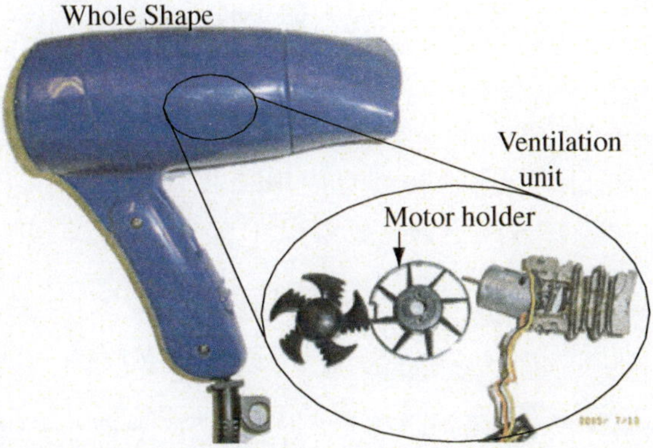

Fig. 8 Execution case: design changing for the motor holder in the ventilation unit of a hair drier

1. Specificity of the structure and the contents of intentional and incidental changes.
2. Specificity of the functions of the elements.
3. Sufficiency of the contents of the points of concern and their technical causes for all types of the designer's error.
4. Sufficiency of the measures corresponding to all causes of the points of concern.
5. Sufficiency of the necessary expertise corresponding to the points of concerns.
6. Process management for measures (charges, limits and progresses of the measures).

In this process, the forms of the sheet are judged (completed) or (not completed) according to the above each factor of evaluation. When the officer in management section considers that one factor is completed, then he checks (completed). If insufficiency or lack of measures are found, the management section orders the *professional adviser* to revise them by taking corrective actions. After the check of the sufficient corrective measures by the management section, the process will be finished.

4 Case Study for System DRBFM

4.1 The Targeting Product

We conducted a case study of the System DRBFM for a design change in the hair drier shown in Fig. 8. The main purpose of this case is to allow readers to understand the process of System DRBFM and its validity of the framework we proposed. We discuss the results of System DRBFM, which involves a discussion of the design changes in one class and its effect on a higher class than the original

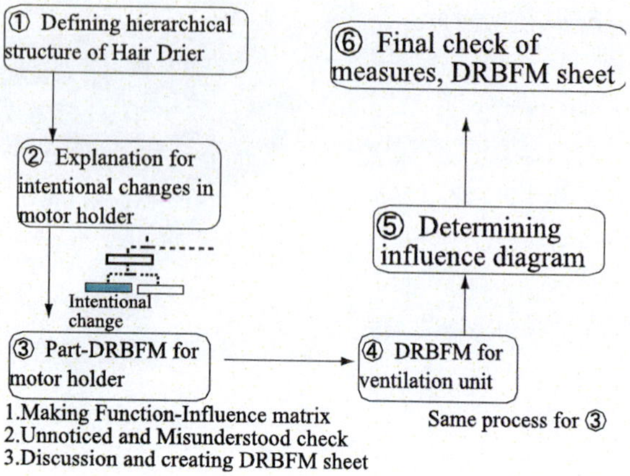

Fig. 9 Execution flow for System DRBFM

class, to simplify the discussion. Furthermore, all points of concern are considered as the possible failure mode to prevent underestimation problems.

4.2 Procedure of the System DRBFM

The entire procedure is shown in Fig. 9. The role of the participants or the designer in charge of hair drier (he will be called only "designer" hereafter) and a *professional advisor* with material strength and mechanical engineering.

1. *Determination of the hierarchical structure of the hair drier.* In this case, the structure of the hair drier is determined previously by the designer as illustrated in Fig. 10.
2. *Explanation of the intentional and incidental changes.* The designer explains the contents of the design change that the means of fixation of the motor holder was changed from screw fixation to direct pressing, as illustrated in Fig. 10. He presents Table 1 in this stage.
3. *Considering the concerns as the result of the changes in the motor holder.* The designer fills a function–effect matrix for the motor holder according to the process illustrated in Fig. 4. The *professional advisor* certifies the unnoticed and misunderstood problems and fills the results in the DRBFM worksheet.
4. *Considering the effects of the points of concern in the motor holder on the ventilation unit.* The designer also creates a function–effect matrix for the ventilation unit.
5. *Summarizing the effect flow diagram.* All of the effects in the function–effects matrices are filled in the effect flow diagram (Fig. 7) by the designer.
6. *Final certification of measures.* The *professional advisor* examines the validity of the measures for all concerns. Finally, he summarizes the found contents of misunderstood problems as a reference to change the attention in operation.

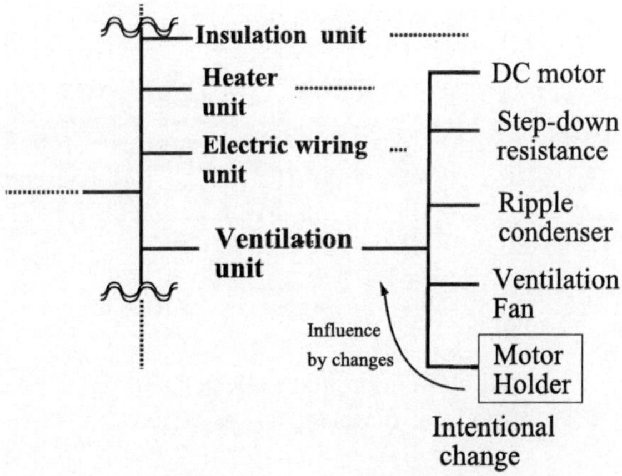

Fig. 10 Hierarchical structure for considered components of a hair drier

Table 1 Comparison of new design with ordinary design (designer-made)

Part name	Comparison list	New design	Ordinary design
Motor holder	Material	PP resin	
	Surface	PP resin	
	Structure	$\phi 27.3 \pm 0.1$ elongation 1.5	$\phi 28$
	Manufacturing	Direct pressure	Screw ×2
	Stress	PP resin	

4.3 Result and Discussion

4.3.1 Considering the Points of Concern Due to the Change in the Motor Holder

Table of Fig. 11 shows the results of consideration of the concerns in the motor holder. The content of the intentional change is shown in Fig. 10. The function of the motor holder is to *fix the motor in place*. The designer presents the points of concern considered and summarizes these into the failure mode using the KJ method [8] under certification by the *professional advisor*.

- *Cracking*
  - Stress concentration caused by too tight fixation
  - Stress concentration by tight R (curvature radius) at the bottom of the holder
  - Cracking due to low temperature shrinkage
  - Stress increase by a wide tightened area
- *Deformation*
  - Deformation by heat stress
  - Loosening during long service
  - Tangential deformation due to moisture absorption

Fig. 11 Function–influence matrix for the motor holder

Part Name	Intentional change	Function of Part				
		Motor Holder	Ventilation fan	Condenser	Resistance	DC motor
		Fixation of Motor	*Rotation*	*Accumulation*	*Current Adjustment*	*Rotation generation*
Motor holder	Inside diameter and fixed method	◎ Cracking				
		△ Transformation				
		△ Dissolution				
		△ Construction failure				

- *Dissolution*
  - Internal dissolution due to heat
  - Adsorption of dissolved materials
- *Manufacturing failure*
  - Cracking by inclined pressure
  - Insufficient deburring in processing
  - Insufficient pressing to normal depth

Figure 12 shows a fault tree example for the cracking mode. The *professional advisor* added creep fracture, fatigue fracture and low temperature shrinkage in the tree as unnoticed concerns. A summary of the concerns is shown in table of Fig. 11.

4.3.2 Considering the Effect of the Concerns in the Motor Holder on the Ventilation Unit

The summary of the concerns is presented in table of Fig. 13. The determined function of the ventilation unit is *to allow air to flow smoothly.* The designer did not consider any concern, so the *professional advisor* pointed out the following viewpoint as the determined effect due to the change in the motor holder.

- *Vibration due to a gap in the position of the motor holder*
  - Resonance vibration
  Therefore, the obtained unnoticed problem was resonance.

4.3.3 Final Certification Using Effect Flow Diagram and DRBFM Work Sheet

The considered effect by the change in the motor holder to the function of the ventilation unit is shown in Fig. 14. The obtained points of concern are written as the follows.

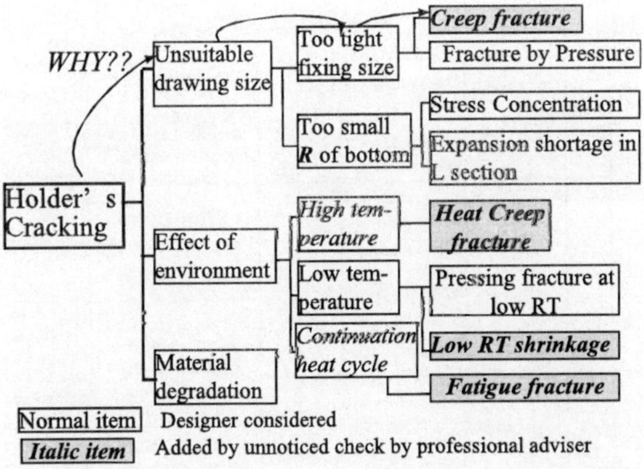

Fig. 12 Root cause analysis of interpreting the failure mode of cracking into specific design aspects

Part Name	Intentional change	Points of concerns for parts		Function of Component: Ventilation Unit	
				air flow generation	
Motor holder	Inside diameter and fixed method	◎	Cracking	◎	Abnormal vibration
		△	Transformation		
		△	Dissolution		
		△	Construction failure		

Fig. 13 Function–influence matrix for ventilation unit

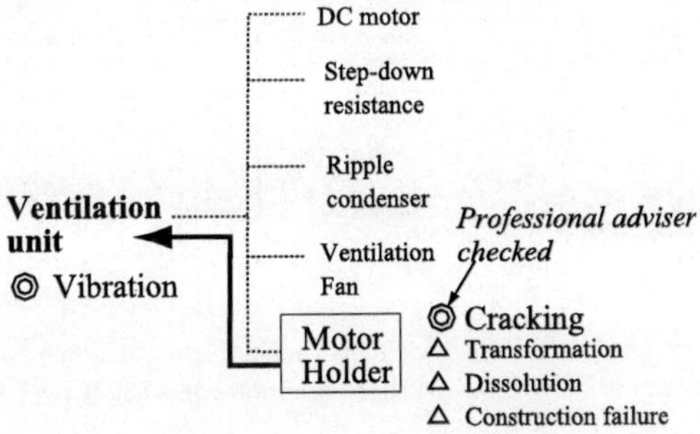

Fig. 14 Influence diagram for a failure mode in a ventilation due to the intentional change in the motor holder

- *Unnoticed problem*; fatigue fracture, creep fracture, resonance vibration
- *Considered problem* (*possibly involving misunderstood problem*); cracking on pressing, cracking due to low temperature shrinkage, falling the motor holder away due to heat deformation, stress concentration in *R*-part and degradation of material.

The obtained DRBFM worksheets illustrated in Fig. 15 and Fig. 15a involves the considerations for the motor holder. All of the unnoticed problems were visualized by the certification by the *professional advisor* and measures were determined after the discussion between them.

(a)

Item Name	Function	Concern / Potential failure mode	Causes of failure mode / Result of Root cause	Effect to customer	Rank	Corresponding Design steps to prevent concern (Included Design rule items, Design standard, Check items)	Recommended actions (results of DRBFM): Solution to be reflected in "Design"	Solution to be reflected in "Evaluation"	Solution to "Production"
Motor holder	Fixiation of motor	Cracking	***Creep Fracture by fixiation***	Im-possible to use	***A***	***unnoticed***		***Creep test at Max tightening tolerance (O mm), (RT O hr)***	
			Creep Fracture by high temperature		***A***	***unnoticed***		***Creep test at Max tightening tolerance (O mm), (O ℃ O hr)***	
			Fatigue fracture		***A***	***unnoticed***	***Heat stress amplitude calculation and Life prediction by S-N diagram***		
			Fracture in the bottom of holder by stress concentration		A	Stress calculation for max tightening width : calculated stress is below the tensile strength	***(no need)***		
			Pressusre fracture, Low temperature shrinkage			Stress calculation for ductile and brittle fracture in low temperature	***(no need)***		
			Softening by high temperature		A	Using same resin materials as before	***Re-calculation and tightening width up***	***(Misunderstood)***	
		Construc-tion failure	Manufacturing error			Manufacturing under the same condition as before	***(no need)***		

(b)

unit Name	Function	Concerns / Potential failure mode due to changes	Causes of failure mode / Result of Root cause	Effect on customer	Rank	Corresponding Design steps to prevent concern (Included Design rule items, Design standard, Check items)	Recommended actions (results of DRBFM): Solution to be reflected in "Design"	Solution to be reflected in "Evaluation"	Solution to be reflected in "Production"
Ventila-tion unit	Flow generation	***Vibration***	***Resonance viblation due to cracking of holder***	Im-possible to use	***A***	***unnoticed***	***Resonance point measurement and vibration endurance test***		

Fig. 15 DRBFM work sheets (normal font. designer filled, bold italic font. professional adviser added) **a** DRBFM worksheet for motor holder, **b** DRBFM worksheet for ventilation unit

Next, the process of finding a misunderstood problem and taking measures are described. The stress distribution in the pressure zone of the motor holder which the designer calculated using elastic theory [18] is determined as follows:

$$\sigma_r = -\frac{r_o^2 - r_i^2}{4 r_i r_o^2} E\delta, \quad \sigma\theta = \frac{r_o^2 + r_i^2}{4 r_i r_o^2} E\delta \qquad (1)$$

E Young's modulus, δ pressing width r_i, inner radius of holder, r_o outer radius of holder.

The condition of brittle fracture by the crack [15] is shown in the next equation.

$$K \geq K_{Ic} \tag{2}$$

K stress intensity factor, K_{Ic} fracture toughness for plane strain.

Furthermore, the fracture toughness under the plane strain [15] is represented by the next equation.

$$K_{Ic} = \sqrt{\frac{Eg}{1 - \nu^2}} \tag{3}$$

g energy release rate, ν Poisson's ratio.

However, the condition of small scale yielding for the crack length a [15] must be satisfied when we apply equations (2), (3).

$$a \geq 2.5 \times \left(\frac{K_{Ic}}{\sigma_{\mathrm{Y}}}\right)^2 \tag{4}$$

σ_Y yielding stress.

The brittle transition temperature of polypropylene resin, which is considered as the material of the motor holder, is about −10°C [5]. Based on this fact, the *professional advisor* asked the designer whether he considered the change in the material constants according to the change in environmental temperature. The designer then noticed that he had misunderstood and decided to calculate again by considering the change in temperature.

At first, the case of pressing at low temperature is considered. Young's modulus E becomes double and the energy release rate g becomes 0.1 times at −10°C [4] which is the lowest temperature considered by the designer. Substituting the calculated values into Eq. 3 shows that K_{Ic} decreases by 0.55 times. However, substituting $K_{Ic} = 1.2\,(\mathrm{MPa}\sqrt{m})$, $\sigma_Y = 39.2\,\mathrm{MPa}$ into Eq. 4 yields $a = 2.3 > 1.35$ mm which Eq. 4 does not satisfy. Therefore, the crack penetrates the entire thickness of the motor holder before the brittle fracture occurs. Furthermore, the condition of plasticity collapse is represented by $\sigma_\theta \geq \sigma_B$. Substituting $r_i = 14$, $r_o = 15$, $\delta = 0.1$ mm into Eq. 1 yields the next value of tangential stress at −10°C

$$\frac{0.014^2 + 0.015^2}{4 \times 0.014 \times 0.015^2} 1960 \times 0.0001 \cong 6.5 < 39.2\,\mathrm{MPa} \tag{5}$$

Therefore, fracture does not occur at the low temperature considered, which means that the misunderstood problem did not cause failure.

Next, the case of falling the motor holder away due to heat deformation at high temperature is discussed. The condition of falling away is represented by $\sigma_r = 0$. E becomes 0.25 times at 80°C [5] when the designer considered the highest temperature in the area of the holder. Substituting this value into Eq. 1 obtains $\sigma_r = -0.15\,\mathrm{MPa}$. When considering that air pressure (0.10 MPa) acts on the outer surface of the holder, the value of σ_r can be zero in the case of vibration and impact. The *professional advisor* ordered the designer to measure the increase in the width of pressing or re-selecting of materials. The remaining problems considered, which is pressing in normal temperature, stress concentration and degradation of materials, involve no misunderstood problems as the result of the checks by the *professional advisor*.

Figure 15b shows the results of discussion on the possible failure in the ventilation unit. The *professional advisor* discusses the measures for resonance, which is the unnoticed problem, with the designer and determines a specific test for this failure mode.

4.3.4 Learning by the Designer and the Organization Using Obtained Errors

For all of the unnoticed and misunderstood problems, the *professional advisor* advised the designer about the error in his design and the rationality of the obtained measures. In the discussion process the lack of designer's knowledge about temperature dependency in the values of material properties in deformation behavior has been visualized. The *professional adviser* then explained the designer why he thought the calculation included a mistake and general theory of elastic deformation. Furthermore, he introduced the designer an appropriate text to understand his explanation. This supports the argument that the design review process we proposed can possess both the role of review process and OJT practices. Furthermore, the organization should consider revising the attention in operation to prevent the recurrence of the misunderstood problems those previously discussed.

4.4 Evaluation of DRBFM Worksheet

The form of the DRBFM worksheet is supposed to evaluate by the mentioned regulation.

1. The structure of the product is specified in Figs. 8 and 10. The details of the intentional change are explicitly shown in Table 1 and Fig. 16.
2. The functions of element or component are described in the prescribed columns in Tables of Figs. 11 and 13.

3 • *Finding unnoticed problems* All of the intersecting columns in Tables of Figs. 11 and 13 are discussed regarding the existence of the concern and the concern is classified by the fault tree in Fig. 12 to specify unnoticed technical causes.
  • *Finding misunderstood problem* All of the concerns considered by the designer are certified whether the contents of measures involve misunderstood problems by *professional advisor*, as shown in Fig. 15b.

4. All of the concerns corresponding to appropriate measures are illustrated in Fig. 15.

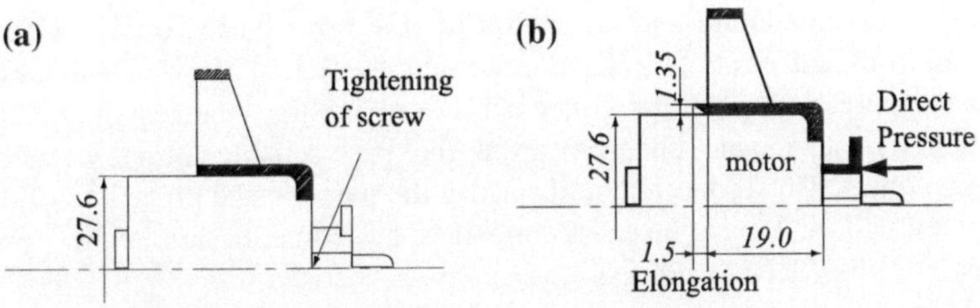

Fig. 16 Intentional change of the motor holder in a hair drier, **a** geometry of motor holder of ordinary design, **b** geometry of motor holder after design change

5. The expertise of the *professional advisor* is in the strength of materials, which corresponds to fracture phenomena such as cracks or fatigue.
6. Specific deadlines and persons in charge are determined. In this case, this process was omitted.

This evaluation then concludes the case study validly finished.

5 Learning Policy by Personal Designers and Organizations

All of the finding process is recorded on a DRBFM worksheet [12]. A management section evaluates the sufficiency of the forms of the DRBFM worksheet. If some insufficiency points are found, probably caused by errors in judgments by a *professional adviser*, the worksheet is then revised again. Remaining errors will be noticed by an unpredicted claim by a manufacturing section or by claims from commercial users. The process of the System DRBFM can specify the designer's unnoticed, underestimated and misunderstood errors. The designer who made the errors can acquire the knowledge he lacks as a result of the coaching by a *professional advisor* and the contents of the design code. Subsequently, the organization summarizes the results of the System DRBFM, which involves the process of finding problems corresponding to the failure mode in the part's function, as *Attention in operation*. In another case of System DRBFM, the *professional adviser* can ask a designer about an already-found error using *Attention in operation* and this makes the designer easily notice his error. This learning policy is then expected to prevent the recurrence of those errors by another designer.

5.1 Importance of Introducing OJT Concepts in Management Systems of the System DRBFM Method

Design review is utilized in various industry fields and its qualitative effectiveness has been perceived [2]. However, the review process involves some problems such as the uncertainty in practices, the insufficient preparation and the considerable load on the participants [7]. Yamada classified [21] these problems into (1) mere formalization in contents, (2) difficulty in gathering talented participants, and (3) difficulty in accumulation and utilization of past error data. To deal with these problems, Shimizu et al. [12] have proposed the DRBFM, which discusses a possible failure (the point of concern) caused by the changes in the design (intentional or incidental changes) from the past reliable design certified by commercial use. This proposal could specify the purpose and procedure of design review and allow all engineers to practice the same design review process according to the DRBFM process.

In this paper, the authors introduced the OJT concept into the DRBFM process to overcome the problem of formalization. The main purpose of an ordinary design review [7] was to discuss reliability problems to improve the product's safety, which only the organization had obtained the benefit by each practice. The participants were then forced to discuss the problem that suffered their motivations for design review. Our proposed DRBFM process introduced OJT which can give the benefits by the practices of DRBFM for both the organization and the participants. Furthermore, transportation of the knowledge or experience in the *professional adviser* to those lacking in the designer can be useful to acknowledge [17]. Consequently, these efforts enables to maintain the enthusiasm for participation of designer with actual profits obtained by each practice that means the improvement of reliability and safety of products simultaneously.

In an ordinary design review, the participants included executives or an unrelated person for the safety of the products [7], who had no specific role for a discussion. We determine the necessary role of participants such as the designer, a *professional advisor* and *reviewers* corresponding to the failures by concerns. The DRBFM manager can select a necessary *professional advisor* from candidates who have been previously authorized for their expertise related to stress-strength model [16] by the organization.

As for the final point of view, the content of utilization policy for accumulated past data on errors have not been obvious. This paper proposes that the necessary data in DRBFM contains worksheet, a design code and attention in operation (the process of finding problems). The proposed data will be used in future case according to the process of DRBFM that means the specification of the use of accumulated data.

5.2 Application of the System DRBFM Method in New Design

The System DRBFM is also applicable in the case of new design. At first we define product structure as follows:

$\text{Product} = \{\text{Sub}_1, \text{Sub}_2 ..., \text{Sub}_i, ...\}$ Sub_i: Subsystem

$\text{Subsystem} = \{C_{i1}, C_{i2} ..., C_{ij}, ...\}$ C_{ij}: Component

$\text{Component} = \{e_{ij1}, e_{ij2} ..., e_{ijk}, ...\}$ e_{ijk}: Element

$\text{Sub}_i \supseteq C_{ij} \supseteq e_{ijk} \longleftrightarrow \boldsymbol{f} \supseteq f_{\text{Sub}_i} \supseteq f_{C_{ij}} \supseteq f_{e_{ijk}}$ f: Function of subsystem, component or element

$\text{Sub}_i \supseteq C_{ij} \supseteq e_{ijk} \longleftrightarrow \boldsymbol{d} \supseteq d_i \supseteq d_{ij} \supseteq d_{ijk} = \{d_1, d_2, ..., d_n,\}$ d_n: design parameter

$\mathrm{Sub}_i, C_{ij}, e_{ijk}$ has one function $f_{\mathrm{Sub}_i}, f_{C_{ij}}, f_{e_{ijk}}$, respectively. This assumption means the mapping between part to its function in each level of the product is bijection (one-to-one correspondence). Moreover, $\mathrm{Sub}_i, C_{ij}, e_{ijk}$ are expressed by the sets of design parameter $\boldsymbol{d}$, respectively. The meaning of "new" design includes the following various cases.

1. Function $\boldsymbol{f}$ changing
   (a) Adding new function f_{new} by adding new part.
   (b) Removing one function f by removing one part.
   (c) Newly composing the function set.
2. Function $\boldsymbol{f}$ unchanging
   (a) Modifying the value of design parameter d_n in Sub_i, C_{ij}, or e_{ijk}.
   (b) Adding new design parameter d_{new} in Sub_i, C_{ij}, or e_{ijk}.
   (c) Changing the link between subsystems, components or elements.

These changes occur in each level of the hierarchical structure of product. Case 1(a), 1(b) can be treated by considering the failure mode in the part and comparing effect flow of these failure modes to neighbor parts before addition/removing and after. In Case 1(c), it is necessary to find same/similar partial functional part as the previous one to prepare comparing target. In discussion, a failure mode list in same functional parts will be helpful. Cases 2(a) can be discussed easily by comparing the values in each parameter and predicting its effect to cause failure mode. Case 2(b) is also considered by comparing empty $\emptyset$ and new value of d_{new}. However, this case is a little difficult in prediction, so the support by using failure mode list and previous trouble data of part which include the parts possessing d_{new} may be needed. In Case 2(c), the moving parts has both the vanishing link to neighbor parts and adding new links. Therefore this case can also be compared connections among neighbor parts in order to predict concerns from changes in connections. In summary, the System DRBFM is also applicable in the above cases of new design by preparing comparison target and supports by failure mode list, past trouble data and effect flow diagram of failure mode. If there is no comparison target, in the case of completely original function, the prediction of failure mode depend on the participant's knowledge and experience, supported by failure mode list and past trouble data. Creation is often defined by the innovative composition of some elements. It means that the participant probably enables to find same or similar functional part as previous one in new product structure partly. Consequently, this case may occupy little part of total design review process, which does not seriously suffer the applicability.

6 Summary

This chapter introduce the framework of the System DRBFM and its procedure to visualize latent unnoticed and misunderstood problems in the System DRBFM process. Unnoticed problems in design can be visualized by examining the entire function–effects matrices in each class of the product's structure. Misunderstood problems can also be visualized by the certification through a *professional advisor* based on the contexts of the design code and its attention in operation. This process will help designers to predict sufficient number of failure modes, which can improve reliability and safety of their products in use.

In the view of management systems,organizations should update the contents of the design code and its attention in operation according to the visualized contents of the designers' problems. Updated codes and attention can help another designer to observe the past problems in his design. *Professional Advisor*s and a design code can compensate for the visualized lack of knowledge by the designer in the System DRBFM process. This concept will improve the quality of management systems of design reviews because the system can provide merits of conducting the DRBFM method for both designers and managers. The stored data will also be useful not only for proactive prevention of failures in service, but also in the cases of safety certification processes which is necessary to export products in global markets.

References

1. Editorial Board (2004) Special report: management techniques for design know–how and knowledge to be applied for practical use. Mach Des (in Japanese) 48(15):7–27
2. Fujita K, Matuo T (2006) Survey and analysis of utilization of tools and methods in product development. Trans Jpn Soc Mech Eng Ser C 72(713):290–297
3. Hatamura Y (1996) Learning from failure. The Nikkan Kogyo Shimbun, Tokyo, pp 467–486 (in Japanese)
4. Itoh K (1980) Plastic data handbook. Kogyo Chosakai, Tokyo, pp 76–163 (in Japanese)
5. Japan Society of High Polymer (1980) Polymer data handbook (in Japanese). Baihuukan, Tokyo, pp 20–27
6. Kletz T (2001) Learning from accidents, 3rd edn. Gulf Professional Publishing, Oxford
7. Makino T (1974) Design review and its problems in practice. Qual Manag 25(2):88–95
8. Munemori J, Nagasawa Y (1991) Development and trial of group ware for organizational design and management. Distributed and cooperative KJ method support system. Inf Softw Technol 33(4):259–264
9. Reason J (1997) Managing the risks of organizational accidents. Ashgate, London
10. Road Transport Bureau (2004) Ministry of land, infrastructure and transport. Analysis report of automotive recalls in fiscal 2004. Government of Japan, Tokyo (in Japanese)
11. Satoh R (1972) Functional design. Japan Management Rationalization Center, Tokyo (in Japanese)
12. Shimizu H, Yoshimura T (2004) Reliability problem prevention method of stimulating creativity and visualizing problems (1st. report). Trans Jpn Soc Mech Eng Ser C 70(689): 243–250
13. Shimizu H, Otsuka Y, Noguchi H (2009) Design review based on failure mode to visualize reliability problems in the development stage of mechanical products. Int J Veh Des 53(3): 149–165
14. Stewart MG, Melchers RE (1996) Probabilistic risk assessment of engineering systems. Chapman & Hall, London
15. Suresh S (1991) Fatigue of materials. Cambridge University Press, Cambridge
16. Suzuki K (2004) Fundamentals and its system of proactive prevention (in Japanese). JUSE, Tokyo

17. Sveiby KE (1996) Transfer of knowledge and the information processing professions. Eur Manag J 14(4):379–388
18. Timoshenko SP, Goodier JN (1970) Theory of elasticity. McGraw-Hill, New York
19. Wageneer WA, Groenewreg J (1987) Accidents at sea: multiple causes and impossible consequences. Int J Man Machine Stud 27:587–598
20. Xijuan L, Yingin W, Shouwei J (2003) A metrics based task analysis model for design review planning. Des Stud 24:375–390
21. Yamada Y (1996) Re-evaluation design review process for enhancing the development process of new products. J Jpn Soc Qual Manag 26(4):356–362

Part III
Applications

Risk-Based Resource Allocation Models for Aviation Security

Laura A. McLay

1 Introduction

The events of September 11, 2001 led to sweeping nationwide changes in aviation security policy and operations. The piecemeal and reactive nature of many of these changes has resulted in large increases in costs and inconvenience to travelers. The August 2006 arrest in London of several suspected terrorists plotting to blow up 10 US-bound transatlantic flights, and the ensuing changes in airport security procedures, serve to further illustrate this point.

Over the past 8 years, there have been numerous changes to all aspects of aviation security systems, all of which have been designed to prevent a reoccurrence of the events on September 11, 2001. Many of the changes implemented have been politically driven. For example, several billion dollars were invested in security devices following September 11, 2001 before any type of systematic analysis of aviation security systems was performed [34]. Coordinated analysis and planning have the potential to determine how taxpayer dollars can be optimally spent and how security system assets can be optimally used.

Next-generation aviation security systems need not merely be makeshift political solutions for mending complex problems; they can be the result of modeling, analysis, and planning. This chapter summarizes analytical approaches for managing risk in aviation security screening systems using operations research methodologies. This chapter focuses on passenger screening problems, an

L. A. McLay (✉)
Department of Statistical Sciences & Operations Research,
Virginia Commonwealth University, 1015 Floyd Avenue,
P.O. Box 843083, Richmond, VA 23284, USA
e-mail: lamclay@vcu.edu

H. Pham (ed.), *Safety and Risk Modeling and Its Applications*,
Springer Series in Reliability Engineering, DOI: 10.1007/978-0-85729-470-8_9,

important and highly visible aspect of aviation security. In addition, it focuses on modeling approaches that seek to optimally use limited resources to manage the risks associated with terrorism.

Before summarizing passenger screening research, first consider that there are two basic approaches to aviation security screening: uniform screening and selective screening. Uniform screening subjects every passenger and their baggage to identical security screening procedures. The argument for uniform screening is that anyone could pose a risk, and hence, all passengers should be screened using the most effective technology and procedures available. The 100% baggage screening mandate, which requires all checked baggage to be screened by a federally certified explosive detection technology (effective December 31, 2002), is a move towards uniform screening [33]. One disadvantage of uniform screening is that it can be very costly to apply expensive new technologies to every passenger or bag.

Selective screening, the alternative to uniform screening, selectively applies security technologies and procedures to a subset of passengers. The argument for selective screening is that most passengers do not pose a risk, and hence, expensive security technologies need not be used on all passengers. Selective screening subjects passengers perceived as high-risk to closer scrutiny by screening them and their baggage with more sensitive and accurate technologies and procedures, while passengers perceived as low-risk are subjected to lower levels of scrutiny. This approach requires that a *prescreening system* perform a risk assessment of each passenger prior to the passenger's arrival. A weakness of selective screening is that it can assign an incorrect degree of risk to a passenger, either by error or through "gaming" of the system by a terrorist. Passenger prescreening is central to managing risks associated with passenger screening, and prescreening systems are discussed throughout this chapter.

In order to put aviation security modeling approaches in context, a brief history of aviation security operations in the United States is presented. Then a survey of modeling approaches for passenger screening is presented. The modeling approaches are subdivided into three categories: (1) checked baggage screening models, (2) passenger screening models, and (3) risk models. The modeling approaches in the first two categories explicitly address resource allocation issues. Modeling approaches in the third category indirectly address resource allocation by providing insight into how risk can be managed by the system. The reader is also referred to several other surveys of operations research methodologies for more information on how to apply analytical methods in the aviation security domain [22, 24, 27, 59].

Designing effective aviation security systems has become a problem of national concern. Since September 11, 2001, numerous changes have been made to aviation security systems. Many of these changes have been politically driven, rather than driven by coordinated, systematic analysis and planning, and as a result, the analysis of how aviation security systems operate continues to lag well behind their actual implementation. Effective passenger screening systems optimally allocate and use scarce security assets and technologies to manage risks associated

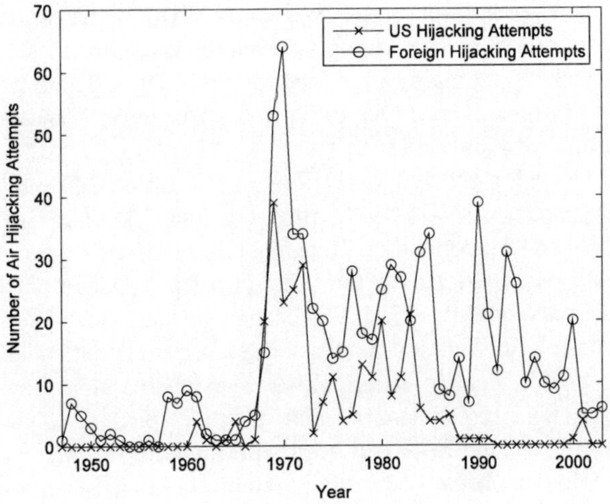

Fig. 1 Foreign and domestic air hijacking attempts, 1947–2003

with terrorism. This chapter summarizes analytical approaches for managing risk in aviation security screening systems using operations research methodologies. It focuses on passenger screening problems, an important and highly visible aspect of aviation security. This chapter shows that next-generation aviation security systems need not merely be makeshift political solutions for mending complex problems; they can be the result of modeling, analysis, and planning. It provides insight into the operation of passenger screening systems and guidance for the design of next-generation aviation security systems.

2 Aviation Security Background

Hijacking attempts were a serious breach of early aviation security in the United States. They were relatively infrequent until 1968, when twenty hijacking attempts occurred on US aircraft and 15 hijacking attempts occurred on foreign aircraft [8]. On September 11, 1970, then President Nixon announced a program of deploying surveillance equipment to the nation's airports to reduce the increased numbers of hijacking attempts. Furthermore, air carriers worked with the Departments of Defense and Transportation to determine whether X-ray devices and metal detectors could be integrated into airports to screen passengers and their carry-on baggage. On February 1, 1972, the Federal Airline Administration (FAA) announced that all passengers were to be screened by at least one approved method, which included a behavioral profile, metal detector, identification check, and physical search. When hijacking attempts persisted, the FAA adopted emergency rules on December 5, 1972, requiring air carriers to use screening procedures to prevent passengers from bringing weapons and explosives onto the aircraft [37]. Figure 1 shows the number of hijacking attempts from 1947 to 2003 [8]. There were at least twenty-three total domestic hijacking attempts during the years from 1969 to 1972, but the number of hijacking attempts plummeted to two in 1973. There were relatively few domestic hijacking attempts after 1972, with six or fewer domestic hijacking attempts in any single year since 1984.

Aviation security operations at US airports did not significantly change from December 1972 to 1996. However, the destruction of Pan Am Flight 103 over Lockerbie, Scotland on December 21, 1988 and the crash of TWA Flight 800 on July 17, 1996 led to the creation of the Commission on Aviation Safety and Security on July 25, 1996, headed by then Vice-President Al Gore. The Commission on Aviation Safety and Security recommended that the aviation industry improve security using existing explosive detection technologies, automated passenger prescreening, and positive passenger-baggage matching (PPBM). Moreover, the FAA had been working with the airlines to annually purchase and deploy explosive detection systems (EDSs) at airports throughout the United States. From 1998 until September 11, 2001, EDSs were only used to screen checked baggage of *selectee* passengers, those who were not cleared by the Computer-Aided Passenger Prescreening System (CAPPS), a computer risk assessment system developed in conjunction with the FAA, Northwest Airlines, and the United States Department of Justice [50]. The checked baggage of *nonselectee* passengers, those who were cleared by such a system, received no additional security attention. There were no further differences between selectee and nonselectee passengers. During this time, approximately 95% of passengers were classified as nonselectees [1].

The terrorist events on September 11, 2001 prompted Congress to enact the Aviation and Transportation Security Act on November 19, 2001, which transferred aviation security from the FAA to the newly created Transportation Security Administration (TSA), then part of the Department of Transportation (now part of the Department of Homeland Security). This act prescribed additional security procedures to be implemented, including screening all checked baggage for explosives by EDSs or alternative techniques [33, 34]. Given such a policy, there would no longer be any distinction between selectee and nonselectee passengers, since all checked baggage would be screened. In order to meet this requirement, 1,200 EDSs and 6,000 explosive trace detection systems (ETDs) were deployed to the nation's 429 commercial airports at a cost of $2.5B [49]. Before these security devices were deployed, some checked baggage was physically screened for explosives by hand searches, and PPBM removed checked baggage from aircraft if the passengers who owned the baggage did not board the aircraft. The Aviation and Transportation Security Act also required that security personnel become federal employees (this was accomplished on November 18, 2002), as well as improved perimeter security and cargo screening.

The TSA responded to these proposals through the development of CAPPS II, an enhanced computer-based system for systematically prescreening passengers. CAPPS II was a revision of CAPPS, which did not classify all of the 19 terrorists on September 11, 2001 as selectees. Although the true number of these terrorists classified as selectees is classified, it has been estimated in the public domain that between 6 and 11 of the terrorists were classified as selectees [2, 35, 43].

On July 14, 2004, the TSA announced that CAPPS II had been dismantled over privacy concerns despite having invested $100M into its development. Shortly thereafter, the TSA announced plans to replace CAPPS II with a revised version of CAPPS in conjunction with *Secure Flight* [55]. Secure Flight identifies passengers who are not permitted to fly based on federal terrorist watch lists. At present, the number of passengers flagged by Secure Flight is extremely small. Those passengers who are cleared by Secure Flight are then prescreened by CAPPS. CAPPS distinguishes selectees and nonselectees by requiring additional screening (such as hand searches) for selectees and their carry-on baggage. How CAPPS operates is considered highly sensitive and may change based on changes in national or international situations, intelligence information, or the risk level of the Homeland Security Advisory System [48].

Several other changes have been made to aviation security since September 11, 2001. The set of items prohibited from being carried onto aircraft have changed several times. Soon after September 11, 2001, box cutters, knives, nail clippers, and lighters were prohibited. Most recently, in December 2005, the TSA announced that small knives, scissors, and tools would be permitted to be carried on to commercial flights [52]. Severe restrictions for traveling with liquids and gels were added in response to the August 2006 plot to blow up ten US-bound transatlantic flights.

Moreover, the TSA implemented the Registered Traveler program to reduce the screening time for travelers who provide personal and biometric information in order to verify their identities [53]. At present, more than 20 airports participate in this program [45]. The TSA has also been deploying explosive detection trace portals and Explosives Detection Canine Teams, as they become available [51, 56]. The Federal Air Marshal Service (FAMS) program began in 1968, when it was called the Sky Marshal Program. FAMS was expanded after September 11, 2001, and the number of federal air marshals increased from fewer than fifty to several thousand during this time [54].

At present, a number of new passenger screening technologies are being developed and deployed, including trace substance detectors (also known as puffers) and backscatter X-rays [44]. New screening technologies and procedures are being developed to screen passenger baggage and air cargo for nuclear and radiological material on incoming international flights (both commercial and private flights) [26, 36, 41].

In addition, the TSA is building a layered aviation security system called Checkpoint Evolution, which constantly adapts to changing threats by utilizing Behavior Detection Officers, law enforcement, and process engineering [10]. In addition, the TSA has embraced transparency, which involves facilitating discussion with travelers and stakeholders as well as maintaining a weblog called *Evolution of Security* (http://www.tsa.gov/blog/).

Most aviation security efforts have been applied to passenger and baggage screening. The Implementing Recommendations of the 9/11 Commission Act of 2007 established a number of security benchmarks for air cargo, including a 100% screening mandate for air cargo transported on passenger aircraft by August 2010 [17]. As a result, a significant amount of resources are being directed to screening air cargo at present.

The primary objective of all these efforts is to improve security operations at the nation's commercial airports while reducing the time required to perform passenger screening. To meet these objectives, the TSA must develop new security system paradigms that can optimally use and simultaneously coordinate several security technologies and procedures. New security procedures put in place by the TSA have the potential to affect a large number of passengers and baggage. Note that in 2005, there were nearly 700 million passengers, with forecasts of over one billion passengers by 2015 [46, 47].

3 Resource Allocation Models For Aviation Security

This section summarizes a number of research efforts that apply analytical methods and operations research methodologies to risk-based resource allocation models for aviation security. The modeling approaches are subdivided into three categories: (1) checked baggage screening models, (2) passenger screening models, and (3) risk models.

3.1 Checked Baggage Screening Models

Screening checked baggage for explosives is a critical component in aviation security system systems. At present, all checked baggage is screened for explosives by a TSA-certified EDS, ETD, or alternative screening procedure. The Aviation and Transportation Security Act required that 100% of checked baggage be screened for explosives. Prior to September 11, 2001, only selectee checked baggage was screened for explosives (approximately 5% of passengers). Increasing the capacity of the checked baggage screening system to meet the 100% screening mandate was achieved at an enormous cost.

One way to improve selective screening systems is to use expensive baggage screening technologies with low throughput to screen passengers perceived as higher-risk. This has the potential to be a more cost-effective approach to screen passengers primarily by increasing throughput. Butler and Poole [4] designed a layered approach to screening passengers and baggage instead of the existing TSA policy of 100% checked baggage screening using EDSs by considering the economic impact of using different screening technologies. They consider three groups of passengers: lower-risk passengers who have volunteered for extensive background checks, lower-risk passengers about whom little is known, and higher-risk passengers. They recommend screening baggage with three layers of baggage screening devices. By sending baggage through three layers of security devices composed of EDSs, high-throughput backscatter and dual-energy X-ray devices, and hand searches, throughput is increased while the overall false clear rate remains at a level comparable to that of the 100% baggage screening mandate. Butler and Poole make similar recommendations for passenger screening. One implication of this screening system is that the resulting improved throughput indirectly decreases space requirements and waiting times in airport lobbies, which is of interest because many airport lobbies were not designed to accommodate extensive screening systems and excessively long waiting lines.

A number of papers perform a cost-benefit analysis for evaluating the impact of using baggage screening devices and technologies to address risk-based issues associated with checked baggage screening. The first of such models by Virta et al. [58] assesses the trade-offs between screening only selectee baggage and screening both selectee and nonselectee baggage by a single security device using the baggage screening paradigm in place prior to September 11, 2001. They provide a cost model that captures the cost of deploying, maintaining, and operating a single baggage screening security device over a 10-year period that is composed of eight cost elements:

1. the annual direct cost of purchasing the security device,
2. the annual direct cost of operating and maintaining the security device,
3. the annual direct cost for inspecting baggage,
4. the annual direct cost of false alarms,
5. the annual direct cost of true alarms,
6. the annual direct cost of true clears,
7. the annual *indirect* cost of false clears,
8. the annual *indirect* cost associated with not fully utilizing the baggage screening capacity.

The direct costs (items 1–6) most accurately characterize the cost of screening. The difference in the risk associated with selectee and nonselectee baggage is captured by the *prscreening multiplier* $\beta = P_{\mathrm{T/S}}/P_{\mathrm{T/NS}}$, the ratio of the proportion of threats T in selectee S versus nonselectee NS checked baggage. Virta et al. [58] evaluate the cost model according to three scenarios. Their analysis indicates that as more nonselectee bags are screened (in addition to selectee bags), the expected annual cost per bag screened decreases, and the expected number of detected threat increases. The marginal increase in security per dollar spent is significantly lower for the 100% baggage screening mandate as compared to only screening selectee bags.

This basic cost-benefit model [58] has been extended several times. Jacobson et al. [12] analyzed the impact of deterrence on checked baggage security screening systems. Deterrence is modeled by reducing the probability that a checked bag contains a threat P_T^{DET},

$$P_T^{\mathrm{DET}} = P_T(1 - \rho\delta^{\eta}),$$

where P_T is the baseline probability that a checked bag contains a threat in the absence of screening, δ is the proportion of all checked bags that are screened

($0 \leq \delta \leq 1$), the deterrence multiplier ρ captures the maximum possible level of deterrence, and the deterrence exponent $\eta > 0$ captures the effect of deterrence on reducing the threat level. Since selectees and nonselectees may be screened at different rates, then the conditional probabilities that selectee or nonselectee bags are threats change according to the proportion screened.

Jacobson et al. [13] perform a cost-benefit analysis for the 100% baggage screening mandate for current and next-generation screening technologies. They evaluate the cost effectiveness of EDSs in both single-device and two-device systems, and then consider the effect of new technologies. The two-device configurations assume a cascading sequence where bags are screened by the second device only if the first device signals an alarm. They find that a risk-based approach, in which selectee and nonselectee bags are screened differently, is shown to significantly improve security.

McLay et al. [30] perform a cost-benefit analysis to compare the hypothetical impact of next-generation baggage screening technology, where the new more effective and expensive screening device is used to screen selectee baggage and the existing security device is used to screen nonselectee baggage. They adapt the cost model provided by [58] and provide a risk model to capture the ability of the prescreening system to correctly identify threat baggage across different levels of β. They report results for a model in which the false clear rate of the next-generation model is reduced by a factor of α ($0 < \alpha \leq 1$) and the purchase, installation, maintenance, and screening costs increase by a factor of $1/\alpha$, $1/\alpha^2$, or $1/\sqrt{\alpha}$. Their results indicate that the accuracy of the prescreening system is more important for reducing the number of successful attacks than the effectiveness of the checked baggage screening devices at detecting threats when few passengers are classified as selectees. They conclude that using expensive and accurate baggage screening technologies on selectees is warranted only if there is an effective prescreening system in place.

Cost-benefit analyses are useful for reporting the performance of baggage screening systems under hypothetical operating conditions to assess the potential impact of changes to baggage screening operations. However, a cost-benefit analysis cannot predict how to optimally use security devices. To address this issue, a number of research papers have applied discrete optimization and integer programming models to determine how to optimally deploy and use baggage screening devices [27].

Several discrete optimization models determine how to optimally deploy and use limited baggage screening devices (such as EDSs) in the baggage screening paradigm in place prior to September 11, 2001, in which only selectee baggage is screened. Jacobson et al. [11] provide a framework for measuring the effectiveness of a baggage screening security device deployment at a particular station. A station is a set of airport facilities that share security resources. There may be several stations in a large, hub airport.

Jacobson et al. [16] propose three performance measures for assessing the effectiveness of security device deployments for screening selectee checked baggage across a set of flights. Note that a flight segment is the flight between takeoff and landing of an aircraft from one airport to another. A flight is *uncovered* if one or more selectee bags on the flight has not been screened, while a flight segment is *covered* if all selectee bags on it have been screened. The performance measures are:

1. Uncovered flight segments (UFS), which captures the total number of uncovered flights.
2. Uncovered passenger segments (UPS), which captures the total number of passengers on uncovered flights.
3. Uncovered baggage segments (UBS), which captures the total number of unscreened selectee bags (regardless of flight).

They find that deploying security devices according to the UFS and UPS performance measures result in very different solutions. However, they note that it is sometimes possible to simultaneously improve both UFS and UPS performance measures.

The concept of coverage was extended to several other research models for baggage screening systems. Jacobson et al. [14] formulate problems that model multiple sets of flights originating from multiple stations subject to a finite amount of security resources. Examples illustrate strategies that may provide more robust device allocations across the UFS and UPS performance measures.

Jacobson et al. [15] and Virta et al. [57] considers the impact of originating and transferring passengers on the effectiveness of baggage screening security systems. In particular, they consider classifying selectees into two types: those at their point of origin and those transferring. Note that a selectee bag screened at its point of origin is covered for two flight segments as opposed to one flight segment for selectee bags screened while transferring. This is noteworthy since at least two of the hijackers on September 11, 2001 were transferring passengers.

3.2 Passenger Screening Models

The Aviation and Transportation Security Act eliminated the distinction between selectee and nonselectee checked baggage and indirectly shifted the distinction between selectee and nonselectee passengers to passenger and carry-on baggage screening. This section focuses on the issues surrounding the screening of passengers and their carry-on baggage at airport security checkpoints.

There are a number of papers that apply discrete optimization, integer programming, and Markov decision process methodologies for analyzing risk-based passenger screening systems. These papers introduce systematic approaches for designing enhanced passenger screening systems by considering multilevel passenger screening strategies. *Multilevel screening* considers two or more levels of security to screen passengers, which generalizes the binary system of CAPPS. These papers show how multilevel screening models can be used to provide insights into the operation and performance of aviation security systems.

McLay et al. [28] are the first to introduce a framework for multilevel passenger screening using discrete optimization methodologies and algorithms. In their framework, each passenger is assigned to a *class* that defines one of the levels of security, which corresponds to a set of procedures using security screening devices and personnel. The integer programming formulation maximizes a security measure subject to a budget constraint. Their model is given by

$$
\begin{aligned}
\max \quad & \sum_{i=1}^{M} L_i R_i \\
\text{subject to} \quad & \sum_{i=1}^{M} \sum_{j=1}^{N} MC_i x_{ij} + \sum_{i=1}^{M} FC_i y_i \leq B \\
& \sum_{i=1}^{M} x_{ij} = 1, \quad j = 1, 2, \ldots, N \\
& \frac{1}{N} \sum_{j=1}^{N} x_{ij} - y_j \leq 0, \quad i = 1, 2, \ldots, M \\
& y_i \in \{0, 1\}, \quad i = 1, 2, \ldots, M \\
& x_{ij} \in \{0, 1\}, \quad i = 1, 2, \ldots, M,\ j = 1, 2, \ldots, N,
\end{aligned}
\tag{1}
$$

where a set of N passengers that must be assigned to one of M classes. Each class is defined by a security level L_i, a fixed cost associated with device purchase and installation costs FC_i, and a marginal cost associated with passenger inspection costs MC_i, $i = 1, 2, \ldots, M$. There is a budget B for screening the N passengers and risk level function R_i that captures the conditional probability that a threat is assigned to class i given that there is a threat, $i = 1, 2, \ldots, M$. The decision variables $y_i = 1(0)$ if there is (not) at least one passenger assigned to class $i = 1, 2, \ldots, M$, and $x_{ij} = 1(0)$ if passenger j is (not) assigned to class i for $i = 1, 2, \ldots, M$, $j = 1, 2, \ldots, N$. Note that in this model, each passenger must be assigned to a class, and hence, all passengers undergo security screening. A polynomial-time greedy heuristic is provided that obtains approximate solutions that use no more than two classes. McLay et al. [28] provide an example in which the objective is interpreted as the conditional true alarm probability. They conclude that using as few as two classes for passenger screening is sufficient for designing effective, risk-based passenger screening systems.

McLay et al. [29] develop a second multilevel screening model that considers how to optimally use security devices once they are in place. The contribution of their model is that it illustrates how to use security devices that are shared by security classes. In this model, each of the M security classes corresponds to several passenger screening device types (there are V device types available), where each device type has an associated capacity (throughput) c_k. The resulting integer programming formulation is given by

$$
\begin{aligned}
\max \quad & \sum_{i=1}^{M} L_i R_i \\
\text{subject to} \quad & \sum_{i=1}^{M}\sum_{j=1}^{N} d_{ik} x_{ij} \leq c_{k}, \quad k = 1, 2, \ldots, V \\
& \sum_{i=1}^{M} x_{ij} = 1, \quad j = 1, 2, \ldots, N \\
& x_{ij} \in \{0, 1\}, \quad i = 1, 2, \ldots, M,\ j = 1, 2, \ldots, N.
\end{aligned} \tag{2}
$$

All parameters and decision variables in this model are the same as those in (1). In addition, $d_{ik} = 1(0)$ if class $i = 1, 2,\ldots,M$ uses (does not use) device type $k = 1, 2,\ldots,V$.

Lazar Babu et al. [20] investigate the possible benefit from using multiple classes for screening passengers using linear programming models. The objective of their model is to minimize the probability of the system giving a false alarm (a proxy for passenger inconvenience), subject to false clear and screening time constraints. They find that using multiple classes are beneficial for security, even when a prescreening system is not used to differentiate passenger risk. Nie et al. [39] extend this model to incorporate the effect of passenger prescreening and the number of screeners at security checkpoints. They evaluate the trade-offs between two performance measures, the probability of a false alarm and the total number of screeners needed.

Nearly all research models assume that each security device or checkpoint yields a binary response, either alarm or clear. Nie et al. [38] extend the research by Lazar Babu et al. [20] to consider checkpoints that yield a magnitude of response. These checkpoint outcomes are used to sequentially "score" passengers. Nie et al. [38] provide a method for accurately grouping passengers according to their risk level after passengers are scored by all checkpoints, using multivariate statistical analysis and optimization.

The models [20, 28, 29, 39] are *static*, meaning that they make passenger screening decisions for many passengers at the same time. In these models, the set of passengers to be screened at a particular station in an airport in a given period of time is assumed to be known, and hence, passenger risk levels are assumed to be known a priori.

A number of models for *dynamic* passenger screening provide insight into how passengers can be optimally screened in real-time. In the models presented, passengers arrive at a security station in a *sequential* manner, with passenger risk levels considered *stochastic* prior to their arrivals. In these problems, passengers enter a security screening sequentially, and each passenger's risk level (assessed by a prescreening system such as CAPPS) becomes known to the TSA upon first entering the screening process. This necessitates a change in the solution methodology.

McLay et al. [32] introduce a sequential, stochastic passenger screening model that determines how to optimally assign passengers (in real-time) to aviation security resources. Passengers are classified as either selectees or nonselectees, with screening procedures in place for passengers with each classification. Passengers arrive sequentially, and a prescreening system determines each passenger's perceived risk level, which becomes known upon arrival. The objective is to use the passengers' perceived risk levels to determine the optimal policy for screening passengers that maximizes the expected number of true alarms, subject to capacity and assignment constraints. Their model is formulated as a Markov decision process, and an optimal policy is found using dynamic programming. Analysis of this is an example suggests that extremely high-risk passengers are almost certainly classified as selectees, regardless of when these passengers arrive. Practically speaking, this means that the screening policies provided the model is not easily gamed by would-be terrorists.

McLay et al. [31] extend this model to consider the impact of multilevel screening. Finding a solution to their Markov decision process optimality equations using dynamic programming is computationally intractable. To find a screening policy, they present a heuristic to provide approximate solutions in real-time. A method is provided to verify when the heuristic yields the optimal policy. Analysis suggests that their heuristic is almost certain to assign extremely high-risk passengers to the most secure class, regardless of when these passengers are screened.

Lee et al. [23] examine the same sequential, stochastic multilevel screening problem introduced in McLay et al. [31] using a different approach. Instead of using Markov decision process and discrete optimization methodologies, Lee et al. [23] apply control theory methodologies to modulate the assignment of passengers to classes. They introduce a real-time sequential binary passenger assignment model as a discrete-time difference equation. Through a probabilistic analysis, a closed-loop policy is presented to achieve desired security class occupancies for a set of passengers anticipated to undergo screening, while maximizing the overall system security. The same closed-loop policy is also shown by applying feedback linearization to the fractional passenger assignment model.

Nikolaev et al. [40] propose a two-stage model for sequential, stochastic multilevel passenger screening problems. The first stage analyzes the purchase of security devices, while the second stage determines the screening assignments of sequentially arriving passengers. Their model is transformed into a deterministic integer program rather than modeled using Markov decision processes. Their model is formulated as

$$\max_{v \in V} E_R\left(\max_{a \in A} G(a, v, R)\right), \tag{3}$$

where the function $G(a,v,R)$ measures the level of security of the passenger screening system with passenger assignment variable a, and security device allocation variable v from the set of all possible assignments, A, and device allocations, V. Since each passenger's perceived risk level R, is unknown prior to check-in, the objective is to maximize the expected total security (or the expected number of threat items detected) overall passenger assignments, A, for a fixed time period, subject to security device capacity, budget, and space constraints. The benefit of this approach is that it balances long-term issue of security device acquisition and deployment with the short-term issue of optimally using the security devices that are in place.

3.3 Risk Models

The resource allocation models surveyed in Sects. 3.1 and 3.2 provide insight into how to optimally use scarce aviation security screening resources to manage risk. This section surveys research that provides guidance for how risk should be modeled in aviation security systems, which in turn influences how aviation security resources are allocated.

Barnett [1] highlights several important issues regarding the tradeoffs between safety and security in the security procedures in place on September 11, 2001. In particular, he notes that the reduction of security scrutiny paid to selectees and their carry-on baggage may have contributed to the events of September 11, 2001. Note that the only practical difference between selectees and nonselectees involved checked baggage, yet the terrorists did not have checked baggage. Barnett also recommends implementing positive passenger baggage matching (PPBM) for all checked baggage. Under PPBM, a passenger's checked bag would only travel on an aircraft if the passenger is known to be aboard the plane. The costs and delays associated with PPBM were studied by Barnett et al. [3], who report the results of a large-scale 2-week experiment at several commercial airports to test which costs and disruptions would arise from using PPBM for all flights. Barnett et al. [3] reports that implementing PPBM would delay domestic flights an average of approximately 1 min per flight (approximately one in seven flights would experience delays that would average 7 min in length), would cost approximately 40 cents per checked bag (in 2001 US dollars), and would not reduce the number of flights.

After September 11, 2001, CAPPS was revised, and it was used to differentiate between passengers and their carry-on baggage. A number of papers applied mathematical models to highlight potential weaknesses with various versions of CAPPS.

A frequently mentioned criticism of any system designed to classify passengers into risk classes, including CAPPS, is that such systems can be gamed through extensive trial and error sampling by a variety of passengers through the system. Carnival Booth is an algorithm that shows how threat passengers can determine a set of circumstances under which they are classified as nonselectees [7]. It shows that a system using prescreening may be less secure than systems that employ random searches. Carnival Booth takes advantage of passengers being aware of whether they are classified as selectees or nonselectees; a system using prescreening is more difficult to defeat if passengers do not know if they are selectees or nonselectees.

Barnett [2] provides simple mathematical models for evaluating the potential impact of passenger prescreening systems such as CAPPS. Barnett [2] posits that the conditional probability that a terrorist attack is detected by passenger screening Q is

$$Q = PH + (1 - P)L, \tag{4}$$

where P is the probability that a terrorist is classified as a selectee, H is the conditional probability that selectee screening detects the attack, and L is the conditional probability that nonselectee screening detects the attack. Barnett notes that the goal of prescreening appears to maximize P, whereas taking a systems perspective suggests that a prescreening system may alter both H and L, reducing the total effectiveness of passenger screening. Ultimately, Barnett suggests that prescreening may only improve aviation security under a particular set of circumstances and recommends that it be transitioned from a security centerpiece to one of many components.

A number of experts agree with Barnett's assessment of prescreening, but are more optimistic of its potential usefulness. Caulkins [6] argues that prescreening may be a worthy investment since it reduces screening costs and uses an extremely small fraction of the budget allocated for passenger screening. Caulkins [6] also suggests several ways that prescreening could be used to increase the overall security even if terrorists are incorrectly classified as nonselectees. Cartensen [5] and Ravid [42] indicate that prescreening could successfully deter attacks on commercial aircraft, while Ravid [42] notes that the objective of prescreening is to deter terrorist events, not necessarily to capture terrorists.

Martonosi and Barnett [25] examine the effectiveness of risk-based passenger screening systems in greater detail. They note that passengers could be classified as selectees because they are perceived as high-risk (with probability P) or are low-risk passengers who are randomly classified as selectees (with probability R). Using the notation introduced in (4), the conditional probability that a single attack is successful is

$$Q = (1 - P)(1 - R)(1 - L) + (P + (1 - P)R)(1 - H).$$

They explore the impact of deterrence, since the events on September 11, 2001 and other terrorist attacks involved terrorists working in tandem. In particular, they define a terrorist group's *deterrence threshold*, the minimum probability that all terrorists succeed in circumventing security in order to proceed with the attack. McLay et al. [32] discuss an alternative way to frame deterrence by noting that if several members of the group of threats are prevented from boarding their flights, a terrorist attack could be canceled. Therefore, it is not necessary to prevent all threats from boarding their flights.

All passenger screening systems implicitly define a set of rules for when a system alarm is sounded. A system alarm determines which passengers and bags may board a flight and which may not. A system alarm is said to be signaled for passengers and bags that are flagged by security procedures. In extreme scenarios, entire airport terminals are shut down so that a bomb squad can examine a bag that has yielded a system alarm. In general, passengers or bags that yield a system alarm are delayed while they are subject to additional security tests and interviewed by security personnel or law enforcement.

Several papers analyze the tradeoffs between the false alarm and false clear rates associated with different system alarm rules. That is, each device yields a response for each passenger. The system yields one of two possible outcomes: alarm or clear, which is a function of the device outcomes and can be defined in several ways. Kobza and Jacobson [18] formally provide a model for defining a system alarm for aviation security procedures. They illustrate how the false alarm and false clear rates can be interpreted as Type I (a false alarm is given) and Type II (a threat is not detected) errors and analyze their relationship for screening systems consisting of multiple devices.

Kobza and Jacobson [19] consider the design of security system architectures using reliability models in the context of aviation security baggage screening systems. Different objects (checked bags) can take different paths through the system, and hence, are screened by varying subsets of screening devices. Their model is analyzed based on Type I and Type II errors, and it is formulated for a series of dependent devices. Kobza and Jacobson [19] note that a system alarm is typically defined in one of two possible ways: at least one device alarm signals a system alarm, or all device alarms signal a system alarm. Their results indicate that multiple-device systems can be more effective than single-device systems, taking into account the probability of errors by each sub-system.

Glässer et al. [9] propose a computational model to evaluate aviation security screening performance using probability models that checks the consistency, coherence, and completeness of security requirements as defined by the FAA guidelines. They combine probabilistic variants of abstract state machines and model checking for analyzing aviation security models. Their model provides a tool for analyzing the effectiveness of security checkpoint screening and to identify potential security deficiencies.

Lee and Jacobson [21] provide alternative performance measures for evaluating the effectiveness of dynamic multilevel passenger screening systems introduced in Sect. 3.2. All of these models are designed to optimize the expected value of a performance measure, where the performance measure reports the expected number of detected threats. Lee and Jacobson [21] provide three alternative measures to evaluate the retrospective performance of a passenger screening system, where the set of passengers is known in hindsight:

1. *The under-screened passenger.* Passengers who are under-screened are assigned to a class that is less effective than their assignments resulting from the retrospective optimal solution.
2. *The over-screened passenger.* Passengers who are over-screened are assigned to a class that is more effective than their assignments resulting from the retrospective optimal solution.

3. *The optimally screened passenger.* Passengers who are optimally screened are assigned to the same class as their assignments resulting from the retrospective optimal solution.

All passenger types are modeled using Bernoulli random variables. Note that the under-screened passengers are of greatest concern, since this represents terrorists who receive low levels of security scrutiny. Lee and Jacobson [21] provide estimators to evaluate the expected number and variance of under-screened passengers, and they provide an upper bound of the probability that a dynamic passenger assignment policy results in zero under-screened passengers.

4 Conclusions

Analytical and operations research methodologies can be used to make a difference in aviation security. New directions in aviation security need not merely be makeshift political solutions for mending complex problems; they can be the result of modeling, analysis, and planning. By illustrating several ways in which operations research has made an impact in passenger prescreening systems, it is shown to have a place in the design and analysis of aviation security systems. However, there are some limitations. When doing operations research modeling (or in fact, mathematical modeling of any type), one must often make assumptions that may limit the applicability of the results obtained. Though such assumptions are often based on reasonable and realistic factors, they may pose difficulties in facilitating the transfer of the operations research analysis to decision-makers, since errors can lead to security breakdowns that may place people at an unnecessary risk. Second, operations research models quite often look at an application's average or mean performance. In aviation security systems, average performance does not always capture the most interesting and salient aspects of such operations, which are often concerned with rare events and events "at the extremes".

The issues discussed here represent but the tip of the iceberg. There are numerous problems in aviation security that can benefit from operations research methodologies, including providing methods for detecting nuclear and radiological weapons, improving perimeter access security with respect to airport employees, designing models for cargo screening, analyzing passenger throughput and space associated with security lines, and modeling secondary screening of passengers and their baggage when screening devices give an alarm response, to name just a few. Analytical methodologies can be used to not only gain insight into ways to improve aviation security system operations and performance, but also to make a lasting impression on our nation's security and well-being.

Acknowledgments This material is based upon work supported by the U.S. Department of Homeland Security under Grant Award Number 2008-DN-077-ARI001-02. The views and conclusions contained in this document are those of the authors and should not be interpreted as necessarily representing the official policies, either expressed or implied, of the U.S. Department of Homeland Security.

References

1. Barnett A (2001) The worst day ever. OR/MS Today 28(6):28–31
2. Barnett A (2004) CAPPS II: the foundation of aviation security? Risk Anal 24(4):909–916
3. Barnett A, Shumsky R, Hansen M, Odoni A, Gosling G (2001) Safe at home? An experiment in domestic airline security. Oper Res 49(2):181–195
4. Butler V, Poole RW Jr. (2002) Rethinking checked-baggage screening. Technical report, Reason Public Policy Institute, Public Policy No. 297, Los Angeles
5. Cartensen P (2004) The lamppost, the wizard and the law: reflections on Professor Barnett's assessment of CAPPS II. Risk Anal 24(4):917–919
6. Caulkins JP (2004) CAPPS II: a risky choice concerning an untested risk detection technology. Risk Anal 24(4):921–924
7. Chakrabarti S, Strauss A (2002) Carnival booth: an algorithm for defeating the computer-assisted passenger screening system. First Monday 7(10). www.firstmonday.org. Accessed 23 Oct 2003
8. Dugan L, Lafree G, Piquero AR (2005) Testing a rational choice model of airline hijackings. Criminology 43(4):1031–1065
9. Glässer U, Rastkar S, Vajihollahi M (2006) Computational modeling and experimental validation of aviation security procedures. Lect Notes Comput Sci 3975:420–431 Springer, Berlin
10. Hawley K (2009) United States Department of Homeland Security, Transportation Security Administration. Oral testimoney before the United States House of Representatives, Committee on Homeland Security, Subcommittee on Transportation Security and Infrastructure Protection, March 18
11. Jacobson SH, Bowman JM, Kobza JE (2001) Modeling and analyzing the performance of aviation security systems using baggage value performance measures. IMA J Manag Math 12(1):3–22
12. Jacobson SH, Karnani T, Kobza JE (2005) Assessing the impact of deterrence on aviation checked baggage screening strategies. Int J Risk Assess Manag 5(1):1–15
13. Jacobson SH, Karnani T, Kobza JE, Ritchie L (2006) A cost-benefit analysis of alternative configurations for aviation-checked baggage security screening. Risk Anal 26(2):297–310
14. Jacobson SH, McLay LA, Kobza JE, Bowman JM (2005) Modeling and analyzing multiple station baggage screening security system performance. Naval Res Logist 52(1):30–45
15. Jacobson SH, McLay LA, Virta JL, Kobza JE (2005) Integer program models for the deployment of airport baggage screening security devices. Optim Eng 6(3):339–359
16. Jacobson SH, Virta JE, Bowman JM, Kobza JE, Nestor JJ (2003) Modeling aviation baggage screening security systems: a case study. IIE Trans 35(3):259–269
17. Kelly E (2009) United States Department of Homeland Security, Transportation Security Administration. Statement before the United States House of Representatives, Committee on Homeland Security, Subcommittee on Transportation Security and Infrastructure Protection, March 18
18. Kobza JE, Jacobson SH (1996) Addressing the dependency problem in access security system architecture design. Risk Anal 16(6):801–812
19. Kobza JE, Jacobson SH (1997) Probability models for access security system architectures. J Oper Res Soc 48(3):255–263
20. Lazar Babu VL, Batta R, Lin L (2006) Passenger grouping under constant threat probability in an airport security system. Eur J Oper Res 168:633–644
21. Lee AJ, Jacobson SH (2009) Performance measures for sequential aviation security screening policies. Technical report, University of Illinois, Urbana
22. Lee AJ, Jacobson SH, Nikolaev AG (2006) Protecting air transportation: a survey of operations research applications to aviation security. J Trans Secur 1:160–184
23. Lee AJ, McLay LA, Jacobson SH (2009) Designing aviation security passenger screening systems using nonlinear control. SIAM J Control Optim 48(4):2085–2105

24. Lee AJ, Nikolaev AG, Jacobson SH (2008) Aviation security management, volume 2, chapter operations research applications in aviation security systems, pp 126–145. Praeger Security International
25. Martonosi SE, Barnett A (2006) How effective is security screening of airline passengers? Interfaces 36(6):545–552
26. McIlvain T (2008) Domestic Nuclear Detection Office, Department of Homeland Security. Personal Interview, Nov 25
27. McLay LA, Jacobson SH, Kobza JE (2005) Making skies safer: applying operations research to aviation passenger prescreening systems. OR/MS Today 32(5):24–31
28. McLay LA, Jacobson SH, Kobza JE (2006) A multilevel passenger prescreening problem for aviation security. Naval Res Logist 53(3):183–197
29. McLay LA, Jacobson SH, Kobza JE (2007) Integer programming models and analysis for a multilevel passenger screening problem. IIE Trans 39(1):73–81
30. McLay LA, Jacobson SH, Kobza JE (2008) The tradeoff between technology and prescreening intelligence in checked baggage screening for aviation security. J Transp Secur 1(2):107–126
31. McLay LA, Jacobson SH, Lee AJ (2010) Risk-based policies for airport security checkpoint screening. Transp Sci 44(3):333–349
32. McLay LA, Jacobson SH, Nikolaev AG (2009) A sequential stochastic passenger screening problem for aviation security. IIE Trans 41(6):575–591
33. Mead KM (2002) Key issues concerning implementation of the aviation and transportation security act. Office of Inspector General, Department of Transportation, Washington, D.C. Report Number CC-2002-098
34. Mead KM (2003) Aviation security costs, Transportation Security Administration. Office of Inspector General, Department of Transportation, Washington, D.C. Report Number CC-2003-066
35. Miller L (2003) Delta to test new airport security plan. Associated Press, 28 February 2003. Available at http://news.yahoo.com. Accessed 28 Feb 2003
36. Mullen M (2009) Welcome address. Academic Research Initiative Grantees Conference, Washington, DC April 6–9
37. National Research Council (1996) Airline passenger security screening. National Academy Press, Washington, DC Publication NMAB-482–1
38. Nie X, Batta R, Drury CG, Lin L (2009) The impact of joint responses of devices in an airport security system. Risk Anal 29(2):298–311
39. Nie X, Batta R, Drury CG, Lin L (2009) Passenger grouping with risk levels in an airport security system. Eur J Oper Res 194(2):574–584
40. Nikolaev A, Jacobson SH, McLay LA (2007) A sequential stochastic security system design problem for aviation security. Transp Sci 41(2):182–194
41. Oxford V (2008) Domestic Nuclear Detection Office, Department of Homeland Security. Personal Interview, May 6, 2008
42. Ravid I (2004) Safety versus defense: comments on "CAPPS II: the foundation of aviation security?". Risk Anal 24(4):929–931
43. Rhodes JD (2004) CAPPS II: red light, green light, or 'mother, may I?'. J Homel Secur. Available at www.homelandsecurity.org. Accessed 7 Sep 2005, 9 March 2004
44. Shepard MR (2007) Stopping the suicide bomber: legal, technical, and operational issues. Invited speaker, National Institute of Justice Conference, Arlington
45. Transportation Security Administration (2008) TSA registered traveler: security privacy and compliance standards for sponsoring entities and service providers, version 3.1, Jan 2008
46. United States Department of Transportation, Research and Innovative Technology Administration, Bureau of Transportation Statistics. National transportation statistics (2008). Available at http://www.bts.gov/publications/national_transportation_statistics/, 2008
47. United States Federal Aviation Administration (2002) FAA aerospace forecasts fiscal years 2006–2017. Report, Office of Policy and Plans, Washington, D.C

48. United States Government Accountability Office (2005) Secure Flight development and testing under way, but risks should be managed as system is further developed. Gao-05-356, Washington, D.C., March
49. United States Government Accountability Office (2005) Systematic planning needed to optimize the deployment of checked baggage screening systems. Gao-05-365, Washington, D.C., March 2005
50. United States House of Representatives (2002) Aviation security with a focus on passenger profiling. Hearing memo, 107th congress, Committee on Transportation & Infrastructure, Subcommittee on Aviation, Washington, D.C., 27 Feb 2002
51. United States Transportation Security Administration (2005) Twelve explosives detection canine teams join TSA/Homeland Security. Press release, Department of Homeland Security, Washington, D.C., 14 Nov 2005
52. United States Transportation Security Administration (2005) TSA unveils enhanced security screening procedures and changes to the prohibited items list. Press release, Department of Homeland Security, Washington, D.C., 2 Dec 2005
53. United States Transportation Security Administration (2006) TSA announces key elements of Registered Traveler Program. Press release, Department of Homeland Security, Washington, D.C., 20 Jan 2006
54. United States Transportation Security Administration (2006) FAMS mission and history. Security & Law Enforcement, Department of Homeland Security, Washington, D.C
55. United States Transportation Security Administration (2004) "Secure Flight" to be tested before year's end. Press release, Department of Homeland Security, Washington, D.C., 26 August 2004
56. United States Transportation Security Administration (2005) TSA continues to roll-out explosives detection trace portals. Press release, Department of Homeland Security, Washington, D.C., 27 July 2005
57. Virta JE, Jacobson SH, Kobza JE (2002) Outgoing selectee rates at hub airports. Reliab Eng Syst Saf 76(2):155–165
58. Virta JE, Jacobson SH, Kobza JE (2003) Analyzing the cost of screening selectee and non-selectee baggage. Risk Anal 23(5):897–908
59. Wright PD, Liberatore MJ, Nydick RL (2006) A survey of operations research models and applications in homeland security. Interfaces 36(6):514–529

Complex Risk and Uncertainty Modeling for Emergent Aviation Systems: An Application

Ahmet Oztekin and James T. Luxhøj

1 Introduction

Modeling complex systems is a very broad area of research where, more often than not, a multi-disciplinary approach is needed to achieve a meaningful representation of the subject matter. The analytical methods employed along the process remain as much an art as science, especially, if the subject matter is safety and risk analysis of a real-world system. One aspect that particularly increases the complexity of modeling is the fact that many real-world systems naturally include both discrete and continuous variables.

Probability theory is the method of choice for dealing with uncertainty in many science and engineering disciplines due to its well understood nature and its ability to model various phenomena in the physical world extremely well. When it comes to building representative probabilistic models of complex real-world systems, discrete Bayesian Networks (BNs) provide a popular and powerful tool. Given a set of discrete random variables, discrete BNs provide a formal framework for representing a joint probability distribution. Fore discrete BNs, exact inferencing solutions exist and are well understood. However, discrete BNs are quite limited for modeling uncertainty and performing probabilistic inferencing about hybrid systems, where entities of continuous and discrete nature co-exist. Many complex real-world applications emerge as hybrid systems and require a more general solution.

A. Oztekin · J. T. Luxhøj (✉)
Department of Industrial and Systems Engineering, Rutgers University,
96 Frelinghuysen Road, Piscataway, NJ 08854–8018, USA
e-mail: jluxhoj@rci.rutgers.edu

A. Oztekin
e-mail: oztekin.ahmet@gmail.com

H. Pham (ed.), *Safety and Risk Modeling and Its Applications*,
Springer Series in Reliability Engineering, DOI: 10.1007/978-0-85729-470-8_10,

General *Hybrid Bayesian Networks* (HBNs), where continuous and discrete variables may appear anywhere within the network topology, are a generalization on discrete BNs. General HBNs are inherently more suitable for modeling complex systems and provide a better modeling representation for the vast majority of real-world applications. However, contrary to the discrete-only case, a universal solution for exact inferencing about general HBNs is yet to be developed. Thus, there exist a real demand among the practitioners of uncertainty analysis for a general operational solution for representation of and inferencing about general HBNs.

In this study, we concentrate on the problem of inferencing in HBNs. Our focus, hence our contributions are three-fold: theoretical, algorithmic and practical. Specifically, our major contributions to the larger research domain of representation and inferencing in general HBNs can be summarized as follows:

- From a theoretical point of view, we complement classical probability theory with Fuzzy set theory to develop a hybrid formalism to understand and model complex uncertainty associated with real-world systems. To that end, we provide a novel framework to implement a hybrid Fuzzy-Bayesian methodology to perform exact inferencing in general HBNs.
- From an algorithmic perspective, we provide a suite of inferencing algorithms for general HBNs. In particular, we introduce two transformations for general HBNs to create *Type-I* and *Type-II* Fuzzy-Bayesian Networks (FBNs) and present formal representation techniques and separate inferencing mechanisms.
- Finally, from a practical perspective, we apply our framework, methodology, and techniques to the task of assessing system safety risk due to the introduction of emergent UASs into the NAS. In this context, we present the UAS Domain Safety Risk Model (DSRM).

The UAS DSRM is a general HBN model representing the conditional interactions of a hybrid set of hazards and causal factors. The outcome of the UAS DSRM is a set of probability distributions for individual hazard elements, which are represented as continuous variables in the model. In this study we used synthetic data to populate the UAS DSRM and to determine these probability distributions.

This chapter describes the development of a generalized hybrid Fuzzy-Bayesian methodology for modeling uncertainty associated with complex real-world systems. Safety risk modeling of the emergent Unmanned Aircraft Systems (UAS) operations in the National Airspace System (NAS) is the domain of application for the developed methodology.

2 Fuzzy Bayesian Networks

In this section, we provide the summary of a hybrid Fuzzy-Bayesian framework developed to model complex uncertainty associated with real-world systems. Consequently, in Sect. 3 we apply the developed hybrid methodology to the UAS domain and present sample results for this application. In particular, we start with *Bayesian Networks* and *Fuzzy Sets* and provide brief overviews of these concepts that serve as the foundation upon which our proposed methodology is developed. Then, as the components of the proposed framework, we introduce the notion of *Probability of a Fuzzy Event* and discuss the concepts of *Fuzzy Evidence* and *Fuzzy Updating*. We conclude the section by introducing *Fuzzy-Bayesian Networks* and present an overview of the proposed methodology for probabilistic inferencing in FBNs. For a detailed discussion on the background and the theory of the developed Fuzzy-Bayesian methodology summarized in this section the reader may refer to Oztekin [1].

2.1 Bayesian Networks

Bayesian Networks (BNs) [2] constitute a class of stochastic models for modeling interactions and dependencies among variables representing a system. Bayesian Networks can be used to represent and solve decision problems under uncertainty [3]. BNs, broadly construed, are a compact way of representing a joint probability distribution imposed by a network structure among a set of random variables **V**.

A BN is commonly represented as a Directed Acyclic Graph (DAG) consisting of a set of nodes and directed edges. In a DAG, there is no directed path that starts and ends on the same node. The nodes represent the variables and the edges represent the conditional dependencies among the variables in the model. The absence of a path between two nodes in a BN indicates that these two variables are conditionally independent. A BN represents a collection of *conditional probability distributions* (CPDs), where each node (i.e., random variable) in the graph is denoted by a conditional distribution given its parent nodes. Figure 1 illustrates a discrete BN, where all nodes are binary discrete random variables.

In Fig. 1, the CPDs of each variable given its parents are presented as tables next to the corresponding nodes. Each entry in the tables represents a conditional probability. For example, the first entry given by $0.95 = P(e_1|d_1,c_1)$ in the CPD of variable E represents the probability of variable E is in state e_1 given that variables D and C are in states d_1 and c_1, respectively. An important feature of Bayesian Networks, which makes inferencing possible, is the fact that *given its parents every node is conditionally independent of the nodes, which are not among its descendants*. In other words, a Bayesian Network represents the joint probability distribution over its set of variables in terms of *conditional independencies*.

Exact probabilistic inferencing solutions for discrete BNs exist and well understood. By exact inferencing we mean that, given a query, an intelligent system produces exact results. In the case of Bayesian Networks, given a set of query variables $\boldsymbol{Q}$ and some evidence $\boldsymbol{E} = e$, an exact inferencing algorithm should produce exact numerical results for the probability distribution $P(\boldsymbol{Q}|\boldsymbol{E} = e)$. We can take advantage of the structure of the discrete BN to perform probabilistic inferencing efficiently. *Junction Tree Algorithm* [1, 4, 5], also known as the *clique*

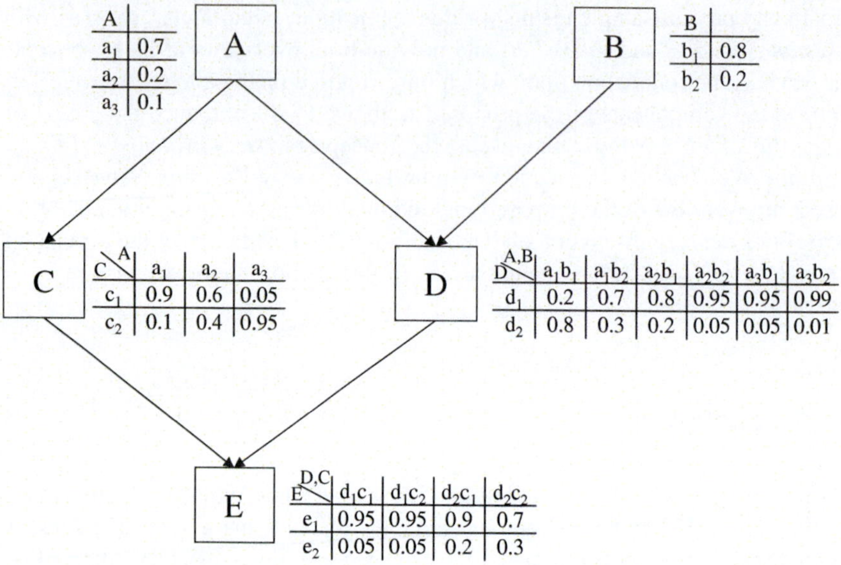

Fig. 1 Example discrete Bayesian network

tree, the *cluster tree*, and the *joint tree* algorithm, is by far the most popular of such methods.

2.1.1 Hybrid Bayesian Networks

Although the inferencing techniques are exact, the modeling ability of discrete BNs is quite limited for representing real-life complex systems, which quite often include continuous variables as well as discrete variables. Shown in Fig. 2, General Hybrid Bayesian Networks (HBN), as the generalization of Bayesian Networks, address this shortcoming inherent to the discrete BNs. However the question of exact inferencing in general HBNs is currently unsolved. For a constrained set of HBNs, Lauritzen proposed an exact method, which makes use of *Conditional Gaussians* (CG) [6, 7]. Today, CG-based models represent the most popular class of hybrid models. However, they have two important restrictions. First, they can only model linear relations between continuous variables. Second, they do not allow discrete nodes to have continuous parents in the network structure. The latter constraint greatly limits the applicability of this model to most real-world situations, where discrete variables may depend on continuous variables.

Lauritzen's approach, from an algorithmic point of view, still represents the state of art. It is based on the *junction tree algorithm* [1, 4, 5], originally developed for discrete BNs. Contrary to the common perception that the extension of the *junction tree algorithm* to hybrid networks is a straightforward implementation, it

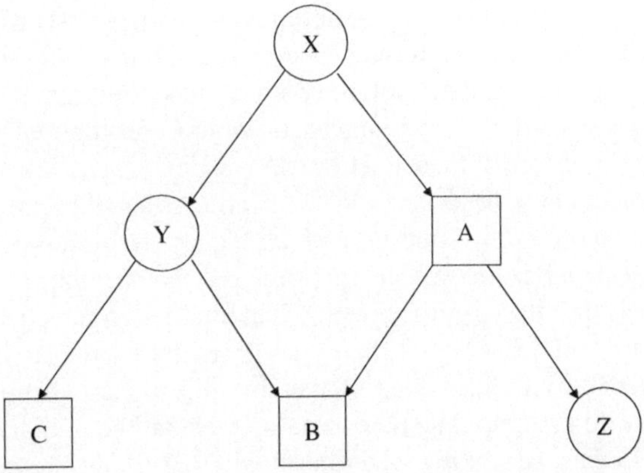

Fig. 2 A general hybrid Bayesian Network, where the sets $\Delta = \{A, B, C\}$ and $\Gamma = \{X, Y, Z\}$ represent the discrete and continuous variables, respectively

has been shown that in many cases the Lauritzen algorithm is intractable even for simple network structures [8].

On the other hand, various approximate algorithms were also introduced for HBNs. The most commonly used framework is based on stochastic sampling. Although, stochastic sampling applies to every class of HBNs (not only CGs), it may take a long time to converge to a reliable answer, and therefore stochastic sampling based approximate algorithms are not suitable for real-time applications.

Within this context, we propose a novel inference algorithm for HBNs. This new framework takes a different view at the whole problem from the vantage point of an analyst whose goal is to assess the safety risk associated with a complex system, for which randomness is only one of the sources of uncertainty. The proposed framework introduces vagueness to the problem in an attempt to bridge the gap between probability and possibility, thereby enabling the application of general HBNs practical for reasoning about complex systems. Next, to facilitate the ensuing discussion we present a concise introduction of the ideas at the crux of Fuzzy Set theory.

2.2 Fuzzy Sets

In classic set theory a certain element can either belong to a set or not, such as in an optimization problem a certain solution can either be feasible or not, which can be represented in mathematical terms by using an indicator function. The nature of membership to classical sets requires precision and assumes that the model parameters represent exactly either our perception of the phenomenon modeled or the real features of the actual system. More importantly, precision implies that our model of the real-world system does not contain any ambiguities. Therefore, an observation about a certain model parameter can only assume the 0-or-1 assignment of an indicator function over its defined domain to determine whether or not it belongs to a collection of mutually exclusive states defining the model parameter. This crisp, deterministic, and precise worldview underlines a whole body of work for formal modeling and reasoning about real-world systems.

A more realistic point of view would admit that the real world is more complex and uncertain. Traditionally, uncertainty in the real world is addressed primarily by probability theory. However, randomness is only one of the many sources of uncertainty in real-world applications. Another major source of uncertainty is *ambiguity*. The question of ambiguity is directly related to the notion of *set membership*. However, this time the membership is represented by a continuous function that can assume any real number on the closed interval [0,1] instead of 0 or 1 only. Fuzzy Set theory proposed by Lotfi Zadeh [9] makes use of this more generally defined idea of membership to formally model the ambiguity in real-world systems. The new idea, here, is that the notion of set membership is the key to decision making when faced with uncertainty in general. In this respect, Fuzzy Set theory could be interpreted as a generalization of the classic Set theory.

Our past research experience on modeling the safety risk in the civil aviation domain indicates that the uncertainty that needs to be quantified originates from two major sources: randomness associated with domain variables and ambiguity associated with their states. Within this context, a Fuzzy-Bayesian hybrid approach provides the most appropriate tools to tackle the problem of modeling the uncertainty associated with a real-world complex system.

2.3 A Hybrid Fuzzy-Bayesian Framework

We propose a hybrid Fuzzy-Bayesian formalism to overcome the complexity and tractability issues of the existing inferencing algorithms with an additional emphasis on improving the representative power of general HBNs for real-world systems. In this section, we provide an overview of the proposed analytical approach complete with its inferencing formalism for a generic Fuzzy-Bayesian Network.

2.3.1 A Fuzzy Bayes Formulation

A very heated debate has been underway since the introduction of Fuzzy logic and possibility theory about their relevance within the scientific community dominated by a world view which believes that probabilistic methods are necessary and sufficient to understand uncertainty in real-world complex systems [10, 11].

We believe that possibility theory based on Fuzzy logic and probability theory are *complementary rather than competitive* [11]. They simply address different sources of uncertainty. Probability deals with randomness, whereas Fuzzy Sets helps us to understand the vagueness.

Within this context, we presented the notion of *probability of a Fuzzy event* and introduced the concept of a *Fuzzy random variable*. In particular, the probability of

Fig. 3 *Fuzzy*-crisp variable pairs in Fuzzy-Bayesian networks

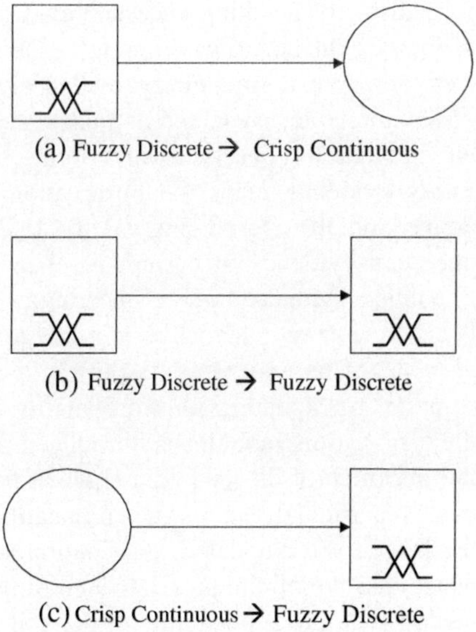

a Fuzzy event is a purely stochastic problem where the event itself is vaguely defined i.e., represented by Fuzzy sets. There are two possible approaches one can adopt: The probability of a Fuzzy event is a scalar (i.e., a crisp real number or a measure) or it can be represented by a Fuzzy set. We conclude that the latter option necessitates the adoption of an inferencing mechanism based solely on Fuzzy logic, which from an algorithmic point of view is considered to be suboptimal. Whereas, if the probability of a Fuzzy event is assumed to be a scalar measure then the inferencing mechanism could capitalize the well-established algorithms developed for probabilistic reasoning methods such as Bayesian Networks. Based on this analysis, we introduce a novel *Fuzzy Bayes formulation* and outline a formalism to determine the conditional probability due to the interactions of Fuzzy and crisp variables. This novel representation of the *hybrid conditional probability* of crisp and Fuzzy variable pairs lies at the crux of inferencing about the proposed FBNs. The major types of such conditional interactions are illustrated in Fig. 3.

Fuzzy discrete nodes in Fig. 3 denote the Fuzzy counterpart of a crisp continuous variable discretized by Fuzzy transformation.

2.3.2 Fuzzy Evidence and Fuzzy Updating

There are two aspects of Bayesian Networks that are still subject to improvement and therefore research: how to represent continuous variables in a general HBN setting and how to deal with uncertain information as evidence? Our Fuzzy Bayes formulation provides a mechanism to represent continuous variables and associated conditional dependencies in a general HBN setting. As part of the theoretical foundation of the proposed Fuzzy-Bayesian framework, we also present a formalism for handling uncertain evidence. In particular, we introduce the notion of Fuzzy evidence to incorporate vague or ambiguous information into a Bayesian Network. We define Fuzzy evidence as a type of uncertain evidence, where observations are presented as Fuzzy sets rather than delta or indicator functions, which place the observed variable in one of the mutually exclusive states. Thus, Fuzzy evidence maps the observation to a set of predetermined Fuzzy states defined on the closed interval [0,1]. We discuss alternative representations of uncertain evidence in current practice and show that, as opposed to virtual or likelihood evidence, where uncertainty is presented as a probability distribution [12, 13]. Fuzzy evidence is suitable to be used in conjunction with continuous variables. Consequently, we present a formal methodology, a new approximate solution, for updating joint probability distributions when Fuzzy evidence about the distribution variables is introduced. We utilize the relative entropy concept of the information theory [12, 14] when formalizing our Fuzzy updating methodology. In particular, we outline an updating scheme for the prior distribution, where the posterior (or updated) distribution satisfying a constraint set (i.e., Fuzzy evidence) has the minimum relative entropy with respect to the prior distribution. We demonstrate the applicability of our solution for updating with single and multiple uncertain evidences with a detailed numerical example and concluded that, when multiple uncertain evidences are present, simultaneous updating should be preferred over consecutive updating especially when dealing with moderate to large size Bayesian Networks for which the inherent complexity is already known to be high.

For a detailed discussion on the Fuzzy Evidence and Fuzzy Updating, including the derivations and numerical examples, the reader may refer to Oztekin [1].

2.3.3 Fuzzy-Bayesian Networks

Historically, uncertainty about real-world systems has been modeled by probabilistic tools, which are crisp, deterministic and precise in character. However, as discussed in earlier sections, different forms of uncertainty exist and probability theory only addresses the randomness aspect of it. Fuzzy Set theory provides a means for representing uncertainty due to vagueness such as the uncertainty in natural language. However, uncertainty due to vagueness (i.e., fuzziness), in fact, exists not only in human cognition and languages, but also in most systems modeled by Bayesian Networks. Consider variables such as *temperature, age* or *speed*, which are inherently continuous but represented as discrete when included in discrete BNs. For such variables an implicit mapping is involved whenever an observation (i.e., evidence) about them needs to be introduced to the model. Axioms of probability dictate that once such mapping is performed on a continuous variable the resulting discrete states must cover the whole domain of the original variable and be mutually exclusive, so that every single observation falls into one and only one state and no two states co-exist at the same space and time. For the purposes of approximation and in cases without a pressing need for

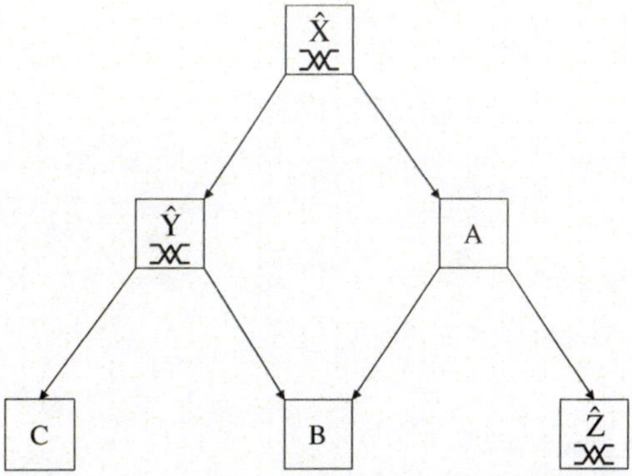

Fig. 4 The general FBN derived from the HBN of Fig. 2

accuracy, such a quantification may be justifiable. However, not every continuous variable behaves sensibly under discretization. Consider *temperature* defined on the frame [0, 40]°C and we decided to use a three state discretization scheme *cold, warm*, and *hot* corresponding to the intervals [0, 10)°C, [10, 25)°C, [25, 40]°C, respectively. A reading of 24.9°C from the thermometer would fall under discrete state *cold,* whereas, 25°C would be labeled as *warm.* We believe that there is no meaningful way of determining a crisp boundary between these states. Hence, using classical sets with crisp boundaries when discretizing a continuous domain may generate some unpredictable results for BNs.

Right at this point, the concept of *fuzziness* becomes very interesting. In fact, Fuzzy Set theory and its implementation of degrees of membership idea to sets provide a structured way to improve on classical discretization techniques. Nevertheless, the distinction between Fuzzy Set theory and probability theory should be made clear. *Fuzziness describes the ambiguity of an event, whereas randomness describes the uncertainty in the occurrence of the event* [15]. Within this context, we see promise in combining the two concepts to complement each other, so that various limitations of classical Bayesian Networks will be overcome by the resulting *hybrid* methodology.

Fuzzy-Bayesian Networks (FBNs) emerge as powerful tools that combine the representation power of Fuzzy Set theory over poorly defined problem domains with the algorithmic strength of Bayesian Networks. Given a general HBN, a general FBNs can be constructed by transforming all continuous variables and associated conditional probability distributions into the Fuzzy domain. For the general HBN in Fig. 2 the result of the transformation to a general FBN is given in Fig. 4.

In Fig. 4, the variable set $\{\hat{X}, \hat{Y}, \hat{Z}\}$ corresponds to the Fuzzy-discrete transformations of the originally continuous variable set $\{X, Y, Z\}$. We provide explicit formulations to perform these transformations using the Fuzzy Bayes formulation outlined in Sect. 2.3.1, which require the Fuzzy sets and corresponding membership functions defined on the frames of all continuous variables in the HBN to be given or constructed first. In the resulting FBN, all originally continuous variables are now replaced by their counterpart Fuzzy-discrete variables whose

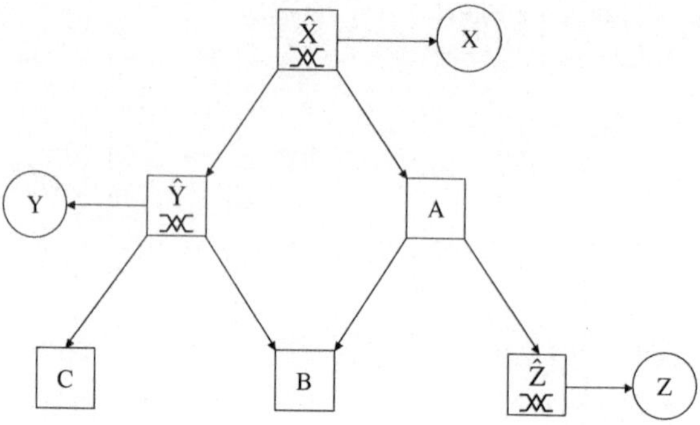

Fig. 5 Type-I FBN derived from the HBN Fig. 2

states correspond to the Fuzzy states identified for the original continuous variable for the purpose of this transformation. Furthermore, after the transformation, all conditional distributions in the FBN can be represented by discrete multinomial distributions. It follows that exact inferencing algorithms such as variable elimination or junction tree algorithm for discrete BNs can be applied to perform probabilistic reasoning about general FBNs.

Although, with general FBNs we achieved practical exact inferencing for general HBNs, since only the Fuzzy-discrete transformations, not the original continuous variables, are present in general FBNs, there is still room for improvement to reach a better approximation of the original hybrid network. Therefore, we introduce two new forms of Fuzzy transformation for general HBNs, namely *Type-I* and *Type-II* FBNs, with increased sophistication in their representation of the original HBN and complexity in inferencing.

A *Type-I* FBN is created in two consecutive steps. First step involves, as in a general FBN transformation, the replacement of all continuous variables in the original general HBN with their Fuzzy-discrete counterparts. In the second step, the original continuous variables are added and connected to the network with a directed link originating from their respective Fuzzy-discrete counterpart. Illustrated in Fig. 5, the resulting form represents an HBN where continuous variables have only discrete parents, for which we showed that exact inferencing solutions exist. We used a Conditional Gaussian (CG) model to represent the conditional distributions of the original continuous variables given their Fuzzy-discrete counterparts and develop a exact inferencing algorithm for *Type-I* FBNs based on this assumption.

As for the second form of transformation for general HBNs, we introduce *Type-II* FBNs which involve a finer approximation when representing the original hybrid network as compared to the *Type-I* transformation and hence, present a greater computational challenge. To construct a *Type-II* FBN, the same fuzzy-transformation, as defined for the *Type-I* case, is applied however this time only to those continuous variables whose descendants in the original HBN include discrete variables. For the transformed part of the resulting hybrid network, similar to *Type-I* FBNs, we use a CG to model the conditional distributions of the original continuous variables given their Fuzzy-discrete counterparts. Whereas, for the conditional distribution of the remaining continuous variables in the *Type-II* FNB,

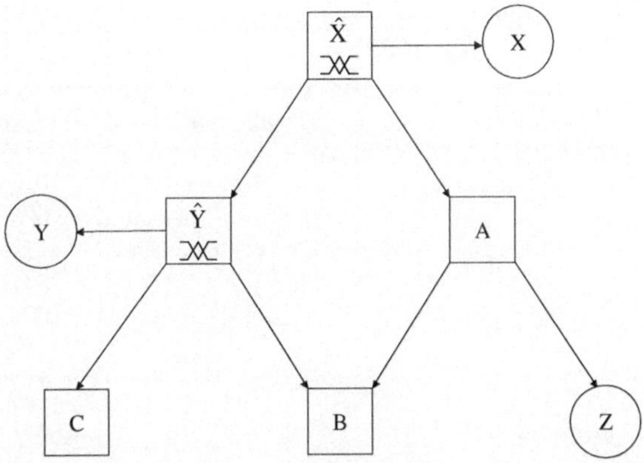

Fig. 6 Type-II FBN derived from the HBN of Fig. 2

we use a Conditional Gaussian Regression (CGR) model. Using these models we developed an algorithm for inferencing in *Type-II* FBNs.

As compared to *Type-I*, *Type-II* FBNs brings about a much accurate approximation of the *original* general HBN, however with the additional cost of increased complexity of the exact inferencing algorithm. The *Type-II* FBN transformation of our example general HBN are depicted in Fig. 6.

A comparative analysis of *Type-I* and *Type-II* FBNs are presented in Table 1.

For a detailed discussion on inferencing in *Type-I* and *Type-II* FBNs and for the developed inferencing algorithms the reader may refer to Oztekin [1].

3 Application of Research Methodology

In this section we focus on the application of the hybrid Fuzzy-Bayesian framework introduced in the preceding section of this report. Unmanned Aircraft Systems (UASs) are selected as our domain of interest. In the following sections we first model the UAS system safety risk as a general HBN using on a regulatory-based approach. The resulting HBN is denoted as the UAS Domain Safety Risk Model (DSRM). Finally, on the UAS DRSM we apply our hybrid Fuzzy-Bayesian methodology outlined in the preceding section and conclude with presenting and analyzing the results.

3.1 Development of a System-level Taxonomy for Categorization of UAS Hazards

UAS having being successfully employed in the last decade by various military applications are, inevitably, making their way into the civilian world. This new frontier in civil aviation adds another dimension to the ever-increasing complexity

Table 1 Comparison of algorithms for *Type-I* and *Type-II* FBNs

	Type-I FBNs	*Type-II*FBNs
Representation of general HBNs	All continuous variables are fuzzified to create Fuzzy-discrete counterparts	A better approximation of a general HBN Only the continuous variables with discrete descendants are fuzzified
	CPDs of continuous variables given Fuzzy-discrete counterparts are assumed to be CG	CPDs of continuous variables given Fuzzy-discrete counterpart are assumed to be CG
		CPDs of the remaining continuous variables including the ones with hybrid parentage are modeled by a CGR model
		The joint distributions of the hybrid cliques presented by CGRs are mixtures of Gaussians
		To perform message passing in the *Type-II* junction tree mixtures are approximated by single Gaussians
Inferencing	Exact	Approximate
Computational complexity	Comparable to discrete BNs	High compared to *Type-I* FBNs

of the current NAS in the United States. The future inclusion of private and commercial operations of the UAS into the NAS, unavoidably, raises safety concerns. As the NAS becomes increasingly more complex and constrained, the associated hazard and safety risk modeling must also mature in sophistication. Thus, there is a need for advanced studies focusing on risk-based system safety analysis of emergent UAS operations.

One of the first steps in the proposed UAS system safety analysis is hazard identification and analysis. To that end, a new hazard taxonomy was developed. This taxonomy, termed the Hazard Classification and Analysis System (HCAS) identifies four main hazard system sources: Airmen, UAS, Operations, and Environment. The basic framework of the proposed taxonomy is based on the FAA regulatory perspective (i.e., Title 14, Code of Federal Regulations (14 CFR) chapters on Aircraft, Airmen, Certification/Airworthiness, Flight Operations, etc.). Such an approach uniquely distinguishes the HCAS taxonomy from all other UAS hazard analyses being performed by the Department of Defense (DoD), the RTCA-Special Committee (SC) 203 [4, 5], etc.

Safety analysis has a fundamental role to play in the identification of hazard source potentials, the understanding of the underlying causal factors, the likelihood assessment of these factors, the severity evaluation of the potential consequence(s) of mishaps, and the prioritization of mitigations.

A sound system-level safety analysis relies heavily on properly identifying the key components of the area of interest. In particular, the identification of potential hazard sources and sub-sources within the systemic structure of the problem domain should be considered as a fundamentally important step in system safety analysis. Furthermore, since semantics play a crucial role while defining the domain variables, a systematic taxonomy that balances fidelity and generalization provides a solid foundation for a meaningful and relevant system safety analysis. Within this context, we present the HCAS taxonomy specifically designed and developed to identify and categorize individual system-level hazard sources for the UAS operations.

HCAS categorizes the UAS hazards consistent with the 14 CFR Sub chapters, thereby establishing the taxonomy on the FAA regulatory framework. The advantage of the proposed approach is to allow direct association of hazards identified with regulatory requirements or vice versa. The system not only provides the FAA as well as the UAS community the tools to determine safety and regulatory implications of UAS operating in the NAS, but also falls in directly under the FAA Safety Management System (SMS) Doctrine.

At the crux of the HCAS taxonomy lie two closely related yet distinct concepts: *hazards* and *hazard sources*. Based on Leveson's definition of hazard [16] we adopted our own definitions for both concepts within the context of this application:

Hazard. A hazard is a state or set of conditions of a system that, together with other conditions in the environment of the system, may lead to an accident (loss event).

Hazard Source. Hazard source are primarily components of the UAS domain; hence a state or set of conditions of these components may lead to hazardous potential of the domain itself. Each hazard source category corresponds to a key component of the domain of interest. Thus, these components do not represent neither individual hazards nor categories of hazards of the UAS domain.

HCAS is a continuously evolving taxonomy. The current version of the taxonomy has been developed in multiple phases as the product of numerous knowledge elicitation sessions with subject matter experts spanning a time period of 2 years.

The idea behind the HCAS development effort is to provide a structured framework to identify and classify both system and sub-system hazard sources for UAS operations. Based on the above hazard and hazard source definitions, in the current HCAS taxonomy four systems-level hazard sources are identified as UAS, Airmen, Operations, Environment. These system-level hazard sources form the four main HCAS *cubes* depicted in Fig. 7 and in tabular form in Table 2.

There is one particular aspect of our approach to taxonomy development that needs to be underlined. Our approach, while originally scenario-based, evolved into a more regulatory-based perspective during the course of the development. In a sense, this focus shift was natural considering the fact that, right from the start, our goal was to develop a generalized taxonomy for system-level UAS hazards that would have applicability across a broad spectrum of FAA regulations. This aspect of our approach uniquely distinguishes the HCAS taxonomy from all other UAS hazard analyses being performed by the Department of Defense (DoD), the RTCA-Special Committee (SC) 203, etc.

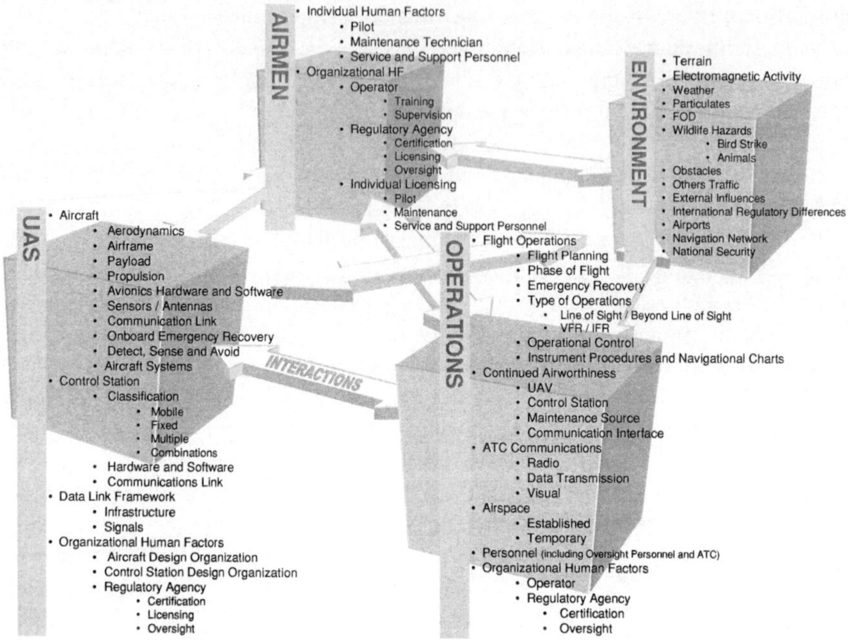

Fig. 7 System and subsystem-level UAS hazard sources—HCAS version 3.5

The HCAS taxonomy may also be used to construct influence/causal factor diagrams representing hypothetical or notional UAS safety risk scenarios. The use of modifiers placed on the HCAS taxonomy elements, such as “inappropriate”, “inadequate”, etc., may be used to create such an influence diagram. These influence diagrams may then be used to study the interactions among various causal factors associated with the hazards. Conceptually, HCAS represents a hierarchical structure for UAS hazard sources. In particular, at the very top, there are system-level hazard sources, which, in lower levels, are decomposed into their subsystem-level hazard sources. Since civil UAS operations are relatively new and emergent, databases of mishaps are not readily available. This idea is further explored in the next section to develop a methodology for modeling safety risk of the UAS domain as a hybrid Bayesian Network.

3.2 UAS Domain Safety Risk Model

The HCAS taxonomy provides a systematic approach to identify hazards associated with UAS operations in the NAS. However, hazards are not causal factors, which are the essential building blocks of influence diagrams (i.e., Bayesian Nets) representing various risk scenarios of the UAS operations. Thus, the decomposition of hazards into their constituent causal factors is another important step in the

Table 2 Outline of the HCAS taxonomy

OUTLINE – HCAS Taxonomy (Version 3.5)

1. UAS (Systems Level)

1.1. Aircraft (Subsystems Level)
- 1.1.1. Aerodynamics
- 1.1.2. Airframe
- 1.1.3. Payload
- 1.1.4. Propulsion
- 1.1.5. Avionics Hardware and Software
- 1.1.6. Sensors / Antennas
- 1.1.7. Communication Link
- 1.1.8. Onboard Emergency Recovery
- 1.1.9. Detect, Sense and Avoid
- 1.1.10. Aircraft Systems

1.2. Control Station
- 1.2.1. Classification
  - 1.2.1.1. Mobile
  - 1.2.1.2. Fixed
  - 1.2.1.3. Multiple
  - 1.2.1.4. Combinations
- 1.2.2. Hardware and Software
- 1.2.3. Communications Link

1.3. Data Link Framework
- 1.3.1. Infrastructure
- 1.3.2. Signals

1.4. Organizational Human Factors
- 1.4.1. Aircraft Design Organization
- 1.4.2. Control Station Design Organization
- 1.4.3. Regulatory Agency
  - 1.4.3.1. Certification (includes authorizations and special approvals, operations specifications, designations, etc.)
  - 1.4.3.2. Licensing
  - 1.4.3.3. Oversight

2. AIRMEN

2.1. Individual Human Factors (HF)
- 2.1.1. Pilot
- 2.1.2. Maintenance Technician
- 2.1.3. Service and Support Personnel

2.2. Organizational HF
- 2.2.1. Operator
  - 2.2.1.1. Training
  - 2.2.1.2. Supervision
- 2.2.2. Regulatory Agency
  - 2.2.2.1. Certification
  - 2.2.2.2. Licensing
  - 2.2.2.3. Oversight
- 2.2.3. Individual Licensing
  - 2.2.3.1. Pilot
  - 2.2.3.2. Maintenance
  - 2.2.3.3. Service and Support Personnel

3. OPERATIONS

3.1. Flight Operations
- 3.1.1. Flight Planning
- 3.1.2. Phases of Flight (include pre and post flight operations by the ground support personnel, see and avoid, right of way-conflict resolution)
- 3.1.3. Emergency Procedures
- 3.1.4. Type of Operations
  - 3.1.4.1. Line of Sight / Beyond Line of Sight
  - 3.1.4.2. VFR / IFR
- 3.1.5. Operational Control
- 3.1.6. Instrument Procedures and Navigation Charts

3.2. Continued Airworthiness
- 3.2.1. UAV
- 3.2.2. Control Station
- 3.2.3. Maintenance Source (Facility and Individual)

3.3. ATC Communications
- 3.3.1. Radio
- 3.3.2. Data Transmission
- 3.3.3. Visual

3.4. Airspace
- 3.4.1. Established
- 3.4.2. Temporary

3.5. Personnel (including Oversight Personnel and ATC)

3.6. Organizational Human Factors
- 3.6.1. Operator
- 3.6.2. Regulatory Agency
  - 3.6.2.1. Certification
  - 3.6.2.2. Licensing
  - 3.6.2.3. Oversight

4. ENVIRONMENT

4.1. Terrain

4.2. Electromagnetic Activity

4.3. Weather

4.4. Particulates (including Volcanic Ash)

4.5. FOD

4.6. Wild Life Hazards
- 4.6.1. Bird Strike
- 4.6.2. Animals

4.7. Obstacles

4.8. Other Traffic

4.9. External Influences (Social, Political)

4.10. International Regulatory Differences

4.11. Airports

4.12. Navigation Network (ground and space based infrastructure and signals)

4.13. National Security

5. INTERACTIONS

5.1. UAS / Environment

5.2. Operations / Environment (includes NAS interconnectivity)

5.3. Airmen / Environment

5.4. UAS / Operations / Environment

5.5. UAS / Airmen / Environment

5.6. Operations / Airmen / Environment

5.7. UAS / Operations / Airmen / Environment

development of a comprehensive scheme for UAS safety risk modeling. Underlying causes of the hazards, such as failure modes, operator and software errors, design flaws, etc., need to be identified in order to eventually determine the mishap risk and the hazard mitigations. However, HCAS is not a taxonomy of causal factors. Although the resulting taxonomy for the UAS hazard sources is intended to be generic and inclusive, it represents an inductive reasoning approach with particular emphasis on a given set of UAS hazard scenarios. Hence, to determine a taxonomy of UAS causal factors, which are, strictly speaking, hierarchically at a lower level than hazard sources, we chose to employ deductive reasoning and based our analysis on the current FAA regulations for commercial civil aviation. Knowledge elicitation sessions with subject matter experts are heavily utilized throughout this process. Subsequently, individual causal factors are mapped to the taxonomy of UAS hazard sources resulting in a seamless analysis that is generic enough to cover most possible UAS operational scenarios yet provides the necessary level of fidelity to map their prominent features into a database.

At the crux of our regulatory-based approach lie the following assumptions:

- UAS integration will impact the entire NAS because of the wide-ranges of UAS size, weight, performance characteristics, airspace access, and unique operation issues.
- There are no sufficient data and proven methods to perform UAS safety analysis with the traditional event-driven approach.
- The regulations provide the essential safety net for the NAS safety.
- There exist a set of causal factors, which can be identified, associated with each relevant regulatory section.
- With proper descriptions of causal factors, the interdependencies (linkages) among themselves can be demonstrated.
- These linkages form the basis to analyze UAS safety risk by applying probabilistic reasoning methodologies such as BNs through the Hazard Classification and Analysis System (HCAS) model.

Within the context of these assumptions, we derive the individual causal factors from existing regulations governing all current aviation-related operations in the NAS. The role of the HCAS taxonomy introduced in the prior section is to provide structure to this regulatory-based casual factor identification process.

The current structure of the Federal Aviation Regulations (FARs) in the US represents a hierarchy. The FARs, as part of Title 14 of the Code of Federal Regulations (CFR), are organized into *Subchapters*. Each subchapter is then organized into *Parts*. Each part deals with a specific type of aviation activity. For example, 14 CFR Part 121 contains rules and requirements for Domestic, Flag, and Supplemental Operations of US registered aircraft. Individual FAR Parts are further divided sequentially into *Subparts*, *Section*, and *Subsections*. The derivation process for the causal factors closely mimic this hierarchical structure. In particular, causal factors are extracted from within the context of a FAR Part keeping possible applicability for UAS operations into consideration. Consequently, each causal factor is categorized under a sub-system level hazard source defined by the HCAS taxonomy, thereby establishing a viable connection between regulations and hazard sources. Figure 8 is a notional diagram depicting the connection between regulations and hazard sources (i.e., HCAS element) through causal factors.

This regulatory-based process has two main objectives; first, for each FAR part, to identify, describe, and define the causal factors; second, to determine interactions and connections among these causal factors. These connections constitute the foundation upon which the Bayesian Networks representing causal dependencies within the context of a UAS risk or hazard source will be constructed.

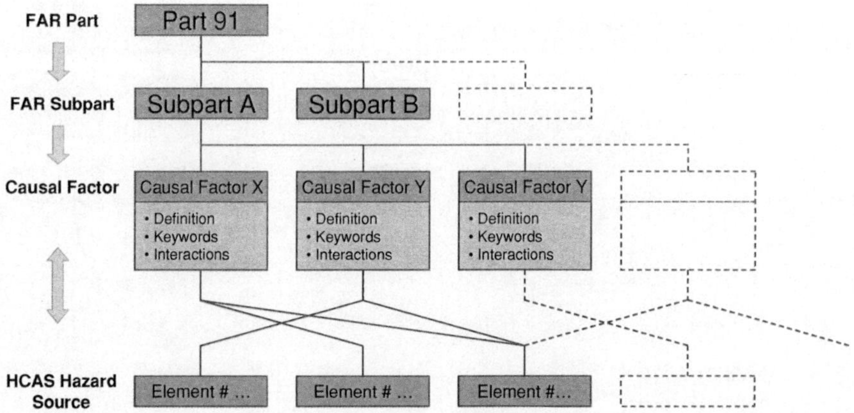

Fig. 8 Causal factors are the link between FARs and HCAS taxonomy

Since individual causal factors are identified and defined solely based on FARs, their derivation, as a creative process relies heavily on knowledge elicitation sessions with subject matter experts and will need some vetting within the aviation community.

Based on the ideas outlined in the preceding section, we introduce a Regulatory-based Causal Factor Framework (RCFF) to study the potential safety impacts of introducing emerging UAS operations into the well-established National Airspace System (NAS). Formally, RCFF is a systematic process for the creation of causal factors that are derived from the regulations to functions to hazards to causal factors [17, 18]. It provides a qualitative means of identifying and assessing hazards controlled by existing regulations. The RCFF is a novel system safety process for analyzing hazards and associated causal factors due to introducing new technology into NAS. Introducing these new technologies to the NAS not only has the potential of impacting the entire system (NAS), but also leads to greater uncertainties of their safety impacts due to the very limited knowledge with no actual operational data in the NAS. Safety risk analyses, essentially events-driven and largely built upon past experience, and vast amount of actual operational data, may not provide adequate technical information for risk controls. The proposed RCFF approach is attempting to overcome some of these uncertainties by utilizing the existing regulations, which provide the minimum safety standards, as a measure to assess whether all potential risk areas are addressed while using the event-driven approach.

Conceptually, the RCFF identifies causal factors based on existing regulatory structure representing a hierarchical framework. At the very top, covering the whole NAS, Federal Aviation Regulations (FARs) provide the minimum requirement for safe operations. Within the context of FAR Subchapters, functional models provide fidelity to conceptualize the risk associated with the proposed UAS operations. Consequently, groups of causal factors are identified to

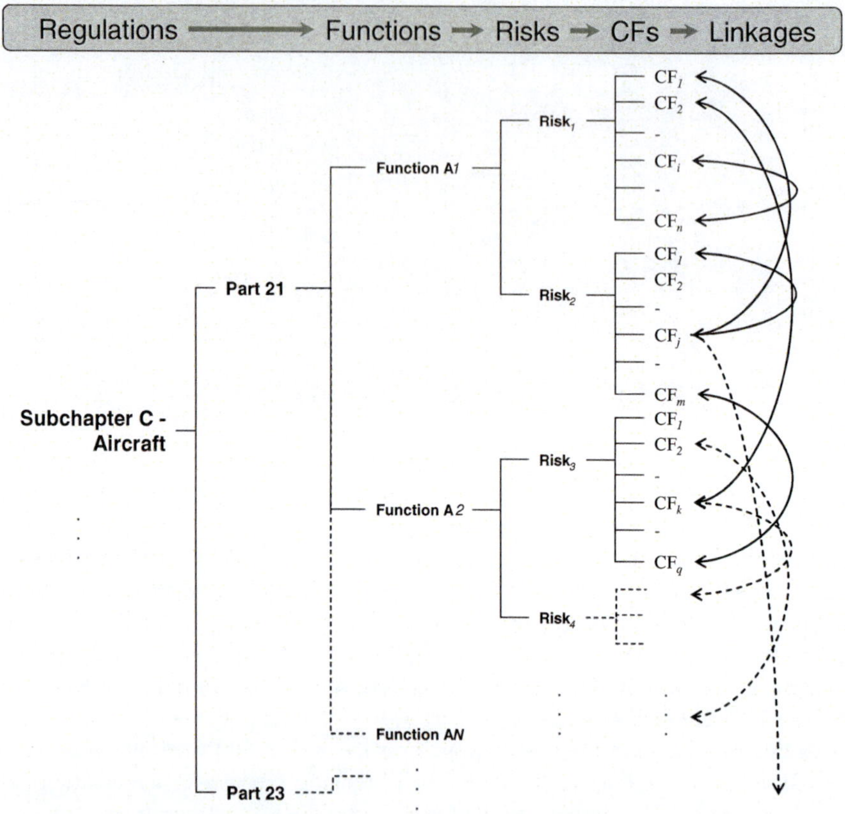

Fig. 9 A sample conceptual RCFF hierarchy for 14 CFR Subchapter C-Aircraft (*CFs* causal factors)

outline the underpinnings of each UAS related risk. However, unlike conventional hierarchical methodologies such as Fault Trees, the proposed framework, illustrated in Fig. 9, also emphasizes the interactions and connectivity among various components and compartments comprising the whole domain.

The process of building a UAS domain safety risk model using the RCFF methodology starts with one of the 14 CFR Subchapters and related FAR parts (i.e., regulations). For each FAR part, generic functions of operational activities are identified. These functions provide context to identify safety risk related to UAS operations and create a domain where the subsequent causal modeling effort becomes conceptually relevant to current UAS safety concerns. The idea of operational functions are successfully employed by [19] and [20] to develop system engineering models for 14 CFR Part 121 air carrier operational and Part 137 oversight activities, respectively.

Determining operational functions is a relatively simple process compared to identifying individual causal factors and determining their interactions. The

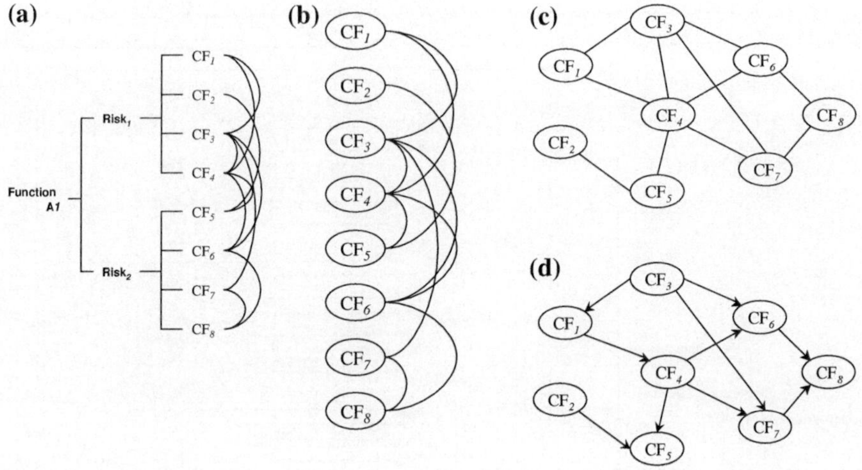

Fig. 10 Transitional steps from a functional domain of the RCFF to a directed acyclic graph. **a** A group of causal factors of tworisks and their interactions withinthe context of a functional domain. **b** A group of causal factors connected through undirected links. **c** A directed acyclic graph

latter, which is essentially a creative process, requires a fairly good understanding of regulations, hence a detailed study of FARs. Thus, heavy involvement of subject matter experts is necessary to identify causal factors and their connections.

Continuing with the development of the methodology, at this stage of the model development, we identify a set of causal factors, for each functional domain and determine possible connections among these causal factors. These connections are *undirected*, i.e., the causal factor pairs are not *ordered* since the interactions do not imply any causality at this point. Nevertheless, the resulting structure exhibits the fundamental attributes of a *graph*, where causal factors are *nodes* or *vertices* and the connections/interactions are *edges*. Figure 10 illustrates this concept.

Recalling the discussions in Sect. 2.1, if this graphical structure can be converted into a *Directed Acyclic Graph* (DAG), a Bayesian Network based on the RCFF emerges. In order to construct a DAG, we need to determine a *direction of causality* for each connected causal factor pair. Consequently, the resulting *directed graph* needs to be revised to identify and eliminate the connections causing *cycling*.

The next step in building the UAS Domain Safety Risk Model is to identify the hazards associated with each operational function governed by the particular FAR part that we focus on at that time and determine the nodes (i.e., causal factors) in the DAG with most immediate and strongest casual impact on these risks. In this sense, the hazard nodes serve as the *terminating node* or *sink* of the DAG. We use the HCAS taxonomy, which represents a systematic approach to identification and

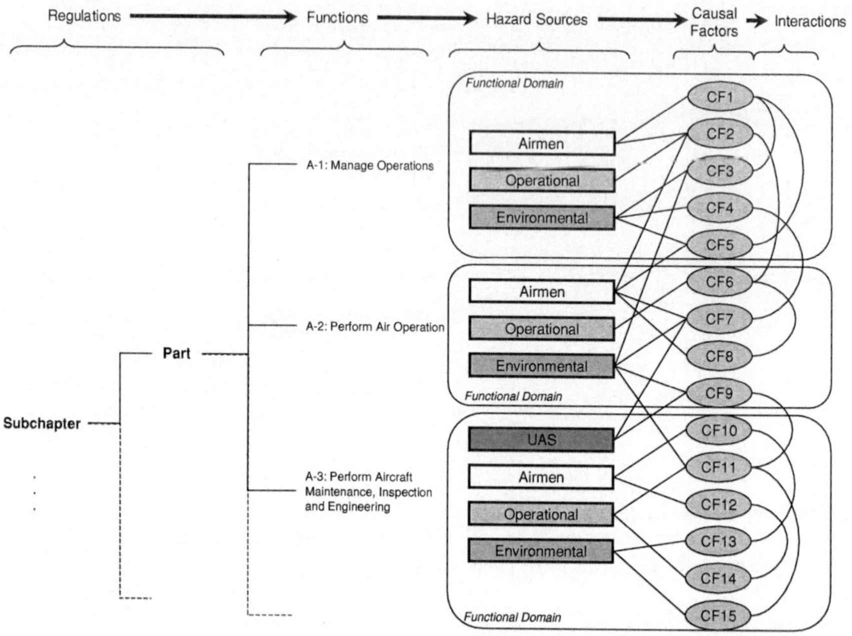

Fig. 11 A conceptual illustration of the UAS DSRM

categorization of *UAS hazard sources* to determine the hazard nodes in the UAS DSRM. A conceptual illustration of the UAS DSRM based on the RCFF methodology outlined above is provided in Fig. 11.

3.2.1 UAS DSRM Based on 14 CFR Part 91

For the purposes of this application, we apply the RCFF methodology to 14 CFR Part 91 "General Operating and Flight Rules" to develop a UAS DSRM introduced in the preceding section. 14 CFR Part 91 of Subchapter F-*Air Traffic and General Operating Rules* is organized into 12 Subparts. Broadly construed, Part 91 is a set of regulations that define the operation of small non-commercial aircraft within the US, however, many other countries defer to these rules. These rules set conditions, such as weather, under which the aircraft may operate, flight operations, equipment, maintenance and alterations, among others.

Although Part 91 covers a large spectrum, subparts B and C constitute its core by outlining fundamental requirements for flight operations, equipment and certification. Thus, our application focuses on these two subparts and individual causal factors are identified accordingly.

The modeling process starts with the identification of individual causal factors of the hazards that Part 91 controls. As an essentially creative process, subject matter experts, during focused sessions under our moderation, derive individual causal

Table 3 Causal factors derived from part 91 subpart B-flight rules

	Causal factor	HCAS element #
1	Inadequate preflight planning	3.1.1, 3.1.2, 3.1.5
2	Inadequate preflight information	3.1.1, 3.1.2, 3.1.5
3	Crewmember not at station	2.1.1, 3.1.2, 3.3
4	Occupants not secured	2.1.1
5	Occupants not informed of use of restraining systems	2.1.1
6	Proximity to other aircraft	1.1.9, 2.1.1, 4.8
7	Right of way rules not followed	2.1.1, 1.1.9,4.8
8	Failure to see and avoid	2.1.1, 4.8
9	Failure to comply with airspace speed limits	2.1.1, 3.3, 3.4
10	Failure to comply with minimum safe altitudes in congested and non-congested areas	4.1, 4.7
11	Inaccurate altimeter setting	2.1.1, 1.1.10, 1.1.6
12	Failure to comply with ATC clearances and instructions	2.1.1, 3.3
13	Failure to comply with ATC light signals	2.1.1, 3.3.3
14	Failure to follow requirements in designated airspace	2.1.1
15	Not following flight restrictions	2.1.1
16	Not complying with fuel requirements	3.1.1, 3.1.2, 3.1.4.2
17	Incomplete VFR flight plan information	2.1.1, 3.1.2
18	Not complying with VFR or special VFR weather minimums	2.1.1, 3.1.1, 3.1.2, 3.1.4.2, 4.3
21	Not complying With VFR/IFR cruising altitude requirements	2.1.1, 3.1.1, 3.1.2, 3.1.4.2
22	Not complying with minimum IFR altitude requirements	2.1.1, 3.1.6, 3.1.4, 3.1.1
23	Flying with VOR equipment that does not meet the check requirements for IFR operations	2.1.1
24	Operating IFR without an ATC clearance and flight plan in controlled Airspace	2.1.1, 3.1.4.
25	Failure to use published instrument approach procedures	2.1.1, 3.1.1, 3.1.4.2
26	Pilot conducts Cat II or Cat III operations without complying with proper training, authorization or procedure requirements (manual)	2.1.1, 3.1.4.2, 2.2.2.2
27	Failure to follow procedure for transitioning from the instrument to the visual portion of an instrument approach using normal maneuvers	2.1.1, 3.1.1, 3.1.4.2
28	Not complying with IFR takeoff procedures or minimums	2.1.1, 3.1.4.2, 4.3, 3.1.6
29	Flying in RVSM airspace without complying to RVSM requirements	2.1.1, 1.1.5, 1.1.10, 1.4.3.2, 2.2.1.1, 2.2.2.2, 3.6.2.2
30	Not complying with IFR course requirements	2.1.1
31	Failure to maintain communications with ATC while flying under IFR	2.1.1
32	Failure to comply with loss of communication procedures	2.1.1, 3.1.3, 1.1.5
33	Failure of an aircraft flying IFR to notify ATC of malfunction of certain required equipment	2.1.1, 3.1.4.2, 1.1.5, 1.1.6, 1.1.10, 3.2.1

factors based on their understanding of the regulatory text and on their expertise on the problem domain. The causal factors for subparts B and C as the result of such knowledge elicitation sessions are given in Tables 3 and 4, respectively.

Table 4 Causal factors derived from part 91 subpart C—equipment, instrument, and certificate requirements

	Causal factor	HCAS element #
1	Fuel tank installed not in accordance per FAA regulatory requirements	1.1.2, 1.1.10
2	Lack of required FAA aircraft certifications	3.2.1, 3.6.2.1
3	Inoperative or missing equipment required for the type of operation	1.1, 3.1.4, 3.2
4	Inoperative or missing emergency locator transmitter(ELT)	1.1.5, 1.1.6
5	Position or anti-collision lights inoperative or not turned on	1.1.10, 1.1.9, 4.8
6	Passengers not provided with sufficient supplemental oxygen	1.1.10, 3.1.1, 3.1.2, 3.1.3
7	Lack of accurate altitude information	1.1.5, 1.1.6, 3.2.1, 3.3.2
8	Lack of or inoperative altitude alert system	1.1.10, 1.2.2
9	Lack of approved Traffic alert and Collision Avoidance System (TCAS)	1.1.10, 1.1.9, 3.2.1, 4.8
10	Inoperative or missing Terrain Awareness and Warning System (TAWS)	1.1.10, 1.1.9, 3.2.1, 4.1, 4.7

Table 5 Possible functional activities of part 91 subpart B and C

Regulation	Functional activity
Subpart B—flight rules	A1.1 Perform flight operations
	A1.2 Perform visual flight operations
	A1.3 Perform instrument flight operations
Subpart C—equipment, instrument, and certificate requirements	A2.1 Manage equipment, instrument, and certificate requirements

Note that the set of causal factors identified in Tables 3 and 4 demonstrates a high level of resolution in terms of the detail that can be achieved using FAR regulations as the sole data source according to the RCFF methodology. For the purposes of the application presented here, we only concentrated or a functional domain representing Part 91 operations, however for a larger functional domain spanning on multiple FAR Parts, this level of detail will result in oversized Bayesian Networks that are unpractical to populate and propagate. Therefore, it is suggested that when larger domains are concerned the level of detail employed for the causal factors identification process should be kept under control to achieve safety risk models that are practically and computationally viable given the resources available.

The *HCAS element #* columns in Tables 3 and 4 indicate the set of specific HCAS taxonomy items related to each causal factor. The functional activities controlled by Subparts B and C is given in Table 5. These functions provide the contextual domain where the UAS domain safety risk model is developed.

Consequently, for each functional domain identified in Table 5 we determine a set of hazard sources that constitute the sink nodes in the Bayesian Network we are about to construct. HCAS taxonomy provides structure during this step of the

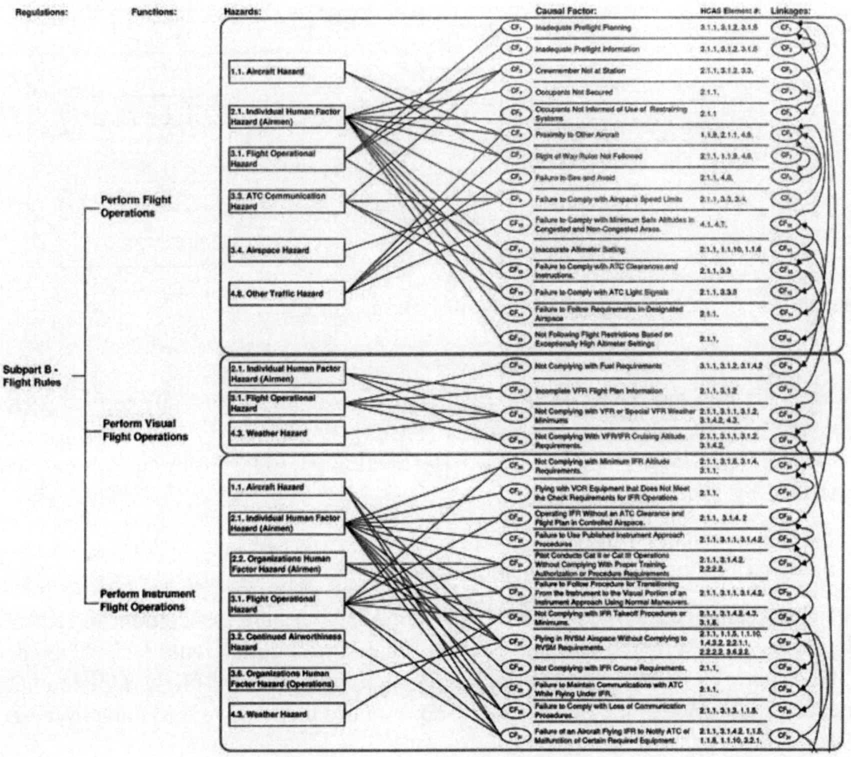

Fig. 12 Part of the initial DSRM (14 CFR Part 91 Subparts B)

process and the information in the *HCAS element #* columns of Tables 3 and 4 determine the set of hazard sources pertinent to each functional activity. Then, we connect each causal factor to the set of hazard sources identified in the HCAS element column with an undirected edge.

We repeat this process for each casual factor under each functional domain. Next, for all causal factors, we determine their interactions including those that connect the causal factors of different functional domains. Consequently, we arrive at an undirected graph structure where on one side of the structure hazard sources accept connections from causal factors and on the other side causal factors are connected among themselves.

This undirected graphical structure is called the initial domain safety risk model (DSRM). The initial DSRM based on the 14 CFR Part 91 Subparts B and C constructed according to the regulatory-based causal factor framework is provided in Figs. 12 and 13. Note that even though the DSRM for Part B and Part C are presented in separate figures, there are links or edges that connect both DSRMs, thereby creating a larger more complex domain model.

In the initial DSRM in Figs. 12 and 13, there are four distinct functional domains, 13 hazards and 37 causal factors. The causal factors, also listed in

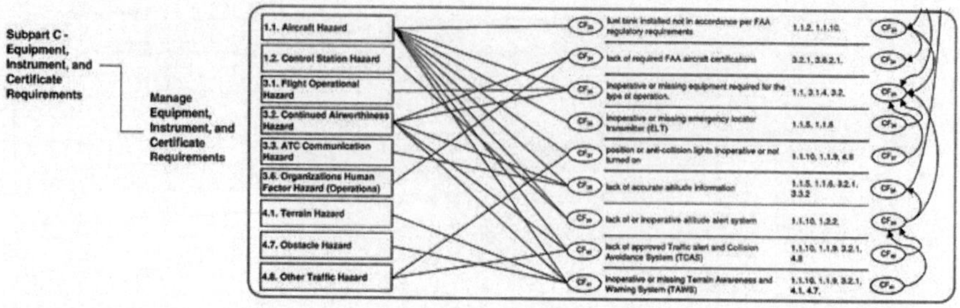

Fig. 13 Part of the initial DSRM (14 CFR Part 91 Subparts C)

Table 3 are derived based on the regulation sections that constitute the subparts with the help of subject matter experts (SMEs). For example, Causal Factor #1 *Inadequate Preflight Planning* based on Section 91.103 "Preflight Action" is described by SMEs as follows

> Pilot fails to follow all necessary steps or makes errors (e.g. calculations, decisions, etc.) during preflight planning based on the available preflight information.

For the same causal factor, SMEs also identified possible associations to HCAS elements 3.1.1 "Flight Planning", 3.1.2 "Phases of Flight", and 3.1.5 "Operational Control". Hence Causal Factor #1 is linked to 3.1 "Flight Operational Hazard" in the HCAS taxonomy, which represents a higher level grouping of these three HCAS elements.

Finally, taking the context of functional domains into consideration, we identify a set of preliminary links between causal factors based on their definitions. At this stage of the process, these links do not imply any causality. The causality of the connections is established as part of the next step where this undirected graph is transferred to a general Hybrid Bayesian Network representing the UAS DSRM.

The process of constructing an initial undirected DSRM can be summarized as follows;

- Identify functional activities, i.e., functional domains, based on the regulatory framework around which the model is to be constructed.
- Identify and describe Causal Factors using the regulatory text.
- Determine associated hazard sources, i.e., HCAS elements for each causal factor.
- Determine hazard groups based on the HCAS elements for each functional domain.
- Connect each causal factor to its hazard group by undirected edges.
- Determine possible prominent interactions between causal factors and depict these interactions by undirected edges.

Starting from the next section, we develop a Hybrid Bayesian Network representing the UAS DSRM and apply our Fuzzy-Bayesian methodology introduced

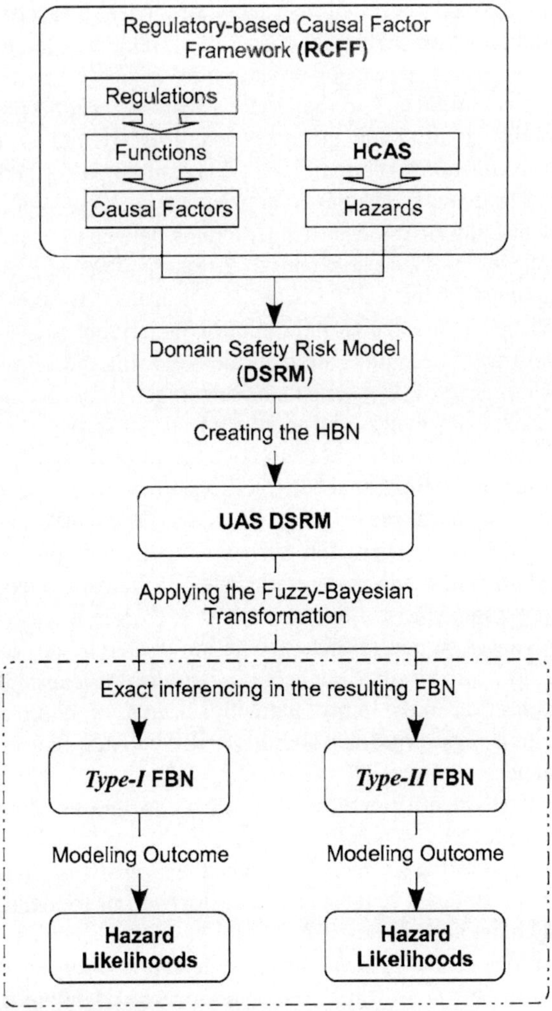

Fig. 14 The flowchart representing the progression of concepts and ideas introduced so far within the context of this application

in Sect. 2 to perform probabilistic inferencing about the resulting Fuzzy Bayesian Network (FBN).

Figure 14 provides a flowchart that outlines this rather lengthy process, which starts with the identification of causal factors and hazards according to the RCCF and ends with the computation of hazard likelihood as the modeling results of the *Type-I* and *Type-II* FBNs using the algorithms developed.

3.2.2 The Hybrid Bayesian Network

The raw model presented in two parts in Figs. 12 and 13 includes 50 nodes comprised of 37 causal factors and 13 hazard elements. Even without taking the linkages between the pairs of nodes into consideration, this preliminary topography of the model constitutes a relatively large network as a BN. However, as discussed above, in order for us to apply our Fuzzy-Bayesian methodology outlined in Sect. 2 to the UAS DSRM, we need to process this preliminary model so that it ultimately becomes a proper Hybrid Bayesian Network about which probabilistic/possibilistic reasoning can be performed.

The process of construction and refinement of the raw model starts with the identifying of the causal interactions between the pairs of causal factors and the linkages between the causal factors and hazard elements. In fact, the actual topology of the UAS DSRM is constructed by these linkages provided that the linkages indicate a casual direction underlying causal dependencies as a Bayesian Network. These links are identified by using the definitions of the causal factors as determined by the subject matter experts based on their interpretation of FAR Part 91. However, to bring structure to this process and facilitate the repeatability of the methodology, we employed a pattern matching approach for the identification of the links between two causal factors. In particular, along with a definition for each causal factor we also identified a set of keywords/phrases to complement the definition. These sets of keywords/phrases provide context to the definitions and emphasize prominent attributes of the causal factors.

We implement a two-step approach while identifying the causal dependencies between two causal factors. The initial step of this approach involves the identification of possible connections among the causal factors. At this stage these connections only imply interactions, thus the undirected edge contains no information regarding the conditional dependency between the two causal factors it connects.

If we identify one or more keywords/phrases shared by two causal factors we connect them by an undirected edge. During this process, to determine meaningful matches between causal factors, the priority is given to searching common or similar phrases. If for a causal factor we fail to identify a key phrase shared with any other causal factor, individual key words become the common patterns to look for to establish a preliminary connection between two causal factors. Practice shows that these initial set of undirected edges need to be reviewed by subject matter experts to identify and remove the connections that cannot be justified within the context of their definitions. At this stage, the SMEs also look for connections that might be overlooked by the pattern matching process.

The second step of our approach in the process of constructing the HBN representing the UAS DSRM is to identify the conditional dependencies among the causal factors. In a Bayesian Network these conditional dependencies are depicted by directed links between causal factors. Throughout this process, once again, SME knowledge can be used to determine the direction of the edges. This direction not only depicts a conditional probability distribution over two random variables but also illustrates a causal interaction between two factors contributing to various functional domain hazards. To identify such a dependency we mainly utilize the definitions of causal factors and the context that the four functional activities listed in Table 5 provide to the overall model domain.

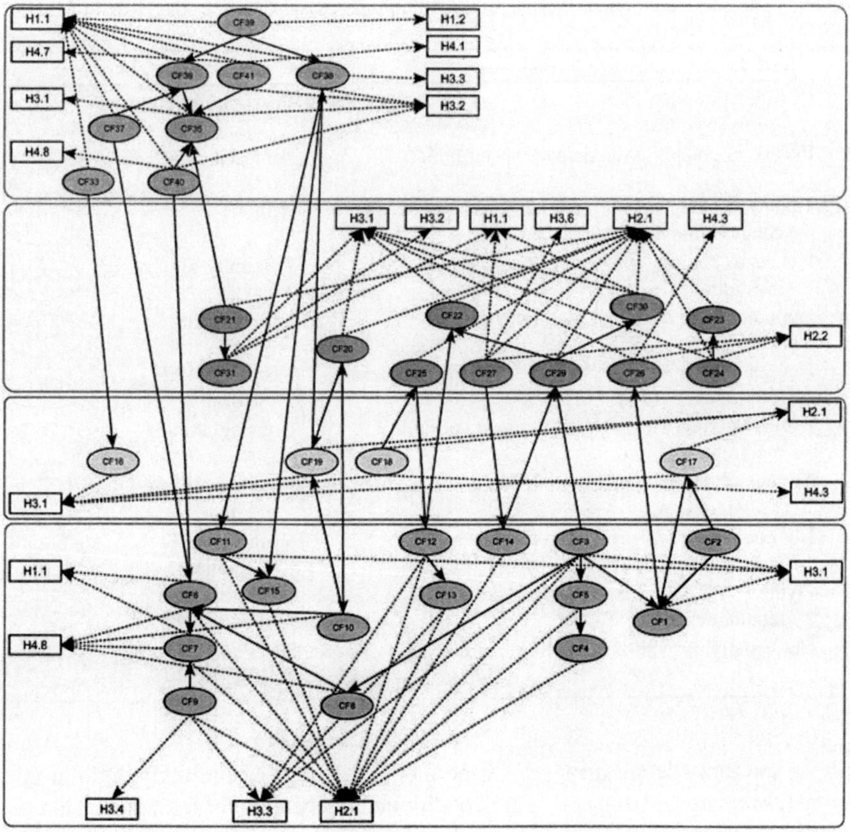

Fig. 15 A DAG of the UAS domain risk model

The Fig. 15 provides the final directed acyclic graph (DAG) depicting the UAS DSRM. While determining the direction of conditional dependencies among the variables of the model, we eliminated some linkages identified during the keyword/phrase matching process outlined above. There are two reasons to perform such a "clean up" in the model. First, the final graph needs to be acyclic. Second, whenever there is a direct connection between two casual factors any secondary conditional dependency through other causal factors only complicates the topology hence the propagation over the final HBN.

In Fig. 15, the connections leading to the Hazard nodes are shown as dashed lines whereas the edges between two causal factors are drawn as solid lines. Although these two connections are depicted differently, strictly speaking, they are the same in nature and represent a conditional dependency between two nodes/variables that they link. Also note that the causal factors and hazards are depicted by ellipse and rectangle shaped nodes, respectively. Finally, Fig. 15 divides the model into four distinct functional domains as outlined in Table 5 and the causal

Table 6 The set of continuous variables representing the causal factors in the HBN of the UAS DSRM

#	Causal Factor	Continuous Variable	
6	Proximity to other aircraft	Proximity to other aircraft	CF_6
9	Failure to comply with airspace speed limits	Airspeed	CF_9
10	Failure to comply with minimum safe altitudes in congested and non-congested areas	Altitude	CF_{10}
12	Failure to comply with ATC clearances and instructions	Proximity to obstacle	CF_{12}
16	Not complying with fuel requirements	Fuel on board	CF_{16}
18	Not complying with VFR or special VFR weather minimums	Visibility (VFR-Cruise)	CF_{18}
19	Not complying with VFR/IFR cruising altitude requirements	Cruising altitude	CF_{19}
20	Not complying with minimum IFR altitude requirements	Minimum IFR altitude	CF_{20}
26	Not complying with IFR takeoff procedures or minimums	Visibility (IFR-takeoff)	CF_{26}
27	Flying in RVSM airspace without complying to RVSM requirements	Vertical separation	CF_{27}
28	Not complying with IFR course requirements	Diversion from IFR course	CF_{28}

factors and hazards are arranged in the model topology accordingly. Within this context, some hazard nodes that are conditionally dependent of (i.e., connected to) the causal factors from multiple functional domains are repeated in these domains. However, the important observation here is that although the boundaries of the functional domains are emphasized by the illustration of the model, there are edges crossing over these boundaries and connecting the four distinct domains into one coherent model representing the system safety risk about one larger domain. This representation of the system safety risk is in line with the thinking and philosophy of the RCFF as well as the DSRM approach outlined in the preceding sections.

Model Variables

In this section we review the types of variables included in the UAS DSRM depicted in Fig. 15. In particular we focus on continuous variable in the model.

A closer look to the variables—both causal factors and hazards of the UAS DSRM reveals that the model contains variables that can be quantitatively expressed through observations. Consider the causal factor "proximity to other aircraft" (CF6), which can be observed and measured in *feet* on a continuous scale, thus it should be represented by a continuous variable to capture a better approximation within the context of a Bayesian Network model such as our UAS

DSRM. Thus, a review of the UAS DSRM is conducted to identify possible quantifiable variables underlying the descriptions/definitions of the causal factors constituting the model. The quantifiable causal factors identified through this process are then replaced by associated continuous variables in the final model. The set of continuous variables representing the causal factors that are quantifiable in the final Hybrid Bayesian Network of the UAS DSRM is listed in Table 6.

Additionally, we consider the hazards identified in the UAS DSRM as continuous variables. In fact, by defining the hazard nodes as continuous entities we are able to assess them quantitatively on a predetermined continuous scale. We believe that the quantitative depiction of individual hazards as the model outcome is a substantial improvement on the qualitative depiction of hazards that the aviation community is accustomed to.

The Final HBN

All other nodes except the causal factors identified as quantifiable in Table 6 and the hazards are considered as qualitative, i.e., discrete variables. Using the notation introduced in Sect. 2.3.3 and in Fig. 2, namely depicting the continuous variables with ellipses and the discrete variables with rectangles, the Fig. 15 illustrates the final Hybrid Bayesian Network of the UAS DSRM. On this HBN we apply in the following section the Fuzzy-Bayesian methodology and present sample results.

In Fig. 16, "CFXX" stands for "Causal Factor XX" and HX.X stands for Hazard Element X.X from within the HCAS Taxonomy. Note also that the Hybrid Bayesian Network depicted above preserves the original domain model of Figs. 12 and 13 which identify four functional models. A Bayesian Network can only be considered complete when the conditional distributions of variables imposed by the network topology are defined and the BN is fully populated. Clemens [21] describes a process that outlines a hierarchy of data sources and their usability. This process indicates an order of preference of data sources for performing risk analysis about complex engineering systems. According to this hierarchy, if the data required to perform risk analysis is provided by preexisting data for the same identical items or components of the system, this preexisting data should be used. Such a perfect case is rarely encountered in real world situations, especially in new technology applications. The next best thing is the preexisting data, however, this time for similar items or components of the systems. If neither of these scenarios is available, published data on similar systems can be used. Finally, if neither preexisting nor published data does not exist, expert knowledge provides a valuable data source to perform the required analysis.

Due to the emergent nature of the UAS operations, historical hard data that can be utilized to populate the UAS DSRM in Fig. 16 is at best limited. Considering this and the proof-of-concept nature proposed safety model, we used synthetic data to populate the developed UAS DSRM.

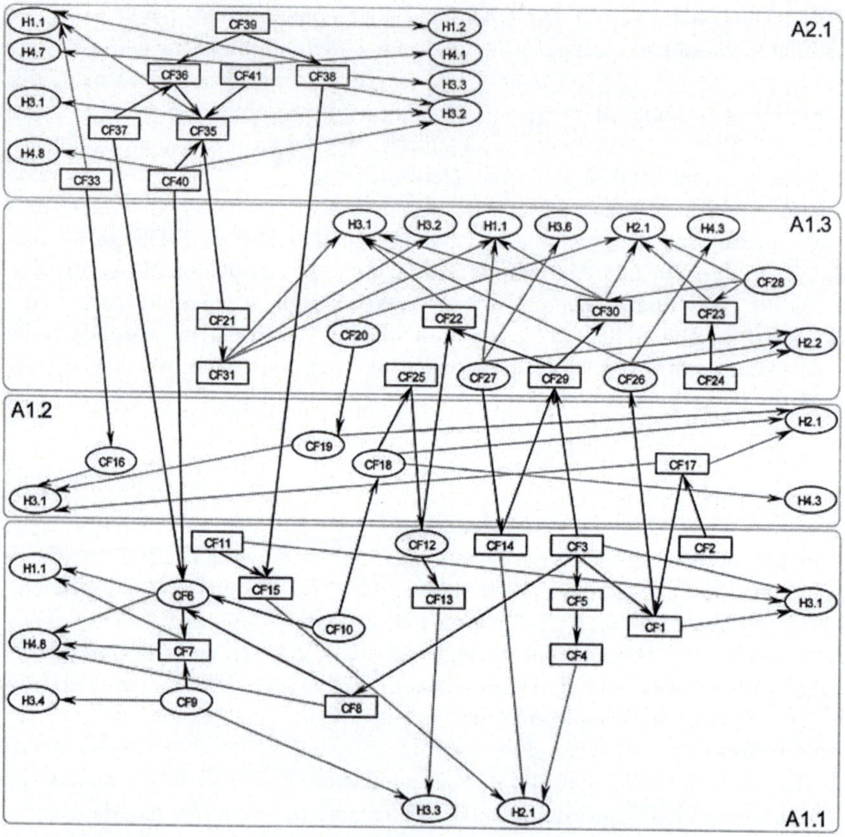

Fig. 16 The final hybrid Bayesian network representing the UAS domain risk model

For a detailed discussion the development of the final HBN representing the UAS DSRM and the classes of synthetic distributions used to populate the hybrid model, the reader may refer to Oztekin [1].

3.2.3 Application of the Fuzzy-Bayesian Methodology to the UAS DSRM

A review of the existing topology of the Hybrid Bayesian Net of Fig. 16 concludes that current popular propagation algorithms could not be used to perform exact inferencing about the UAS DSRM.

Hence, we can use the Fuzzy Bayesian-Methodology developed in this study to perform probabilistic reasoning on the UAS DSRM and calculate marginal distributions of various hazard identified as the result of the modeling process.

Within this context, using the proposed propagation algorithms summarized in Sect. 2, we perform the *Type-I* and *Type-II* transformations of the final HBN illustrated in Fig. 16 and present sample results of the marginal probability distributions of the hazard nodes, which are calculated as the outcome of the UAS DSRM.

Sample Results

Next, we present sample results of our exact inferencing algorithms for *Type-I* and *Type-II* FBN applied on the UAS DSRM for a selected hazard element. Reiterating the concept illustrated in Fig. 14, the marginal probability distributions of the 13 hazards identified in the final HBN of the UAS DSRM of Fig. 16 are the outcome of the application of our research methodology. Therefore, we concentrate solely on hazard nodes while determining the results of applying the *Type-I* and *Type-II* inferencing algorithms to the UAS DSRM. As the results of *Type-I* and *Type-II* FBN application, we present two sets of marginal distributions for the Hazard nodes:

- The marginal distributions after the HBN of the UAS DSRM is initiated with synthetic data. This set of marginal distributions determine a baseline probability distribution for the hazard variables.
- The marginal distributions of hazards when evidence is introduced to the model. These marginals could then be compared to the baseline distributions of the same hazards to reveal the impact of the evidence as an increase or decrease in the hazard likelihoods.

In the latter case, a set of observations regarding some selected continuous and discrete variables constitutes evidence, thereby outlining a scenario about the problem domain modeled by the Bayesian Network. Therefore, for the purposes of this application, a scenario is simply a collection of variables with known values. Within the same context, the scenario depicted in Table 7 is used as evidence to generate associated marginals.

The results of the *Type-I* and *Type-II* inferencing on UAS DSRM constitute 26 plots of marginal probability distributions for 13 hazard elements. These results are provided in [1] and we are going to present them here. However, to elaborate on the information that these plots provide, the results for the hazard element $H_{1.1}$ "*Aircraft Design Related Hazards*" are given in Figs. 17 and 18.

Table 7 The set of the causal factors and their values used as the synthetic scenario

CF_i	Definition	Value
CF_1	Inadequate preflight training	True
CF_9	Airspeed	150 kts
CF_{10}	Altitude	3,000 ft
CF_{14}	Failure to follow requirements in designated airspace	True
CF_{16}	Remaining fuel on board	40
CF_{17}	Incomplete VFR flight plan information	True
CF_{30}	Failure to comply with loss of communication procedures	True

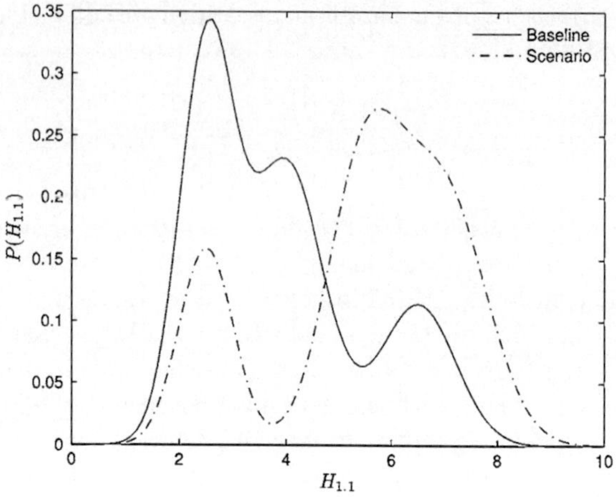

Fig. 17 Marginal probability density functions of hazard H1.1 as the result of the Type-I inferencing

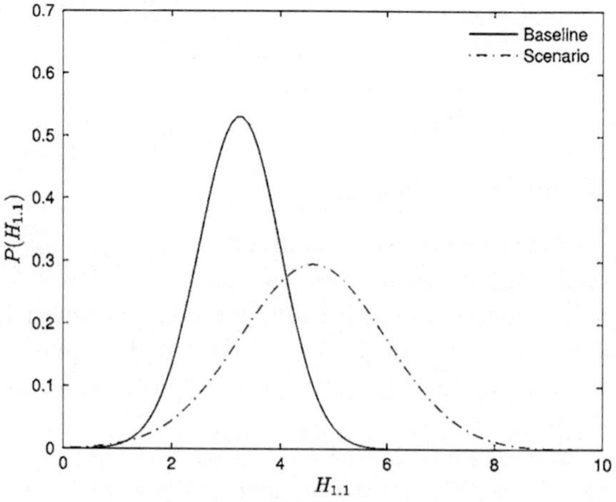

Fig. 18 Marginal probability density functions of hazard H1.1 as the result of the Type-II inferencing

We present, in Fig. 17, the probability distribution of Hazard $H_{1.1}$ as the result of the *Type-I* FBN transformation and inferencing. In the figure, the dotted line represents the baseline marginal distribution for Hazard $H_{1.1}$ "*Aircraft Design Related Hazard*", whereas the solid line represents the marginal probability distribution of the hazard after the evidence associated with the scenario outlined in Table 7 is introduced to the model.

Juxtaposing the baseline and scenario plots of the marginal distributions make it possible to visualize the relative change in the likelihood of individual hazards. For example, in Fig. 17, one can observe a shift in the probability density towards a higher hazard value, which is depicted on a continuous scale [0,10], after the evidence is introduced to the network as compared to the baseline density for the same hazard. Thus, we can deduce that the scenario outlined in Table 7 has a negative impact on likelihood of occurrence of the individual hazard element $H_{1.1}$ *Aircraft Design Related Hazards*.

Furthermore, one can also apply fuzzy transformation on these continuous probability density functions to determine the membership values associated with the states of Fuzzy-discrete counterpart hazards. For example, using sample membership functions for *low*, *medium*, and *high* Fuzzy states defining a Fuzzy hazard, we can determine the Fuzzy hazard $\hat{H}_{1.1}$ for the baseline and scenario cases as follows:

$$\hat{H}_{1.1\,\text{Baseline}} = \{\text{Low} = 0.51, \text{Moderate} = 0.32, \text{High} = 0.17\}$$
$$\hat{H}_{1.1\,\text{Scenario}} = \{\text{Low} = 0.23, \text{Moderate} = 0.34, \text{High} = 0.43\}$$

These Fuzzy hazard values also verify the shift towards a higher value as the result of the scenario introduced to the model.

The results for the *Type-II* inferencing on the UAS DSRM entails two marginal densities presented in a similar fashion: a marginal density for the hazard associated with the Baseline *Type-II* model and a marginal density determined after evidence is introduced to the network. The scenario outlined in Table 7 is used as the evidence.

3.3 A Discussion on Validation

Broadly speaking, any attempt to model a complex system or a real phenomenon results in an approximation of reality with varying degrees of veracity. Thus, a study to compare and contrast the results of the model and the reality is generally considered a important aspect of the modeling methodology. In essence, the RCFF and the UAS DSRM are decision support tools and, as a general practice, decision support tools are validated by comparing the results of test runs against preexisting data or expert judgment. Decision support systems can be validated against known results as well as against expert knowledge [22].

However, due to the emergent nature of UAS, as far as the information on UAS related causal factors and hazards are concerned, data on UAS operations is practically nonexistent. Coupled with our usage of synthetic data to populate the UAS DSRM, this lack of historical data renders the applicability of quantitative validation techniques impossible. Therefore, unless populated by real data, including both actuarial data and expert judgments, any further discussion on validation is premature. However, we can present a concise argument on the repeatability of the regulatory-based modeling results presented as the UAS DSRM. A closer look to the backgrounds of the SMEs whose knowledge has been primarily utilized to construct the UAS domain safety risk model indicates a rather wide coverage in terms of expertise in and understanding of the problem domain. We believe this diversity of experience resulted in a UAS DSRM model with sufficient representative power so that the modeling results should be considered repeatable provided that a similarly diverse group of SMEs is tasked with the developed modeling framework.

From our experience with much smaller discrete only BN models of aviation accidents [22–24], we can foresee that populating the UAS DSRM with real data will require numerous knowledge elicitation sessions as well as an extensive effort to collect field data on UAS accidents/incidents. Such a study, which requires time, resources, and most importantly access to UAS operations, is considered as a possible avenue of future work to improve upon this study. Additional background material on the analytical methods used to support the development of the UAS DSRM are provided in [25–29].

4 Conclusions

It is our hope that this study presents a convincing formal argument on the usefulness of Fuzzy-Bayesian Networks in understanding and modeling complex uncertainty associated with real-world applications, such as the emergent UAS operations.

In this study, we concentrate on the problem of inferencing in general Hybrid Bayesian Networks. In particular, we try to understand and tackle the issues that exact inferencing in general HBNs faces. Our contributions to the larger research domain of representation and inferencing in general HBNs are three-fold: theoretical, algorithmic and practical. Specifically, our major contributions can be outlined as follows:

- From a theoretical point of view, we complement classical probability theory with Fuzzy set theory to develop a hybrid formalism to understand and model complex uncertainty associated with real-world systems. To that end, we provide a novel framework to implement a hybrid Fuzzy-Bayesian methodology to perform exact inferencing in general HBNs where continuous and discrete variables may appear anywhere within the network topology.
- From an algorithmic perspective, we provide a suite of inferencing algorithms for general Hybrid Bayesian Networks. In particular, we introduce two transformations for general HBNs to create *Type-I* and *Type-II* Fuzzy-Bayesian Networks and present formal representation techniques and separate inferencing mechanisms for *Type-I* and *Type-II* FBNs.
- Finally, from a practical perspective, we apply our framework, methodology, and techniques to the task of assessing system safety risk due to the introduction of emergent UASs into the NAS.

Acknowledgments This research is supported by Federal Aviation Administration grant number 08-G-002. The contents of this paper reflect the views of the authors who are solely responsible for the accuracy of the facts, analyses, conclusions, and recommendations represented herein, and do not necessarily reflect the official view or policy of the Federal Aviation Administration.

References

1. Oztekin A (2009) A generalized hybrid Fuzzy-Bayesian methodology for modeling complex uncertainty. Dissertation, Department of Industrial and Systems Engineering, Rutgers University, Piscataway
2. Pearl J (1988) Probabilistic reasoning in intelligent systems: networks of plausible inference. Morgan Kaufmann, San Francisco
3. Lauritzen SL, Spiegelhalter DJ (1988) Local computations with probabilities on graphical structures and their applications to expert systems. J Roy Stat Soc B 50(2):157–224
4. Shenoy PP, Shafer GR (1990) Axioms for probability and belief-function propagation. In: Proceedings of the 6th annual conference on uncertainty in AI (UAI), pp 169–198
5. Huang C, Darwiche A (1996) Inference in belief networks: a procedural guide. Int J Approx Reason 15:225–263
6. Lauritzen SL (1992) Propagation of probabilities, means, and variances in mixed graphical association models. JASA 87(420):1089–1108
7. Lauritzen SL, Jensen F (2001) Stable local computation with conditional Gaussian distributions. Stat Comput 11:113–203
8. Lerner UN (2002) Hybrid Bayesian networks for reasoning about complex systems. Stanford University, Dissertation
9. Zadeh L (1965) Fuzzy sets. Inf Control 8:338–353
10. Klir GL (1989) Is there more to uncertainty than some probability theorist might have us believe? Int J Gen Syst 15:347–378
11. Zadeh LA (1995) Discussion: probability theory and fuzzy logic are complementary rather than competitive. Technometrics 37(3), August
12. Di Tomaso E, Baldwin JF (2008) An approach to hybrid probabilistic models. Int J Approx Reason 47:202–218
13. Korb KB, Nicholson AE (2004) Bayesian artificial intelligence. Chapmen & Hall/CRC, Boca Raton
14. Baldwin JF, Di Tomaso E (2003) Inference and learning in fuzzy-Bayesian network, fuzzy system, FUZZ'03. The IEEE international conference on fuzzy systems, vol 1, pp 630–635, 25–28
15. Ross TJ (1995) Fuzzy logic with engineering applications. McGraw-Hill, New York
16. Leveson NG (1995) Safeware: system safety and computers. Addison-Wesley, Reading
17. Luxhoj JT (2008) Safety risk analysis of unmanned aircraft systems (UAS) integration into the national airspace system (NAS): phase 1 final report. Department of transportation. Federal aviation administration, January 31
18. Luxhoj JT (2009) Safety risk analysis of unmanned aircraft systems (UAS) integration into the national airspace system (NAS): phase 2 final report, Department of transportation. Federal aviation administration, December
19. FAA (2001) Air carrier operations system model, Final Report. DOT/FAA/AR-00/45, Office of Aviation Research, Washington, D.C., March
20. FAA (2007) 14 CFR Part 137 Oversight Model, Final Report. DOT/FAA/AR-06/51, Office of Aviation Research, Washington, D.C
21. Clemens PL (2002) Making component failure probability estimates. www.sverdrup.com/safety/failprob.pdf, December
22. Oztekin A (2005) A case-based reasoning (CBR) approach for accident scenario knowledge management, M.S. Thesis, Department of Industrial and Systems Engineering, Rutgers University, Piscataway
23. Luxhoj JT (2005) Aviation safety in practice: applying principles and tools to measure risk reduction, Safety across high-consequence industries, Saint Louis University, St. Louis, Sep 20–22
24. Luxhoj JT (2005) Model-based reasoning for aviation safety risk assessments, SAE world aerospace congress, Dallas/Fort Worth, Oct 3–6

25. Oztekin A, Luxhoj JT (2009) A regulatory-based approach to safety analysis of unmanned aircraft systems. HCI international conference proceedings (lecture notes in computer science (LNCS) series). Springer, San Diego, pp 19–24, July
26. Oztekin A, Luxhoj JT, Allocco M (2007) A general framework for risk-based system safety analysis of the introduction of emergent aeronautical operations into the national airspace system. Proceedings of the 25th international system safety conference, Baltimore, August 13–17
27. Oztekin A, Luxhoj JT (2008) Hazard, safety risk, and uncertainty modeling of the integration of unmanned aircraft systems into the national airspace. 26th congress of international council of the aeronautical sciences. Anchorage, Sep 14–19
28. Anoll R (2006) Safety checklist, federal aviation administration, unmanned aircraft system (UAS) program office. Washington, DC, December
29. Oztekin A, Luxhoj JT (2009) A hybrid framework for modeling complex risk and uncertainty, IERC. Institute of industrial engineering, Miami, May 30–June 3

Sensor Management Problems of Nuclear Detection

Tamra Carpenter, Jerry Cheng, Fred Roberts and Minge Xie

1 Introduction

Terrorist nuclear attack is a potentially devastating threat to homeland security. It is increasingly important to have the capability to intercept illicit nuclear materials entering the country and to monitor for nuclear threats emerging from within. The effective use of sensors for nuclear and radiological detection requires choosing the right type of sensor, putting it in the right place and activating it at the right time. It also involves interpreting the results of sensor alarms and making decisions that balance various types of risk and uncertainty based on those results. This article describes a variety of approaches to sensor management for nuclear detection that revolve around formulating the related problems using precise mathematical language and then developing tools of the mathematical sciences to solve them. It emphasizes a variety of approaches to sensor management in a multi-institution project on nuclear detection, which is based at Rutgers University and includes Princeton University and Texas State University-San Marcos.

The nuclear and radiological materials whose detection is of particular concern are radiation dispersion devices (RDDs)—more commonly known as dirty bombs—and special nuclear materials (particularly highly enriched uranium and

T. Carpenter (✉) · J. Cheng · F. Roberts
DIMACS, Rutgers University, Piscataway, NJ 08854, USA
e-mail: tcar@dimacs.rutgers.edu

M. Xie
Department of Statistics and Biostatistics Rutgers,
The State University of New Jersey, 501, Hill Center,
Busch Campus, 110 Frelinghuysen Rd, Piscataway, NJ 08854, USA
e-mail: mxie@stat.rutgers.edu

H. Pham (ed.), *Safety and Risk Modeling and Its Applications*,
Springer Series in Reliability Engineering, DOI: 10.1007/978-0-85729-470-8_11,

weapons-grade plutonium) that could provide the fissile material for a nuclear weapon. Throughout this chapter, we will use the generic term "nuclear detection" to include detection of any radiation-emitting material of concern. RDDs could potentially contain a number of different radionuclides, some of which are used commercially, so the RDD threat is less specific than special nuclear materials and harder to differentiate from benign sources of radiation such as those from medical procedures or naturally occurring radioactive materials like the clay found in pottery and kitty litter. The need to distinguish true threats from commonly occurring benign sources and background sources of radiation is a particular challenge in nuclear detection [24].

Nuclear detection arises in a variety of different contexts that pose overlapping, but sometimes quite different, research challenges. This chapter aims to provide an overview of several problems arising in the following settings for nuclear and radiological detection:

Border crossings: At borders, vehicles move through radiation portal monitors (RPMs) that provide passive, non-intrusive screening for the presence of nuclear and radiological materials. In this setting, we have brief contact with all entering vehicles and can detain them for further inspection when alarms occur. Here, the emphasis is on preventing nuclear materials from entering the country, so detection is the main priority and false alarms are tolerated as a natural consequence [24]. Typically, inspection at borders is a layered process and all alarming vehicles are subject to further scrutiny in subsequent layers which can include passing the vehicle through another RPM, screening with handheld radioactive isotope identification devices (RIIDs), or manual inspection of the vehicle. Each progressive level of scrutiny introduces additional delay and inspection cost. Since all RPM alarms are followed with further inspection, new methods can potentially reduce false alarm rates through enhanced analysis of sensor data at each layer and by optimizing the choice of which inspection to perform next.

Ports of entry: At seaports and other ports of entry, we have huge numbers of shipping containers that must be screened, and this must be done in a way that mitigates risk without introducing excessive delays and the ensuing disruptions to commerce. As at borders, we have a layered inspection process that includes passive radiation monitoring and manual inspection of containers, but we also have more stringent testing capabilities using gamma radiography, as well as pre-port information on arriving ships and the cargo they contain. As containers arrive, we must decide which containers to inspect more carefully, and we need to do this without causing excessive disruption to port operations. For a particular detection technology, we have to identify the best method for assessing the risk of a container and perhaps the sequence of screening tests to apply.

Special events: At special events such as a major concert in a city park, a political rally, or a large festival or parade, there may be no existing infrastructure and possibly no restricted points of entry. In such cases, there is a need to locate a system of sensors to provide maximal protection. Here, we can consider the development of methods to determine where to locate sensors so as to optimize detection, as well as routing strategies to efficiently patrol an area or venue.

Urban settings: Major metropolitan areas present attractive targets, but cover large geographic areas that may be difficult to monitor and/or patrol. They present many of the same challenges as special events, but on a grander scale—both in terms of geographic area and duration. Here we discuss methods to use either stationary or moving detectors (e.g., in vehicles) with fixed or random routes. At special events and in urban settings, radiological sources may be contained within moving vehicles or carried by people, introducing the added challenges of detecting devices and materials in transit, typically without the ability to choose whom to screen. If an event is detected, there will be a delay while response measures are enacted and this introduces the problem of identifying and tracking the source vehicle or person.

This chapter describes a variety of approaches to sensor management in a multi-institution project on nuclear detection, which is based at Rutgers University and supported by the US Department of Homeland Security Domestic Nuclear Detection Office in collaboration with the US National Science Foundation. The chapter provides an overview and summary of the project. In so doing, it touches on four themes that have emerged as the areas of greatest emphasis in the project: (1) methods to exploit data from radiation sensors and shipping manifests for classification and decision making; (2) ways to optimize sequential decisions in layered inspection processes; (3) detection using a fleet of mobile radiation sensors; and (4) data sampling strategies for nuclear detection.

2 Analysis of Manifest and Sensor Data

Detection of nuclear materials entering the US currently relies on two important sources of data. One is the radiation sensors that are deployed at all major border crossings and ports of entry to scan containers entering the US, and the other is documents submitted to US Customs and Border Protection (CBP) prior to a shipping container entering the US. Challenges in our project have included obtaining and understanding the information that these sources provide; identifying how we might make greater use of these data throughout the inspection process; and building models to support enhanced decision making based on these data.

2.1 Manifest Data

Customs information is collected at overseas points of embarkation using a variety of custom forms, including a ship's manifest and bill of lading. Recently, there has been increasing emphasis by US CBP on improving the quality of customs data resulting in new, more stringent requirements that accurate manifest information be submitted at least 24 hours before cargo is loaded onto a US-bound vessel. Prior to arrival at US ports, CBP does screening based on such data to determine whether the shipment poses a risk. Identifying mislabeled or anomalous shipments through scrutiny of manifest data is one step in a multi-layer inspection process for containers arriving at ports [58]. Enhancing capabilities to extract information from such data may prove useful in screening for radioactive and nuclear materials.

Early in 2008, we obtained manifest data that provides information on cargo entering US ports over several days. The raw data include over 30,000 records, which were parsed to create a database for use by various teams on our project. Each manifest contains 120 attributes that include information about the shipper, consignee, notify party, and the shipment itself. Shipment data include size, weight, export codes (for hazardous materials, products covered by tariffs, etc.), and a physical description of the cargo manually entered by an inspector. Some of the manifest information is contained in free-form text while the rest is categorical or numeric. It is our observation that there is considerable leeway in the level of information provided, as well as little structure or commonality in the text fields. These issues present challenges and introduce uncertainty that must be dealt with when using this data for screening.

As noted in Wein et al. [58], the Automated Targeting System (ATS) of US CBP is already using manifest information to classify containers as being either "trusted" or "untrusted", but it may have difficulty in correctly classifying a container transporting nuclear material via a trusted shipper. The work of McLay et al. [41] suggests that effective prescreening, such as that based on manifest data, can be an important component of cargo screening when there are limited screening resources.

The Canada Border Services Agency (CBSA) is also using information available in shipping documents to classify containers according to risk [30] and to improve their inspection process [29]. Like the US CBP, CBSA has an automated system that assigns risk scores to indicate the likelihood that a container entering the country has undesirable contents. In recent work, Hoshino et al. [30] have looked for ways to improve the existing system by explicitly taking into account the facts that (1) the problem is inherently unbalanced with only a very small fraction (roughly 2%) of containers being deemed dangerous; and (2) the likelihood of finding a dangerous container seems to vary with time. To deal with these issues they propose a two-stage approach that first fits a baseline classifier without considering time and then applies a time adjustment factor that results in substantially improved performance. In other work [29], Hoshino and CBSA colleagues have developed user-friendly classification methods to predict the presence of fumigants to reduce the time and expense of chemical testing of every marine container referred for further inspection.

Our project includes several studies that leverage classification algorithms and apply them in the context of shipping manifest data. We describe some relevant activities in the remainder of this subsection.

Bayesian Binary Regression: Our project is leveraging anomaly detection methods developed for the intelligence community to look for anomalies and general trends in manifest data. We have begun applying Bayesian LASSO logistic regression using the Bayesian binary regression (BBR) software developed at DIMACS [25, 40] to help analyze manifest data. In particular, we used the following logistic regression model to analyze and discover the potential associations between the risk status (Y) and the origination/destination and contents of the cargo, as well as the history of the shipping company, etc. (covariates X's):

$$\log\left(\frac{P(Y_i = 1)}{1 - P(Y_i = 1)}\right) = \sum_{j=1}^{m} \alpha_j X_{i,j}.$$

In this model, $Y_i = 1$ if the ith container is selected for further inspection and $Y_i = 0$ if it is not. The $X_{i,j}$'s represent the values of covariates, as well as possible interactions among covariates, associated with the ith container. Bayesian logistic regression finds the maximum a posteriori (MAP) estimate of the parameter vector $\alpha = (\alpha_1, \ldots, \alpha_m)^t$ under a Laplace prior distribution. This approach can effectively deal with the large number of covariates/fields extracted from manifest data, as well as the sparsity of the data. From the manifest data, we identify the covariates and interactions among them with statistical significance. This information can help us build a predictive model to assign risk scores to incoming containers.

The response variable of risk status is not available in the manifest data. Nevertheless, we performed a variety of simulation studies based on the manifest data to determine the effectiveness of Bayesian LASSO regression and the BBR software in such an application. Specifically, we selected a small set of covariates from the manifest data (such as port of origin, cargo contents, etc.) and hypothesized a regression relationship between the risk status and the selected set of covariates. We assumed this relationship to be the "true" model and based on this assumed relationship, we simulated the risk status (Y) for each container. Now, pretending that we did not know the assumed relationship, we applied the BBR algorithm and LASSO regression using the simulated risk status and all covariates associated with the containers from the manifest data. Most of the time, the BBR and LASSO regression could identify the selected set of covariates as significant contributors to container risk status, and the predicted risk scores were consistent with simulated risk score from the assumed "true" model [8]. This suggests that the proposed logistic model/BBR approach could indeed provide an effective tool for processing information in the manifest data.

Higher-order Naïve Bayes: In another area of research, project member Bill Pottenger and his students are applying a higher-order na Bayes (HONB) algorithm [21] to classify shipments in the manifest data using the hazardous material export code as the class. In theoretical work, they are developing a novel approach to learning that exploits relationships between attribute/feature values across different shipment manifests. In empirical work, they are selecting nominal classes within the manifest data that result in the most useful models.

The underlying assumption in many traditional machine-learning algorithms is that the instances are independent and identically distributed (i.i.d.). Such models are called "first-order" because in general they only leverage relationships between attributes within instances (e.g., co-occurrence relationships), and do not leverage connections that link attributes from different instances. These critical independence assumptions that are made in traditional machine-learning algorithms prevent them from going beyond instance boundaries to exploit latent "higher-order" relations between instances. Work in our project moves beyond instance boundaries to exploit the latent information captured in higher-order co-occurrence paths between instances within a dataset. The algorithms being developed leverage implicit co-occurrence relationships between attributes in different instances or manifests. Pottenger and his team believe that algorithms leveraging higher-order associations between different attributes of each shipment will allow for more precise identification of anomalous shipment data, especially when such algorithms are part of an online learning environment. The related work in the project assumes that descriptions of products such as "IKEA home furnishings" will be more likely to match with certain container types or ports of departure than will other products; thus, anomalies may be discoverable in manifests that do not observe similar associations. Higher-order na Bayes is especially useful considering the online nature of the manifest data, which implies sparsity during initial model learning.

Results obtained on benchmark corpora [23] show that higher-order na Bayes generates more accurate models on sparse data than first-order na Bayes classifiers, especially with small training sets. Extensive experiments on several data sets from different domains support this conclusion. However, it remains to be determined work whether the classification models point to anomalous shipments that would be identified as "high risk cargo" by inspection domain experts.

2.2 *Radiation Sensor Data*

Detecting nuclear materials at borders and seaports relies on data from the radiation portal monitors that are deployed at all major ports to scan vehicles and containers entering the country [24]. At present, there are roughly 1,100 radiation portal monitors installed and they inspect approximately 90% of the containers and vehicles entering the country [55]. Much of the nation's commercial life depends on the contents of these containers that are carried on the roughly 57,000 trucks, 2,500 aircraft, and 580 sea vessels entering the US each day. Given this volume of cargo, there are two competing priorities in inspection: (1) to process cargo quickly so as not to cause congestion and resulting disruptions to commerce, and

(2) to prevent entry of any illicit nuclear or radiological material. To meet these dual objectives, CBP has adopted a multi-layer approach for inspection, which consists of a "routine" inspection followed by a more stringent inspection of containers that were identified as suspicious during the routine inspection. Wein et al. [58] describe the current layered approach and consider how to optimize an 11-layer security system that includes shipper certification, container seals, the ATS system, passive and active radiation testing, and manual inspection to improve detection. Our project is developing approaches for making decisions during routine screening.

Statistical Learning: As trucks at border crossings move though portal radiation sensors, the portal captures the energy spectrum every tenth of a second across a range of channels going from low frequency to high frequency. It can be 256 channels or coarser bands consisting of frequency counts in only 5 non-overlapping, exhaustive bands corresponding to channels 0–5, 6–10, 11–40, 21–80 and 81–256, respectively [33]. Project member Siddhartha Dalal formulated a Bayesian learning approach for modeling the energy emitted by an unknown source and classifying it as belonging to one of K defined classes [12]. These classes would include radioactive materials of high concern, such as high energy Uranium, depleted Uranium, Plutonium, Cobalt-57, and Barium-133, as well as benign materials that may be sources of emission such as medical waste or kitty litter.

Denote by $Z = (R_1,\ldots,R_5)$ the observed radiation counts in the five non-overlapping channels from the training set of the portal radiation sensors data. Dalal and Han [12] assumed that the total count $N = R_1 + R_2 + \cdots + R_5$ given a class $C = c$, $c = 1, 2,\ldots,K$, follows a Poisson distribution. Furthermore, they assumed that, given total count $N = n$ in class $C = c$, the observed radiation counts $Z = (R_1,\ldots, R_5)$ follows a multinomial distribution. Based on these Poisson process and multinomial models—both of which are conventional assumptions for this type of count data—they were able to derive the following classification rule from Bayes formula:

$$P\{C = c|Z = (r_1,\ldots, r_5)\} \propto P\{N = n|C = c\}$$
$$P\{R_1 = r_1,\ldots, R_5 = r_5)\,|N = n, C = c\}P\{C = c\}.$$

Here, $(r_1,\ldots,r_5)$ are observed radiation counts in class c in the training data, $n = r_1 + r_2 + \cdots + r_5$ and $P\{C = c\}$ can be estimated by the fraction of containers of class c in the training set.

This classification model can be used to develop a scoring model that assigns a risk score to each new container. A potential risk scoring model [7] could be

$$s^* = \sum a_c P\{C = c|Z^* = (r_1^*,\ldots,r_5^*)\},$$

where: $(r_1^*,\ldots, r_5^*)$ is the radiation sensor reading of a new container; a_c is the average risk score of the class c computed from the training data and classification model; and the sum $\sum$ is over all $c = 1,\ldots, K$ classes. We can use this predictive model to assign a risk score to each incoming container. This will give us a likelihood that the new container has nuclear or illicit material so further investigation can be conducted accordingly.

Machine Learning: In a second study using radiation sensor data, project members Bill Pottenger, Jason Perry, Christopher Janneck and Christie Nelson applied machine learning techniques to analyze gamma-ray spectra generated by CZT-based handheld detectors to see whether they could distinguish non-threat sources of radiation from possible threat materials. One way to cast this problem as a machine learning problem is to train a set of classifiers to identify the presence of any gamma-emitting radioisotopes from a predetermined library, as above. However, another approach may be necessary to build a robust real-world solution for distinguishing threat-from non-threat isotopes. For instance, in security applications, very specific types of accuracy may yield practical advantages for reducing false alarms. One example would be a mechanism to provide a very high confidence level in detecting known non-threat isotopes such as Technetium-99m (Tc99m)—the most common medical isotope and one which would generally not present a threat. A system to accurately discern Tc99m as the radiation source could safely indicate when no further inquiry is necessary. At the same time, the system must be sensitive to a wide variety of other known and unknown radioisotopes, so that no potential sources of threat are missed. These, in turn, must be distinguished from fluctuations in the natural radioactive background, in order to minimize false alarms. This requires both an optimal framing of the machine learning problem and a very finely tuned classification system which takes advantage of all available data.

In initial investigations using data obtained from a CZT-based hand held detector, Perry [46] formulated a three-class classification problem with classes corresponding to: (1) presence of Tc99m; (2) presence of other known and unknown isotopes; and (3) all natural background radiation conditions. Using Support Vector Machines (SVMs) for classification, his experiments showed that the single-isotope Tc99m class is distinguishable from the other classes of isotopes with near-perfect accuracy. However, separating the class with all other isotopes from the normal background noise and detector anomalies is much more difficult and requires further study.

Statistical Change Detection and Identification: Project members Savas Dayanik, Warren Powell and Kazutoshi Yamazaki have developed new online statistical change detection and identification rules to identify pattern changes in sensor readings that indicate the presence of either hazardous materials or a malfunctioning of the sensor [15, 47]. We envision these procedures operating in real time as vehicles or containers are scanned through portals or other sensing devices. The algorithms are designed to make diagnoses within a prescribed low level of false alarms and to work with sensing devices that have only a small amount of associated computational capability. In general, the method models a set of potential "disruptions" that include a variety of sensor failure modes and detection events corresponding to detection of different materials. The methods analyze sensor readings and both determine when to sound an alarm and identify the suspected alarm trigger (i.e., the failure mode or material detected).

A typical change detection and identification rule consists of a pair of an alarm delay (the difference between the time that some disruption first occurs and the time that the algorithm declares an alarm) and a diagnosis rule. Because observations are collected sequentially over time, optimal rules are typically the solutions of a dynamic program in a Bayesian framework [13, 14]. Unfortunately, optimal solutions are often not in closed form, and due to the curse of dimensionality of state space, their numerical implementations require large computing power and memory. Therefore, a classical dynamic programming approach to the change detection and identification problem does not easily lead to online optimal rules that can be run on devices with limited associated computational capability. However, it is possible to develop simple nearly optimal decision rules by combining dynamic programming and renewal theory for stochastic processes, which is the methodology adopted by Dayanik et al. [15].

Their methods [15, 60] seek to find an alarm time and an identification rule such that the decision risk associated with any potential "disruption" is below a specified threshold (which is an adjustable model parameter) and the detection delay is minimized. The specific types of risk considered include the risk associated with a real nuclear event whose detection is missed or delayed, risks from investigating false alarms, and misdiagnoses of detected real disruptions. Precise solution of this problem would require solving a constrained stochastic optimization problem to find a rule that would concurrently minimize for each of the potential disruptions, and it is unclear whether this would even have a solution. Therefore, Dayanik et al. [15] studied the optimal asymptotic performance as the bounds on allowable decision risk converge to zero. Results show that, for small allowable decision risk, the minimum expected detection delay over all admissible rules can be attained by a common admissible rule that can exploit recursive equations to minimize the need for computational power that may be lacking with small sensing devices. The result is a simple, computable policy that determines when a signal change has occurred and determines the cause of the change (i.e., the sensor failure mode or material detected) in the presence of noisy readings from sensors that may fail due to aging or operational stress. This policy closely approximates an optimal policy [60], but is much easier to compute, potentially allowing it to be used in real time.

2.3 Combining Data Sources

Together, sensor data and manifest data provide terabytes of data on millions of containers and their contents. While methods have emerged to analyze each set of data separately (including the methods that we have described), efforts to combine these data for more powerful capabilities for detecting illicit nuclear material are still relatively new. Methods for combining such data hold the promise of considerable improvement in detection. Recently, we obtained an additional month of manifest data that are coordinated with radiation detection data that we also hope to attain. Such data—linking radiation sources, manifest data, and radiation portal readings—would be used to provide a rich source of training data for building a classifier for specific ports of entry. Our future work will study how to combine manifest data with sensor data in our statistical and machine learning methods. For instance, we will study how to compare materials claimed on the manifest with those identified in classification using radiation sensor data to identify potential anomalies. In addition, we will investigate how we can use manifest information to inform the classification task itself in order to improve accuracy. For instance, such methods could include information from the manifest to determine which materials to consider when defining the K classes in Dalal and Han's model [12]. Our aim is to use these combined methods for more powerful decision-making capabilities during the routine screening process, enabling us to more definitively identify suspicious containers and reduce delay of benign containers. Along these lines, researchers at Lawrence Livermore National Laboratory [37] have developed a Context-Aware Nuclear Evaluation System (CANES), which combines data from multiple types of sensors (such as RPMs and RIIDs) with context data that includes information on distance from source and type of conveyance and applies machine-learning algorithms for threat assessment.

3 Optimizing Sequential Decision-Making Strategies for Inspection

In addition to analyzing the sensor data itself, another aspect of sensor management is deciding which test to apply to incoming cargo and in which order to apply them in light of practical considerations such as budgets on inspection time and/or cost. To date, several researchers have studied paradigms for modeling and optimizing container inspection at ports [1, 5, 11, 19, 38, 39, 49, 53, 58]. Rather than focusing on making a decision based on given sets of data (as in Sect. 2), the emphasis here is on determining the sequence of tests to apply to optimize the inspection process.

At ports of entry, we envision a stream of entities arriving for inspection and a decision maker having to decide how to inspect each one. This includes deciding which to subject to further inspection and which to pass through with only minimal levels of inspection. Viewed this way, the process becomes a sequential decision-making problem. Sequential decision making is an old subject, but one that has become increasingly important as traditional methods for making sequential decisions fail to keep pace with problem scale. Enumerative algorithms for optimizing port-of-entry inspection rapidly come up against the combinatorial explosion caused by the many possible alternative inspection strategies. Moreover, methods must incorporate practical considerations—such as sensor error—which introduce uncertainty into the models. Work on these topics is being conducted as part of several other projects that are closely aligned with ours and have some overlapping participants. In particular there was another nuclear detection project based at Rutgers and led by Endre Boros and Paul Kantor, and there have been several projects on port-of-entry inspection also based at Rutgers. These projects have developed approaches that bring into the analysis many of the complications—such as sensor error—that arise from practical considerations.

In the port-of-entry inspection projects (see [5] for an overview), the project teams have built on the initial approach to the port-of-entry inspection problem taken by Stroud and Saeger [53], who studied a case that involved different potential tests (we will call them sensors) for deciding whether a cargo contains illicit material. Four such tests currently in use are evaluation of ships manifests, passive radiation signatures, radiographic images, and induced fission. All of the tests have costs associated with them, including the cost of a reading indicating illicit material when there is none (a false positive), the cost of a reading indicating there is no illicit material when there is (a false negative), time costs of using the sensor, delay costs of waiting for the sensor, and fixed cost of equipment, labor, etc. For each sensor the readings for cargo containing illicit material (positives) and readings for cargo not containing illicit material (negatives) are random variables. The model Stroud and Saeger created assigns an output of 0 (absence of illicit material) or 1 (presence of illicit material) for each sensor and defined cost using some of the above costs. In general, n sensors will yield a string (vector) of 0's and 1's of length n, and can be modeled with a binary decision tree (BDT). Stroud and Saeger developed enumerative methods to find the binary decision tree of sensors that would minimize total cost of inspection. The problem becomes intractable already for $n = 4$ if one relies on brute force methods since the number of possible trees expands rapidly, but Stroud and Saeger were able to extend their method to $n = 4$ by making some assumptions about the types of decision functions captured by the BDT. However, their method is not feasible for higher values of n.

The project teams built on this initial approach, making the models and algorithms better suited to address inspection issues that might arise in practice. For instance, the heuristics developed will need to be able to scale up to perhaps 20 or more sensors. Furthermore, there are several interdependent aspects of port-of-entry inspection that need to be explored in tandem [5]. Some of them are:

- Developing simulation models of inspection stations as one part of an operating port. These models can be used to assess the efficiency and effectiveness of security field operations, aid decision makers in quantifying the tradeoff between security goals and their attendant costs, provide feedback for devising improved operations, as well as to provide estimates for some of the cost parameters (such as delays) used in some of the optimization models.
- Studying the sensitivity of optimal and near-optimal trees to the input parameters [1]. As input parameters such as the costs of false positives and false negatives, the costs of delays, etc., are estimated with more or less accuracy, one wants solutions whose sensitivity to changes in these parameters is known and tolerable. Team studies led by Saket Anand et al. [1] show that the optimal

inspection strategy is remarkably insensitive to variations in the parameters needed to apply the Stroud-Saeger method. An important research challenge is to understand why.
- Developing new computational approaches that are inexpensive, scalable, and able to incorporate various cost factors with enough flexibility to include future technologies. Such approaches are based on efficient search heuristics [5, 38, 39], linear programming [6], and dynamic programming [26] and are now able to address problems involving many more sensors in very little time. In related research, Concho and Ramirez-Marquez [11, 49] have used evolutionary algorithms to optimize a decision tree formulation of the inspection process. Their approach is based on the assumption that readings r_j by the jth sensor are normally distributed, with a different distribution depending on whether the container in question is "bad" or "good." Thresholds t_j are used to determine outcomes of inspections, with a container declared suspicious by the jth sensor if $r_j > t_j$. Here, the cost function used depends upon the number of sensors used and the cost of opening a container for manual inspection if needed, but does not take into account the cost of false positives or false negatives, which is a key feature of the work in Stroud and Saeger [53], Anand et al. [1], and Madigan et al. [38] and [39].
- Investigating the optimum threshold levels for sensor alarms so as to minimize overall cost as well as minimize the probability of not detecting hazardous material [1, 5, 19, 38, 39]. Zhu et al. [61], in work extending that of Elsayed et al. [19], consider sensor measurement error independently from the natural variation in the container attribute values. They model situations when measurement errors exist (and are embedded) in the readings obtained by the inspection devices and use a threshold model to identify containers at risk for misclassification. They study optimization of container inspection policies if repeated inspection of at-risk containers is part of the process.
- Exploring use of sensors with many possible output categories. Boros, Kantor and colleagues. [6], in a parallel nuclear detection project, used a large-scale linear programming model approach and considered more container classifications than just the bad or good. They demonstrated the value of a mixed strategy applied to a fraction of the containers. They then added budget constraints to the problem in Goldberg et al. [24].

Several other authors have also considered ways to optimize and improve layered screening systems that include some of the above aspects. Wein et al. [58] consider several of the above issues in a detailed study on how to optimize the inspection strategy for detecting nuclear weapons (or their building blocks) at ports. In so doing, they develop operational models and make specific recommendations on which key uncertainties are most important to resolve, how to improve the existing screening process, as well as how to most effectively utilize new technologies. McLay et al. [41] develop a linear programming model for screening cargo for nuclear materials at ports of entry. Their approach defines a framework for determining alarms when there are limited screening resources. Jacobson et al. [31] look at baggage screening at airports and compare 100%

screening with one type of screening device with screening with a second device when the first device says a bag is suspicious. They calculate costs and benefits of the two methods.

4 Managing Static and Mobile Sensors

In some cases, such as ports and border crossings, entering vehicles are funneled through checkpoints that provide natural locations for radiation sensors. Even in these cases, practical considerations arise because of differing sensor operating characteristics—different sensors have different capacities for inspection over a given time and vary in cost and performance. Wein et al. [57] consider the spatial location of radiation portal monitors at overseas ports to improve detection (which depends on scan time) without creating bottlenecks that would create excessive congestion in the port. Jacobson et al. [32] consider the problem of deployment of baggage screening devices at airports, formulating it as an integer programming problem that takes into account various practical complications.

In less structured settings, such as urban environments, desirable locations are less obvious and need to be determined in other ways. In our project we have explored two different scenarios. The first is a variant of more traditional static sensor location problems that require locating sensors to respond to a set of uncertain events. In this case, we assume that we are placing sensors in a set of fixed locations to minimize the risk of missing a threat. Typically, an implicit assumption is that locations for these sensors must be chosen judiciously because they are too expensive to locate "densely" over the area to be covered. In this way, sensor placement problems are closely related to well-studied facility location problems in the optimization literature [34, 35, 42]. However, the sensor placement problem has sources of uncertainty that are not part of the traditional facility location problem. In sensor placement, it makes sense to consider environments where events to be monitored occur with low probability. Thus, the locations that we need to "cover" have considerable uncertainty yielding stochastic versions of more traditional problems. These stochastic variants appear to be significantly more complex and are not yet well-studied.

Dimitrov et al. [18] locate sensors along a transportation network using a stochastic interdiction model. Here the scenario is that a smuggler needs to get from a given origin to a given destination in the network and the "interdictor" needs to locate sensors to minimize the probability that the smuggler can reach the destination without detection. They envision this problem arising in border protection. Another complication that arises in sensor networks is sensor error. Neidhardt et al. [43] consider optimizing the positioning of error-prone sensors to monitor an area. Their placement strategy seeks to reduce error by having areas "covered" by multiple sensors in an "equitable" fashion through use of a minimax objective.

Wein and Atkinson [54], Atkinson and Wein [2] and Atkinson et al. [3] develop a detection-interdiction model to assess the efficacy of deploying a ring of sensors to protect an urban area from attack with various types of nuclear devices. Their models assume that an adversary is attempting to drive an already-assembled nuclear device into an urban area to maximize expected damage, while a defender combines use of a ring of radiation sensors and a fleet of interdiction vehicles to prevent penetration into the city. Their studies consider sensor errors resulting in missed detections and false alarms that occupy interdiction vehicles.

The second scenario that we explore in our project assumes that sensors are cheap and mobile. Under this basic paradigm, we examine the utility of locating sensors on vehicles such as taxis and police cars in an urban setting. Here the problem is no longer locating sensors; it is developing the statistical capabilities to reconcile readings from multiple sensors that are moving and may be prone to errors. Our project to date has emphasized this second scenario [9, 10], with a focus on managing a "fleet" of mobile sensors. We are examining the viability of fleet-based sensing and addressing related statistical challenges in detection. Hochbaum and Fishbain [27, 28] have considered a similar fleet-based surveillance scenario, while Neidhardt et al. [43] have considered a two-level sensing system that includes a static network of sensors augmented with a mobile pool of opportunistic sensors, such as those on cell phones.

4.1 Opportunistic Surveillance with Mobile Sensors

We envision "opportunistic sensing" as one possible paradigm for sensing with vehicles. In this case, we imagine vehicles (whether taxis, police cars, or some other "fleet") that contain radiation sensors, but their movement is determined by activities other than surveillance, such as routine taxi pickups or police patrolling. This paradigm features a network of mobile sensors operating relatively independently to provide surveillance of an area as an artifact of movement in performing other duties. Such networks can operate in tandem with smaller, more carefully designed, static networks to provide additional coverage and corroboration of alarms (such as in Neidhardt et al. [43]), or they can operate independently.

To illustrate the concept, we envision installing small radiation detection devices, communication capabilities (through cellular networks) and global positioning systems (GPS) in taxis, police vehicles, fire trucks, and/or public transit vehicles to provide surveillance in major urban areas. Such networks aim to leverage technological advances in sensors and positioning systems, miniaturization of devices for sensing and communication, and the pervasiveness of human activity in dense urban areas. Recent advances have made communication infrastructure nearly ubiquitous, while detection devices and positioning systems have become both economical and portable. Thus, large-scale deployment of a mobile sensor network is becoming feasible and affordable. The New York City police department is already using small sensors in vehicles and on officers [36, 51] for radiological detection. The idea of using massive networks of mobile sensors has been adopted and tested by the Radiation Laboratory at Purdue University, where they use a network of cell phones with GPS capabilities to detect and track radiation [48].

The movement and extensive coverage afforded by sensors in taxis is appealing because it could provide pervasive surveillance in dense urban areas, while devices placed on emergency response vehicles or police cars could offer greater control to investigate suspicious regions and allow the possibility of including more powerful (and expensive) sensing capabilities on some vehicles. When vehicles equipped with sensors move within a certain range of a nuclear source, the radiation energy from the source will trigger the sensing devices to send an alarm notification and a GPS position to a central command center over a wireless network. This basic sensing paradigm has many attractive features. First, the random movement and extensive coverage of the vehicles provides constant surveillance for nuclear materials. Second, the mobile sensors do not need to be of high accuracy, since the failure of a small portion of them will not significantly hamper the effectiveness of surveillance coverage because of the sensors' random movements. Next, the movements of the sensors will (in most cases) be difficult to predict by an adversary, and because of the number of sensors, difficult to tamper with. Finally, because the sensors are mounted on vehicles, there are fewer size constraints and power consumption requirements.

Such a mobile sensor network would likely be supplemented by stationary sensors to cover locations with sparse traffic, such as a large park in the city. The methods we have developed can easily be envisioned for such mixed networks by simply viewing the stationary sensors as parked vehicles. While our algorithms can be readily adapted to a variety of settings, we work under the following basic assumptions:

- Nuclear sensors and Global Position System (GPS) tracking devices are installed on a large number of vehicles.
- The sensors and GPS devices constantly send detection and location information to a central surveillance center.
- Real-time tracking signals can be geolocated on a map of the area under surveillance.
- Real-time analysis is done at the surveillance center using sophisticated statistical algorithms to identify potential locations of nuclear sources that appear as clusters of positive sensor readings.

Because sensors are not always 100% accurate, there will potentially be false alarms and missed detections. Statistical methodologies have proven to be effective tools for detecting true signals against random errors. Thus, a challenge that we began to address early in the project was that of processing sensor network information to identify "positive clusters" among the sensor signals that are not due to either random chance or known background sources. Multi-cluster spatial classification methods are ideal for such tasks. Our research has explored two innovative multiple spatial clustering methods. The first method (due to Demattei et al. [16, 17]) is based on data transformation and a step regression model. It provides a formal statistical test of significance against background noise based on the premise that points within a "cluster" should be spatially closer to each other than positive signals outside the cluster that are due to random chance/error. The second approach is based on the recent Ph.D. thesis of Lynette Sun [52]. By mimicking the process of typical sample data generation, Sun [52] and Xie et al. [59] developed an intuitive procedure that introduces a latent modeling structure and uses formal likelihood inference to detect multiple clusters occurring simultaneously within a defined region or time window. They apply model selection techniques to determine the number of clusters, and develop likelihood inference and Expectation Maximization/Markov Chain Monte Carlo algorithms to estimate model parameters, detect clusters and identify cluster locations. Their new method differs from the classical scan statistic in that it can simultaneously detect multiple clusters of varying sizes. This work is readily applicable to identifying clusters of vehicles with "positive" sensor readings for radiation.

This latent model approach was adapted by team members Jerry Cheng and Minge Xie to the context of nuclear surveillance. The method is flexible and able to accommodate a variety of extensions that make it well suited to the nuclear detection problem. The key idea is to use statistical notions of clustering, where a "cluster" involves an unusually large number of events/alarms clumping within a small region of time, space, or locations in a contiguous sequence (suggesting a moving source). Our methods are using modified versions of the traditional statistical method using scan statistics. The idea is to scan the entire study area and try to locate region(s) with unusually high likelihood of incidence. For example, one would use the maximum number of cases in a fixed-size moving window or identify the diameter of the smallest window that contains a fixed number of cases. Early work in our project demonstrated the applicability of this method to detecting clusters of positive radiation sensor readings from taxis. Cheng and Xie also performed simulations for both spatial classification methods under scenarios that include stationary and moving sources. Results of these preliminary simulations suggested that the proposed approach can effectively filter noise and background radiation sources to detect nuclear materials placed in a metropolitan area. For some details of the approach, see Cheng et al. [9].

In the first phase of our project, we emphasized use of taxis in radiation detection. Our subsequent discussions with law enforcement suggested reluctance to depend on the private sector (e.g., taxis) in surveillance. As the project progressed, our emphasis therefore shifted from considering taxis as the primary type of sensing vehicle to police cars or a combination of taxis and police cars. This concept employs the police vehicles in a manner similar to our initial ideas about taxis, but it explores use of smaller fleets, with possibly less random movement, and perhaps higher-quality sensing equipment. A central focus of more recent work has been to compare taxi-based "coverage" to police car "coverage" through simulation studies. This line of investigation aims to determine how many police cars might be enough to get sufficient "coverage" of a region when the police cars contain sensing devices but, as with taxis, their movement is directed toward normal police activity, not radiation detection.

As part of our efforts, project members Jerry Cheng, Fred Roberts, and Minge Xie applied statistical power analysis to determine the number of vehicles required to provide adequate coverage for surveillance of a given network [9]. They developed a model and carried out a large number of simulations to gain intuition and assess detection power under a variety of different assumptions.

The early simulations followed the same basic paradigm but systematically varied one or more of the parameters of interest. The surveillance area in our testing consisted of a 4,000 ft by 10,000 ft area, roughly equal to the area of the roads and sidewalks of Mid/Downtown Manhattan. In this phase of the work, we disregarded the street network and simply considered that a specified number of vehicles are randomly located in the area at a particular "snapshot" in time. At the next time period, the vehicles were again randomly located in the region to correspond with a new "snapshot" in time. The parameters that we could adjust from one experimental run to the next included: the number of vehicles; the effective range of a sensor; and the rates for false positives and false negatives. In these experimental runs, we considered a stationary radiation source, placed randomly in the surveillance area.

We conducted a large number of experiments using this basic framework. For example, in one model, we assumed the effective range for a detector to be150 ft., a false positive rate of 2%, and a false negative rate of 5%, and we varied the number of vehicles (i.e., sensors). (We realize that this range is beyond most presently used detectors, but wanted to concentrate on methodology and relative comparisons, and we experimented with shorter ranges as well, as noted below.) We then ran at least 200 simulations for each number of vehicles and determined whether the source was detected. For each number of vehicles, this gave us an estimate of the "power," which is defined to be the probability of detecting a source for a single random placement of the vehicles (i.e., a single time period). In this model, we found that 4,000 vehicles were needed to get even 75% power. With 2,000 vehicles, the power was about 30%. To give this some perspective, we note that the New York City Police Department has 3,000 plus vehicles in 76 precincts in 5 boroughs, but at any given time only about 500 to 750 would be in the streets of Mid/Downtown Manhattan.

Of course, in practice, we would monitor the alarms over a period of time, not just at a single instant. To reconcile readings from several "snapshots" over time we may wish to use some decision rule such as: detection if a majority of the times there is an alarm; detection if at least once there is an alarm. The number of time periods is another variable that needs to be considered. It is not hard to show that if the statistical power is sufficiently high and majority rule detection is used, then with sufficiently many time periods, the detection probability can be increased significantly. We are currently exploring various rules for detection over time.

We also conducted studies in which we varied the effective range of a sensor. Given current technology, the range of a sensor may actually be closer to 25 ft than it is to the 150 ft assumed in the experiment that we just described. However, since our project is intended to look beyond today's capabilities, we wondered: what would happen if we had a better detector, say with an effective range of

250 ft.? In an experiment similar to the previous one but with a sensor range of 250 ft., 2,000 vehicles yielded 93% power.

There are other aspects of our model that need to be modified when the sensing vehicles are police cars as opposed to taxis. In particular, the assumption of random movement is less appropriate for police cars, since they will tend to remain in their own region/precinct, and they won't move around as randomly or as frequently as taxis. We did a simple study that attempted to make movement slightly more realistic by dividing the region into 20 equal-sized precincts. (There are 22 police precincts in Manhattan.) Next, we placed police cars randomly within each precinct. When we assumed that the number of police vehicles in each precinct is 25—making for a total of 500 police vehicles—and assumed that each detector has a 250 ft. range, then our simulations estimated the power at 35%. This is not very good and is not significantly different than when the same number of vehicles was allowed to roam throughout the region. In other studies, we varied other parameters such as the false positive and false negative rates. We have also investigated hybrid models that involve a mixture of police cars and taxis.

We are implementing a variety of extensions to make our models more realistic, including: more complex hybrid models of taxis and police vehicles with different movement models; hybrid models that include some stationary detectors; hybrid models with more powerful detectors in police vehicles; more realistic movement models; moving sources; multiple sources; fusing information from multiple time periods. We are especially interested in exploring hybrid models that include police cars and taxis that have different models for movement and possibly sensors of differing capability. In work with graduate students Tsvetan Asamov and Adam Marszalek, we also introduced a street network with more realistic models for movement of vehicles, and we plan to use this in future studies.

A mobile sensor nuclear threat detection problem is also studied in Hochbaum [28]. Here, the goal is to identify a small area in the region of interest that has a high concentration of alarms. The paper separates the two goals of small area and high concentration of alarms, which can be conflicting, and introduces a weighing factor for balancing the contribution of the two goals. In contrast to our early work, this work has a specific model of a region as a network with streets and assumes vehicles move along streets; the paper formulates the problem of finding an "optimal" area as a mathematical programming problem and presents a polynomial time algorithm for solving it. The study is extended in Hochbaum and Fishbain [27] with discussions of false alarms, simulations, and methods of aggregating results over time to improve the algorithm's performance.

4.2 Randomized Surveillance Routing

The previous "opportunistic" model considers a surveillance scenario in which we gain surveillance capability by exploiting the random movement of a large vehicle fleet. Another line of research is to consider the case in which our fleet is not large enough to provide sufficient surveillance coverage by purely undirected movement. In this case, the idea is to equip a certain number of vehicles with sensors and dedicate them to the task of performing surveillance within an area. Unlike the previous case, the movement of these vehicles will be prescribed by some "controller". This controller would like to find routes for the vehicles that are "efficient" in the sense that they cover the entire region quickly but also appear "unpredictable" to an adversary [44]. In this case, we represent the region by a graph, where links correspond with streets and nodes correspond with locations. For each vehicle, we would like to create a patrol route that begins and ends at a designated location (e.g., police headquarters) and cannot be predicted by an adversary; yet, taken together, these routes cover the entire region efficiently. These two properties of efficiency and unpredictability are seemingly at odds with each other. Suppose for the moment that we have just one patrolman or patrol vehicle. An extremely efficient route would be a traveling salesman tour of the graph. However, given that it is very efficient, an observer knows that once a location is visited, it will not be visited again. Moreover, if the vehicle drives the same route each day, an observer could predict exactly where it will be at any given time. On the other hand, the vehicle could be very unpredictable by moving totally at random. In this case, it may take a long time to cover the entire region, but an attacker would not be "safe" just because a node was recently patrolled.

Clearly there is a tradeoff between efficiency and unpredictability, and Alantha Newman, a researcher on our project, is trying to formalize this tradeoff by developing formal definitions for "unpredictability" and then defining a route optimization problem for selecting efficient routes that satisfy these formal definitions [44]. Another approach that does not apply a strict definition for unpredictability but does exploit notions of randomness in surveillance [48] is the basis for surveillance at LAX airport. Here the approach is to consider surveillance to be a Bayesian game against an unknown adversary. Nonetheless, the ideas are similar: to find a tradeoff between an efficient assignment of inspection stations each day and one that is unpredictable. The problem is formulated as a game between an inspector and an attacker and mixed strategies are used to find good solutions. The methods have recently been extended to the Pittsburgh airport and are also being used to randomly assign federal air marshals to some of the flights between the US and Europe.

5 Data Sampling Strategies for Sensor Data

Sometimes when we have limited time or budget for data collection, it is advantageous to adjust our data sampling strategy in response to previously collected data. The specific settings that we consider involve considerable uncertainty, where the underlying probability distributions are either unknown or changing through time. Although we do not know the underlying distributions, we nonetheless have the ability to collect information through measurements to help us learn about the environment. In this sense, we may view information collection as a sequential decision problem in which our objective is to learn about our environment. There are a wide variety of practical decision-making settings in which a decision maker has the ability to collect a finite amount of information before he or she must render a decision. Examples of applications in nuclear detection include determining when to subject people and containers to additional scrutiny, positioning sensors, and deciding when to introduce a new sensing technology.

A number of practical adaptations of dynamic programming [4] techniques exist for finding near-optimal solutions to these types of sequential decision problems. One such adaptation that project members Peter Frazier, Warren Powell and Savas Dayanik have explored is the knowledge gradient policy which makes sampling decisions by maximizing the expected value of the sampled information according to a simple heuristic metric [21]. The problems addressed typically involve three dimensions:

1. the decision about what information to measure or collect;
2. the information that is observed when a measurement is made;
3. the decision that is made after observing the new information.

We refer to the first decision as the measurement decision w. We represent knowledge about a problem using a vector μ^n which captures the distribution of belief about a set of parameters after n measurements. After a measurement decision is made, we make an observation (such as on the level of nuclear radiation) which was uncertain before the measurement. Finally, we seek to make an economic decision which we denote by x.

Letting μ^{n+1} represent knowledge after $n + 1$ measurements (which is a random variable before we have made our last observation), measurement decisions will have the fundamental structure of

$$\max_w E\left\{\max_x F\left(x|\mu^{n+1}(w)\right)\right\}$$

We could avoid the measurement and take an action now that requires solving

$$\max_x F(x|\mu^n)$$

The value of a measurement is called the *knowledge gradient*, and is given by

$$\nu^{\mathrm{KG}} = \max_w E\left\{\max_x F\left(x|\mu^{n+1}(w)\right)\right\} - \max_x F(x|\mu^n).$$

For the problem of deciding what to measure, we have been exploring a class of measurement policies we call knowledge gradient policies. With this strategy, we choose the measurement w that yields the largest value of ν_w^{KG} where

$$\nu_w^{\mathrm{KG}} = E\left\{\max_x F\left(x|\mu^{n+1}(w)\right)\right\} - \max_x F(x|\mu^n).$$

This policy chooses the measurement that would be best if you were going to make a single measurement, but it has also been shown to be asymptotically optimal. Not surprisingly, it has been found to work better than any other competing approaches for intermediate measurement budgets. The knowledge gradient (KG) policy [21] offers an easily implemented rule that tells us how to sample information on competing alternatives to learn which is best. The initial version of the knowledge gradient policy applies in settings where the measurements are independent. The team later developed the Correlated Knowledge Gradient (CKG) [22] for determining how to collect information when measurements are correlated, as would occur when detecting nuclear radiation. KG and CKG are provably optimal in certain special cases; have provable bounds on suboptimality in other cases; and are very easy to implement and use.

As the project progressed, the team has considered applying the KG method in more complicated settings, such as collecting information on a graph [50], possibly for moving sensors around a street network. In sensor management decisions, our choice of what to measure (which changes our knowledge state) depends on where we are (our physical state). A measurement at one physical state can affect decisions at other physical states. The result is what we call "the information collecting shortest path problem". The shortest path problem is fundamental to a wide range of optimization problems [52]. For example, these might arise in the process of designing emergency response measures (e.g., how to evacuate New York City, or how to guide emergency response forces). Often, the state of these networks is uncertain (e.g., travel times on links may vary), and it is necessary to collect information which may involve the time-consuming process of dispatching people to collect it. Given this, we would like to collect the most valuable information first.

The immediate goal in this research is sensor management, where we have to move a physical sensor around the network to collect information. We found that an important stepping stone for this larger problem is the problem of determining which link in a network we should measure (without regard to the physical location of a sensor). For example, in the network depicted below, we show a probability density function on each link to describe our belief about the cost on that link. The dark path represents what we currently believe is the shortest path, while the dotted link is one that we might consider measuring. If we do measure this link, we may change our distribution of belief about the link based on the observed measurement. The question is: which link should we measure?

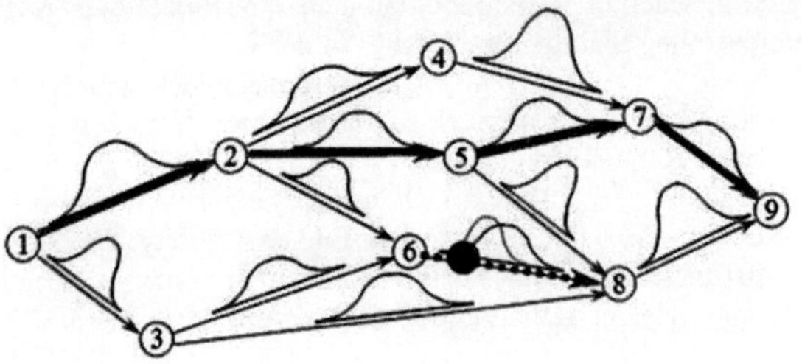

Project members Warren Powell and Ilya Rhyzov have explored this problem of information collection on a graph and have found that they can adapt an existing knowledge gradient method [18] to this new problem. In computational testing they have found that for networks where the paths are not too short, the knowledge gradient works consistently quite well relative to competing techniques [50].

Acknowledgments We thank the members of our nuclear detection and port security project teams, especially those whose work is represented in this chapter: James Abello, Saket Anand, Tsvetan Asamov, Endre Boros, Xueying Chen, Siddhartha Dalal, Savas Dayanik, Elsayed Elsayed, Peter Frazier, Emilie Hogan, Paul Kantor, Mingyu Li, David Madigan, Sushil Mittal, Alantha Newman, Jason Perry, William Pottenger, Warren Powell, Ilya Rhyzov, Warren Scott, Kazutoshi Yamazaki, Christina Young, and Yada Zhu, Christopher Janneck, Adam Marzsalek, Christie Nelson. We also gratefully acknowledge support from the National Science Foundation under grants SES 05-18543, DMS 09-15139, and CBET 07-36134, from NSA under grant H98230-08-1-0104, ONR under grant N00014-07-1-0299, and the US Department of Homeland Security Domestic Nuclear Detection Office under grant 2008-DN-077-ARI012-02.

References

1. Anand S, Madigan D, Mammone R, Pathak S, Roberts F (2006) Experimental analysis of sequential decision making algorithms for port of entry inspection procedures. In: Mehrotra S, Zeng D, Chen H, Thuraisingham B, Wang F-X (eds) Intelligence and security informatics, Proceedings of ISI-2006, Lecture notes in Computer Science 3975. Springer-Verlag, New York, pp 319–330
2. Atkinson M, Wein L (2008) Spatial queueing analysis of an interdiction system to protect cities from a nuclear terrorist attack. Oper Res 56:247–254
3. Atkinson M, Cao Z, Wein L (2008) Optimal stopping analysis of a radiation detection system to protect cities from a nuclear terrorist attack. Risk Anal 28:353–371
4. Bellman R (1957) Dynamic programming. Princeton University Press, Princeton
5. Boros E, Elsayed E, Kantor P, Roberts F, Xie M (2008) Optimization problems for port-of-entry detection systems. In: Chen H, Yang CC (eds) Intelligence and security informatics: techniques and applications. Springer, New York, pp 319–335
6. Boros E, Fedzhora L, Kantor P, Saeger K, Stroud P (2009) Large scale LP model for finding optimal container inspection strategies. Naval Res Logist Q 56:389–486
7. Chen X, Xie M (2010) Enhancing inspection process in nuclear detection by combing information from different sources. Working paper, DIMACS Center, Rutgers University
8. Chen, X, Xie, M, Zhang, CH (2010). A statistical approach for analyzing manifest data in pre-portal intelligence. Working paper, DIMACS Center, Rutgers University
9. Cheng J, Xie M, Roberts F (2009) Design and deployment of a mobile sensor network in surveillance of nuclear materials in metropolitan areas. In: Proceedings of the 15th ISSAT international conference on reliability and quality in design
10. Cheng J, Xie M, Chen R, Roberts FS (2011) A latent model to detect multiple spatial clusters with application in a mobile sensor network for surveillance of nuclear materials. Working paper, DIMACS Center, Rutgers University.
11. Concho A, Ramirez-Marquez JE (2010). An evolutionary algorithm for port-of-entry security optimization considering sensor threshold. Reliab Eng Syst Safety, 95:225–226
12. Dalal S, Han B (2010). Detection of nuclear material in containers entering the US: a Bayesian approach for analyzing radiation portal data. Ann Appl Stat 4:1256-1271

13. Dayanik S, Goulding C (2009). Detection and identification of an unobservable change in the distribution of a Markov-modulated random sequence. IEEE Trans Inform Theory 55: 3323-3345
14. Dayanik S, Goulding C, Poor HV (2008) Bayesian sequential change diagnosis. Math Oper Res 33:475–496
15. Dayanik S, Powell W, Yamazaki K (2008) Asymptotic analysis of sequential change diagnosis problem. Proceedings of the international workshop on applied probability
16. Demattei C, Molinari N, Daures J-P (2007) Arbitrarily shaped multiple spatial cluster detection for case event data. Comput Stat Data Anal 51:3931–3945
17. Demattei C, Molinari N, Daures J-P (2006) SPATCLAS: an R package for arbitrarily shaped multiple spatial cluster detection for case event data. Comput Methods Programs Biomed 84:42–49
18. Dimitrov NB, Gonzalez MA, Michalopoulos DP, Morton DP, Nehme MV, Popova E, Schneider A, Thoreson GG (2008) Interdiction modeling for smuggled nuclear material. Proceedings of the 49th annual meeting of the Institute of Nuclear Materials Management
19. Elsayed E, Schroepfer C, Xie M, Zhang H, Zhu Y (2009) Port-of-entry inspection: sensor deployment policy and optimization. IEEE Trans Autom Sci Eng 6:265–277
20. Frazier P, Powell WB (2008) The knowledge gradient stopping rule for ranking and selection. Proceedings of the Winter Simulation Conference, 2008
21. Frazier P, Powell W, Dayanik S (2008) A knowledge gradient policy for sequential information collection. SIAM J Control Optim 47:2410–2439
22. Frazier P, Powell WB, Dayanik S (2009) The knowledge gradient policy for correlated rewards. INFORMS J Comput 21(4):585–598
23. Ganiz MC, Lytkin NI, Pottenger WM (2009) Leveraging higher order dependencies between features for text classification. In: Buntine et al. (eds.) Machine learning and knowledge discovery in databases, Lecture notes in computer science, vol 5781, pp 375–390
24. Geelhood BD, Ely J, Hansen R, Kouzes R, Schweppe J, Warner R (2003) Overview of portal monitoring at border crossings. IEEE Nucl Sci Symp Conf Rec 1:513–517
25. Genkin A, Lewis D, Madigan D (2007) Large-scale Bayesian logistic regression for text categorization. Technometrics 49:291–304
26. Goldberg N, Word J, Boros E, Kantor P (2011). Optimal sequential inspection policies. Ann Oper Res, (in press) doi:10.1007/s10479-010-0799-6
27. Hochbaum D (2009) The multi-sensor nuclear threat detection problem. In: Chinneck JW, Kristjansson B, Saltzman MJ (eds) Operations research and cyber-infrastructure, operations research/computer science interfaces series, vol 47. Springer, New York, pp 389–399
28. Hochbaum D, Fishbain B (2009). Nuclear threat detection with mobile distributed sensor networks. Ann Oper Res (in press), doi 10.1007/s10479-009-0643-z
29. Hoshino R, Coughtry D, Sivaraja S, Volnyansky I, Auer S, Trichtchenko A. Application and extension of cost curves to marine container inspection. Ann Oper Res (in press), doi 10.1007/s10479-009-0669-2
30. Hoshino R, Oldford W, Zhu M. Two-stage approach for unbalanced classification with time-varying decision boundary: application to marine container inspection, ACM SIGKDD workshop on intelligence and security informatics, ACM: New York
31. Jacobson SH, Karnani T, Kobza JE, Ritchie L (2006) A cost-benefit analysis of alternative device configurations for aviation checked baggage security screenings. Risk Anal 26:297–310
32. Jacobson SH, McLay LA, Virta JL, Kobza JE (2005) Integer program models for the deployment of airport baggage screening security services. Optim Eng 6:339–359
33. Kapoor S, Ramamurthy V (1986) Nuclear radiation detectors. New Age Publishers, India
34. Kariv O, Hakimi S (1979) An algorithmic approach to network location problems. I: the p-centers. SIAM J Appl Math 37:513–538
35. Kariv O, Hakimi S (1979) An algorithmic approach to network location problems. II: the p-medians. SIAM J Appl Math 37:539–560

36. Kearney C (2009) New York police expand dirty bomb security, Reuters News, http://www.reuters.com/article/domesticNews/idUSTRE56067720090702, July 1, 2009
37. Lawrence Livermore National Laboratory (2007) Computing Directorate, Annual Report, p 38, 2007
38. Madigan D, Mittal S, Roberts F (2007) Sequential decision making algorithms for port of entry inspection: overcoming computational challenges. In: Proceedings of the international conference on intelligence and security informatics, pp 1–7
39. Madigan D, Mittal S, Roberts F (2009). Efficient sequential decision-making algorithms for container inspection operations. working paper, DIMACS Center, Rutgers University
40. Madigan D, Genkin A, Lewis DD, Fradkin D (2005) Bayesian multinomial logistic regression for author identification. In: Proceedings of the 25th international workshop on Bayesian inference and maximum entropy methods in science and engineering (MaxEnt 05), pp 509–516
41. McLay L, Lloyd J, Niman E. Interdicting nuclear material on cargo containers using knapsack problem models. Ann Oper Res (in press), doi 10.1007/s10479-009-0667-4
42. Mirchandani B (1990) Discrete location theory. Wiley-Interscience, New York
43. Neidhardt A, Luss H, Krishnan K (2008) Data fusion and optimal placement of fixed and mobile sensors. Proceedings of the IEEE sensors applications symposium (SAS2008)
44. Newman A (2009) Using globally random walks to patrol a network, working paper, DIMACS, Rutgers University Center
45. Paruchuri P, Pearce J, Tambe M, Ordonez F, Karus S (2007) An efficient heuristic approach for security against multiple adversaries. Proceedings of the 6th international joint conference on autonomous agents and multiagent systems, Honolulu
46. Perry J (2009) Clustering and machine learning for gamma ray spectroscopy. Rutgers University, Department of Computer Science, Class Project, May 2009
47. Powell WB (2007) Approximate dynamic programming: solving the curses of dimensionality. Wiley, New York
48. Purdue University: cell phone sensors detect radiation to thwart nuclear terrorism. Online: http://news.uns.purdue.edu/x/2008a/080122FischbachNuclear.html
49. Ramirez-Marquez J (2008) Port-of-entry safety via the reliability optimization of container inspection strategy through an evolutionary approach. Reliab Eng Syst Saf 93:1698–1709
50. Rhyzov I, Powell WB. Optimal learning on a graph (under review)
51. Riggio M (2009) Status report on federal and local efforts to secure radiological sources. Prepared statement of testimony before the United States House of Representatives Committee on Homeland Security, Subcommittee on Emerging Threats, Cybersecurity, and Science and Technology, Sep 14 2009. http://homeland.house.gov/SiteDocuments/20090914150055-28907.pdf
52. Roberts FS, Tesman B (2009) Applied combinatorics, 2nd edn. Chapman&Hall/CRC, Boca Raton, an imprint of Taylor & Francis
53. Stroud PD, Saeger KJ (2003) Enumeration of increasing Boolean expressions and alternative digraph implementations for diagnostic applications. In: Chu H, Ferrer J, Nguyen T, Yu Y (eds) Proceedings volume IV, computer, communication and control technologies: I, pp 328–333
54. Sun QK (2008) Statistical modeling and inference for multiple temporal or spatial cluster detection. Ph.D. thesis, Department of Statistics, Rutgers University
55. Weier D (2008) Radiation detection for DHS applications and the radiation portal monitor project. Talk given at the DIMACS/DyDAn/LPS Workshop on Port Security/Safety, Inspection, Risk Analysis and Modeling, Nov 2008
56. Wein L, Atkinson M (2007) The last line of defense: designing radiation detection-interdiction systems to protect cities from a nuclear terrorist attack. IEEE Trans Nucl Sci 54:654–669
57. Wein L, Liu Y, Cao Z, Flynn S (2007) The optimal spatiotemporal deployment of radiation portal monitors can improve nuclear detection at overseas ports. Sci Glob Secur 15:211–233

58. Wein L, Wilkins A, Bajeva M, Flynn S (2006) Preventing the importation of illicit nuclear materials in shipping containers. Risk Anal 26:1377–1393
59. Xie M, Sun Q, Naus J (2009) A latent model to detect multiple clusters of varying sizes. Biometrics 65:1011–1020
60. Yamazaki K (2009) Essays on sequential analysis: multi-armed bandit with availability constraints and sequential change detection and identification. Ph.D. thesis, Department of Operations Research and Financial Engineering, Princeton University
61. Zhu Y, Li M, Young CM, Xie M, Elsayed E (2009) Port of entry inspection policies: incorporation of measurement errors. Ann Oper Res doi: 10 .1007/510479-010-0681-6

Risk-Informed Decision Making in Nuclear Power Plants

A. K. Verma, Ajit Srividya, Vinod Gopika and Karanki Durga Rao

1 Introduction

Probabilistic Safety Assessment (PSA), also called Probabilistic Risk Assessment (PRA), is currently being widely applied to many fields, viz., nuclear facilities, chemical and process plants, aerospace, and even to financial management. PSA has been accepted all over the world as an important tool to assess the safety of a facility and to aid in ranking safety issues by order of importance. PSA essentially aims at identifying the events and their combination(s) that can lead to severe accidents, assessing the probability of occurrence of each combination, and evaluating the consequences. The main benefit of PSA is to provide insights into design, performance, and environmental impacts, including the identification of dominant risk contributors and the comparison of options for reducing risk. PSA provides the quantitative estimate of risk which is useful for comparison of alternatives in different design and engineering areas. Furthermore, PSA is a conceptual and mathematical tool for deriving numerical estimate of risk and quantifying the uncertainties in these estimates.

PSA studies not only evaluate risk/safety of systems but also their results are very useful in safe, economical, and effective design and operation of NPPs. The latter application is popularly known as "Risk-Informed Decision Making". In this, it provides inputs to decisions on design and back fitting, plant operation, safety

A. K. Verma (✉) · A. Srividya
Indian Institute of Technology Bombay, Mumbai, India
e-mail: akv@ee.iitb.ac.in

V. Gopika
Bhabha Atomic Research Centre, Mumbai, India

K. D. Rao
Paul Scherrer Institut, Villigen PSI, Switzerland

H. Pham (ed.), *Safety and Risk Modeling and Its Applications*,
Springer Series in Reliability Engineering, DOI: 10.1007/978-0-85729-470-8_12,

analysis, and on regulatory issues. PSA offers a consistent and integrated framework for safety related decision making. So worldwide, utilities are performing the PSA of their plants and many regulatory bodies are using it as a risk-informed approach in decision making and some even following it as a risk-based approach in decision making. Over the years, the PSA methodology has matured and even new applications like technical specification optimization, risk-informed in-service inspection, living PSA/Risk Monitor, and Reliability Centered Maintenance (RCM) have emerged. The chapter focuses mainly on the first two applications of RIDM.

Decision making related to regulation of the design, operation, and maintenance of nuclear power plants is a quite challenging problem. Risk informed approaches complements the traditional deterministic engineering analysis methods to provide useful information for the decision making. The emphasis is both on effective risk control and effective resource use. This chapter presents two applications of risk informed decision making, namely, optimization of surveillance test interval and optimization of in-service inspection interval.

2 Technical Specification Optimization

The criterion for regulation of the design and operation of NPP has been derived from deterministic engineering analysis methods. This traditional defence-in-depth philosophy continues to assure a safe condition of the plant following a number of postulated design basis accidents and also achieving several levels of safety. During recent years, both the nuclear utility and nuclear regulatory bodies have recognized that PSA has evolved to the point that it can be used increasingly as a tool in decision making. The key to this risk-informed approach to decision making is that it is complementary to the defence-in-depth philosophy. This has given rise to the advent of various methodologies for optimizing activities related to NPP operation and maintenance. Thus the risk-informed applications emphasize both effective risk control and effective resource expenditures at NPPs by making use of PSA results to focus better on what is critical to safety.

Several studies have emphasized the potential of risk-informed approach and its application to nuclear as well as non-nuclear/chemical industries also. The specific activities related for their resource effectiveness in risk-informed applications are evaluation of technical specifications, in-service inspection, and preventive maintenance. Evaluation of technical specifications is one of the important applications of Risk-Informed decision making. Technical specifications represent a set of parameters according to which systems should be operated, tested, maintained, and repaired. Deciding Test Interval (TI), one of the important technical specifications, with the given resources and risk effectiveness is an optimization problem. Nowadays, special attention is being paid on the use of PSA for risk-informed decision making on plant-specific changes to test intervals in technical specifications.

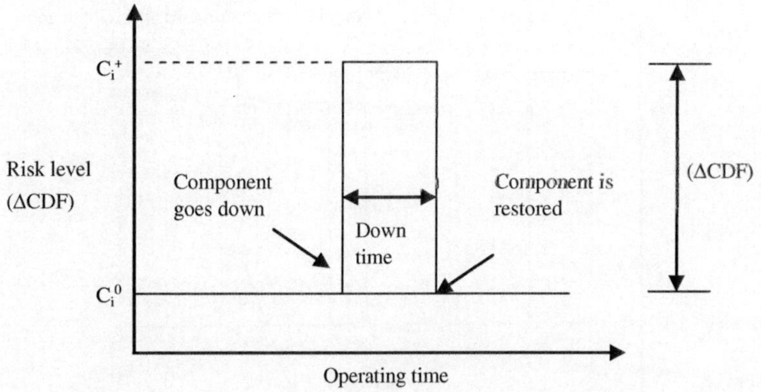

Fig. 1 Increase in risk associated with component outage

2.1 Traditional Approaches for Technical Specification Optimization

The various risk measures and methodology for TS modifications related to Allowed Outage Times (AOTs) and Surveillance Test Intervals (STIs) are discussed here [1]. The steps include the following: (a) identify the STIs and AOTs to be evaluated for consideration of changes, (b) determine the risk contribution associated with the subject STIs and AOT, (c) determine the risk impact from the change of proposed AOTs and STIs by evaluating risk measures of SSCs for which change in AOT/STI is sought, (d) ascertain the acceptability or otherwise of the risk impact (e.g., change in system unavailability, CDF, release frequency, etc.) from target value established for risk-informed decision, and (e) perform sensitivity and uncertainty evaluations to address uncertainties associated with the STI and AOT evaluation.

2.1.1 Measures Applicable for AOT Evaluations

(a) Conditional Risk Given the Limiting Condition of Operation (LCO)

Increase in risk (ΔCDF or ΔLERF) associated with component outage is shown in Fig. 1

(b) Incremental Conditional Core Damage Probability (ICCDP) or Single Downtime Risk

Increase in risk (e.g., single downtime risk r_i of ith component is obtained by multiplying the increase in CDF by the duration of the configuration for the occurrence of a given configuration, i.e., outage of ith component only).

$$r_i = \Delta C_i \times d = (C_i^+ - C_i^0) \times d_i \tag{1}$$

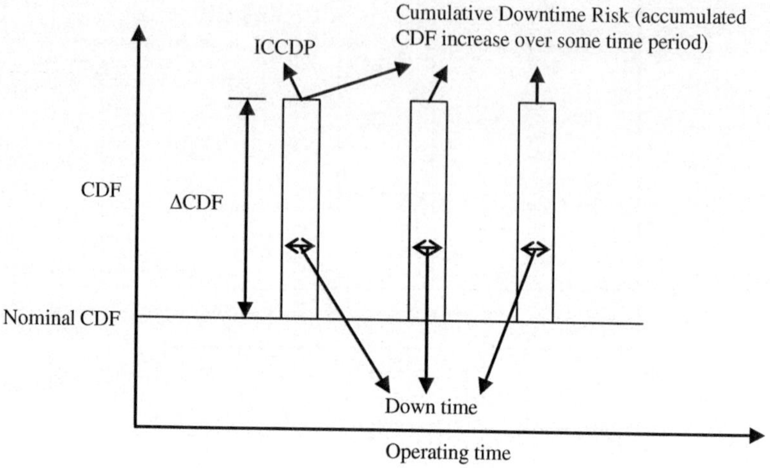

Fig. 2 Illustration of the different risks associated with downtimes

r_i is the single downtime risk of the ith component, C_i^+ is the CDF when component is known down including reconfigurations, C_i^0 is the CDF when component is known up, d_i is the downtime

By imposing an acceptable limit (i.e., target or reference value for risk-informed decision process) to the risk contribution of an AOT, a risk-based AOT can be calculated, $d_{max} = r_{max}/\Delta R$, where ΔR is the change in risk (change in system unavailability, change in CDF (ΔC_i) or change in LERF). Then the risk-based AOT can be compared to the real-time duration of maintenance and to the AOT established in the TS.

(c) Yearly AOT Risk

Risk increase from the projected (or expected) number of downtimes over one-year period is yearly AOT risk. Figure 2 shows the single downtime risk and cumulative downtime risk over some time period.

$$R_i = N_i r_i \tag{2}$$

R_i is the yearly downtime risk for ith component, N is the expected number of downtime occurrences in a year $= wT$, w is the downtime or maintenance frequency $= k\lambda$, where, $k =$ maintenance factor, $\lambda =$ failure rate, and $T =$ time period, 1 year.

Maintenance frequency includes failure frequency and the frequency of maintenance due to degraded or incipient conditions.

When comparing the risk of shutting down with the risk of continuing power operation for a given LCO, the applicable measures are:

- risk of continued power operation for a given downtime, similar to ICCDP and
- risk of shutting down for the same downtime.

The risk associated with simultaneous outages of multiple components, called configuration risk, is calculated as part of AOT changes. The applicable measures are similar to the AOT measures stated above.

2.1.2 Measures Applicable for STI Evaluations

(a) Test-Limited Risk

The analysis of STIs is based on the risk contributions arising from failures occurring between tests and detected at the moment of the test. The STI risk contribution of a component is given by

$$R_D = {}^1\!/_2 \lambda_s T \cdot \Delta R, \tag{3}$$

where ΔR is the risk increase when the component is found failed at the moment of the test, λ_s is the standby constant failure rate, and T is the STI. Similar to the AOT risk contributors, the STIs can be classified and set to a limiting value to the risk contribution,

$$T_{\max} = (2R_{D\max})/(\lambda_s \Delta R) \tag{4}$$

(b) Test-Caused Risk

To evaluate and identify the test-caused risk, events should be analyzed and those caused by a test should be identified. These could be due to failure in human interactions or component wear-out on testing. Failure due to Human Error Probability can be modelled and quantified from detailed Human Reliability Analysis. Component wear-out can be addressed by aging risk analysis. However an integrated approach to work out such test-caused risk is a developing subject and presently is beyond the scope of this chapter.

2.2 Advanced Techniques for Technical Specification Optimization

The issue of risk effectiveness versus resource utilization is an optimization problem where the resources, viz., number of tests conducted, working hours required, costs incurred, radiation exposure, etc., are to be minimized while the performance or unavailability is constrained to be at a given level. As mentioned by Martorell [2], in optimizing test intervals based on risk (or unavailability) and cost, one normally faces multi-modal and non-linear objective functions and a variety of both linear and non-linear constraints. In addition, requirements such as continuity and differentiability of objective and constraints functions add yet another conflicting element to the decision process. Resolution of such complex optimization problems requires numerical methods. However, as traditional approaches usually give poor results under these circumstances, new methods based on Genetic Algorithms (GAs) were investigated in order to try to solve this kind of complex optimization problems [2–5]. This section presents a solution to test interval optimization problem with genetic algorithm along with a case study of a safety system for Pressurized Heavy Water Reactor (PHWR).

2.2.1 Mathematical Modeling of Problem

Notations:

T	Surveillance test interval
M	Mean time to preventive maintenance
t	Mean time to test
M	Maintenance interval
λ	Standby failure rate
c_{ht}	Testing cost per hour
ρ	Per-demand failure probability
C_{hm}	Preventive maintenance cost per hour
d	Mean time to repair
c_{hc}	Corrective maintenance cost per hour

System unavailability model in the PRA is adopted to represent the risk function. It is obvious that by optimizing test intervals based on minimizing the corresponding safety system unavailability one can improve the safety level of NPP. Unavailability function of the system is generally derived from fault tree analysis, which is a logical and graphical description of various combinations of failure events. Minimal cut-sets are obtained from fault tree analysis which represents minimal combinations of basic events (components) leading to unavailability of system. Thus, system unavailability is expressed as a function of unavailability of components. As safety system is considered for case studies and normally all the components in a safety system are in standby mode, the following model (refer Eq. 5) as explained in Martorel et al. [2], Vaurio [6] represents the unavailability of component. It is a function of unavailability arising from random failure during standby mode, surveillance testing, preventive maintenance activity, and corrective maintenance due to observed failure.

$$u(x) = u_r(x) + u_t(x) + u_m(x) + u_c(x) \tag{5}$$

$u(x)$	Represents unavailability of component that depends on the vector of decision variables x
$u_r(x)$	Contribution from random failures $\approx \rho + \lambda T/2$
$u_t(x)$	Contribution from testing $\approx t/T$
$u_m(x)$	Contribution from preventive maintenance $\approx m/M$

$u_c(x)$ Contribution from corrective maintenance $\approx (\rho + \lambda T)d/T$
thus

$$u(x) = \rho + \lambda T/2 + t/T + m/M + (\rho + \lambda T)d/T \tag{6}$$

System unavailability is sum of j number of minimal cut-sets and the product k extents to the number of basic events in the jth cut-set as given in Eq. 7:

$$U(x) \approx \sum_j \prod_k u_{jk}(x) \tag{7}$$

u_{jk} represents the unavailability associated with the basic event k belonging to minimal cut-set number j. Similarly the cost model is given as follows:

$$c(x) = c_t(x) + c_m(x) + c_c(x) \tag{8}$$

The total cost $c(x)$ of the component (yearwise contribution) includes costs due to testing $c_t(x)$, preventive maintenance $c_m(x)$, and corrective maintenance $c_c(x)$.

$$c(x) = \frac{t}{T}c_{ht} + \frac{m}{M}c_{hm} + \frac{1}{T}(\rho + \lambda T)dc_{hc} \tag{9}$$

The total yearly cost of the system having i number of components is given by:

$$C(x) = \sum_i c_i(x) \tag{10}$$

Both risk and cost functions are important to decision making in effective, efficient, and economical safety management of NPPs. In the first case, constraints are applied over one of the two objective functions, risk or cost function. These are referred to as implicit constraints, where, for example, if the selected objective function to be minimized is the risk, $U(x)$, then the constraint is a restriction over the maximum allowed value to its corresponding cost. In the second case, the selected objective function to be minimized is the cost, $C(x)$, and the constraint is stated through the maximum allowed value for the risk. One can also impose constraints directly over the values the decision variables in vector x can take, which are referred to as explicit constraints.

2.2.2 Genetic Algorithm (GA) as Optimization Method

The GA is a stochastic global search method that mimics the metaphor of natural biological evolution. GA operates on a population of potential solutions applying the principle of survival of the fittest to produce better and better approximations to a solution. At each generation, a new set of approximations is created by the process of selecting individuals according to their level of fitness in the problem domain and breeding them together using operators borrowed from natural genetics. This process leads to the evolution of populations of individuals that are

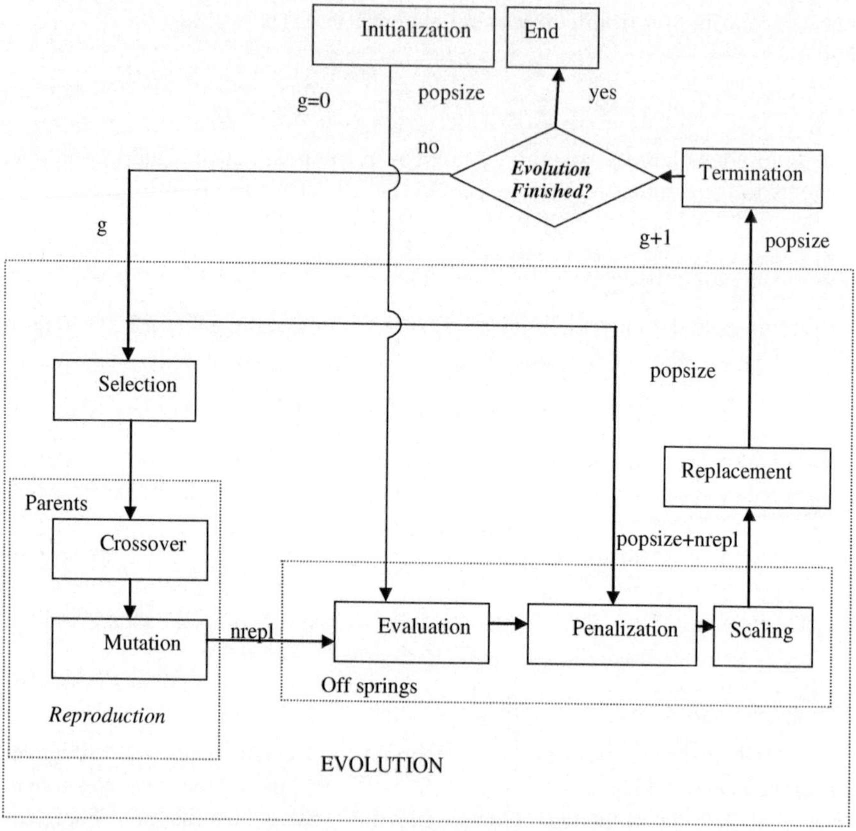

Fig. 3 Steady-state genetic algorithm scheme

better suited to their environment than the individuals that they were created from, just as in natural adaptation. Individuals, or current approximations, are encoded as strings, chromosomes, composed over some alphabet(s), so that the genotypes (chromosome values) are uniquely mapped onto the decision variable (phenotypic) domain. The most commonly used representation in GAs is the binary alphabet {0, 1} although other representations can be used, e.g., ternary, integer, real-valued, etc.

The main feature of the SSGA is the utilization of overlapping populations, as it can be observed in Fig. 3. The SSGA starts with an initial population of a given size. The number of individuals that constitute this base population, denoted by popsize, is selected by the user. This algorithm generates an auxiliary population, of size nrepl, constituted by the offspring obtained after the reproduction of certain individuals selected from the base population. Newly generated offspring is evaluated and then added to the base population. Each individual of the resulting population, composed by popsize + nrepl individuals, is penalized and then scaled to derive a ranking of individuals based on their fitness score. After scaling, the nrepl worst individuals in the ranking are removed in order to return the population to its original size (popsize). Therefore, after replacement, the best individuals remain in the new population constituting the new generation, generically denoted by $g + 1$, which descends from previous one, g. The number of individuals to be replaced, nrepl, is fixed as 6 in the present problem. Once the new population is generated, the algorithm checks if the termination criterion is satisfied. In case the criterion is not satisfied, then the evolution continues to produce new generation as described previously. The best fit of the population that satisfied termination criteria gives the optimum solution to the problem.

The binary encoding scheme of the decision variables is used for the current problem, test interval optimization, due to its simplicity in mutation operation and the range constraint is automatically implicit in the encoding. The roulette-wheel method, which is a stochastic sampling method that picks the individuals by simulating the roulette-wheel, is used in for the process of selection. The one-point crossover has been chosen for the crossover operation, which is a very simple method widely used that provides good results. Population size of 100 (popsize) and auxiliary population size of 6 (nrepl) are taken. Crossover and mutation probabilities of 0.7 and 0.1 are assumed in the calculations. More details about steady-state genetic algorithm can be found in Martorell et al. [2] and Goldberg [7].

2.2.3 Case Studies: Test Interval Optimization for Emergency Core Cooling System of PHWR

Emergency core cooling system (ECCS), one of the important safety systems in a Nuclear power Plant, is designed to remove the decay heat from the fuel following a Loss of Coolant Accident (LOCA) and provides means of transferring decay heat to the ultimate heat sink under all credible modes of failure of the Primary Heat Transport System (PHTS) pressure boundary. The operation of ECCS consists of two phases, viz., injection phase and recirculation phase. The surveillance testing is focused here on only recirculation part. This consists of four pumps that draw water from suppression pool and inject it into the PHT system header after the light water accumulator level becomes low. Upon the occurrence of LOCA, as sensed by low inlet header and header differential pressure signals, ECCS is initiated depending upon the location of LOCA as sensed by header differential pressure. The schematic diagram of ECCS (only recirculation part) in a typical PHWR is shown in Fig. 4.

In this problem, the system components are grouped into three different test strategies. Strategy 1 covers the four motor operated suction valves, namely, SV1, SV2, SV3, and SV4. Strategy 2 covers the four motor operated discharge valves, DV1, DV2, DV3, and DV4. Finally, four pumps, P1, P2, P3, and P4 are placed in the third strategy. It is assumed that all the components in the same group will have the same test interval. Further, test strategies must satisfy the following relationship in our particular case of application:

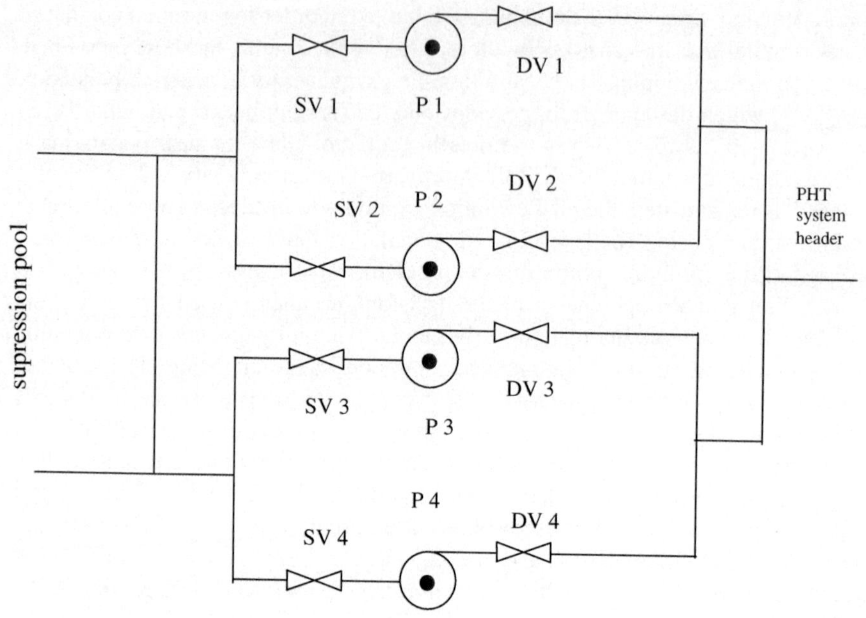

Fig. 4 Schematic diagram of ECCS recirculation

Table 1 Unavailability and cost parameters

S. no	Name	λ (per h)	ρ (failure/demand)	T (h)	T (h)	D (h)	c_{ht} (Rs/h)	c_{hc} (Rs/h)
1	P	3.89e-6	5.3e-4	4	2,190	24	250	200
2	SV	5.83e-6	1.82e-3	1	2,190	2.6	250	200
3	DV	5.83e-6	1.82e-3	1	2,190	2.6	250	200

$$T_2 = k_2 T_1 \text{ and } T_3 = k_3 T_2, \tag{11}$$

where T_1, T_2, and T_3 are test interval for strategy 1, 2 and 3, respectively, where k_2 and k_3 are integers that must lie in between 1 and 10. T_1 must lie between [0, 8,760]. The current practice recommends 1 month for all the components and the cost of test and maintenance for the current practice is Rs. 74082.6 (in Indian Rupees (Rs.)) when it is calculated keeping the failure and repair parameters at their nominal values. It is to be noted that cost of maintenance is a function of failure rate, demand failure probability, and repair time (refer Eq. 9). The unavailability parameters of pumps and valves, and the cost parameters are shown in Table 1.

In developing the cost function, costs of only repairs and testing are considered. Computer coding for the genetic algorithm based optimization has been used to solve the problem [8]. The parameters adopted for genetic algorithm and generic operators are shown in Tables 2 and 3, respectively.

Table 2 Genetic algorithm parameters

S. No.	Parameter	Values
1	Encoding	Binary
2	Chromosome size	22
3	Population size	100
4	Crossover probability	0.7
5	Mutation probability	0.3
6	Replacement	10
7	Generations	5,000
8	Conv. prob.	0.99
9	Diversity	0.01

Table 3 Genetic operators

S. No.	Operator	Method
1	Selection	Roulette-wheel
2	Crossover	One point
3	Mutation	Flip mutator
4	Scaling	Linear

Table 4 Optimized values

Variable	Initial values	Optimized values	
		Unavailability as objective function	Cost as objective function
T_1(h), k_2, k_3	720, 1, 1	480, 1, 2	575, 1, 2
Unavailability	3.86e-6	2.86e-6	3.86e-6
Cost (Rs.)	74082.6	74,082	61998.7

The initial population in SSGA implementation is normally generated using a random method. However, it cannot guarantee the criteria of satisfying constraints, therefore the actual test intervals implemented in the plant are considered for initial population. A generation dependent dynamic penalization model and termination criteria have been used in SSGA.

In the first case, the unavailability of the system has been considered as objective function and cost per year (Rs. 74082.6) as the constraint apart from satisfying above-said intervals for decision variables T_1, k_2, and k_3. In the second case, cost per year has been considered as objective function and unavailability (3.86e-6) as the constraint. The results achieved for the optimized values of unavailability/cost, the cost/unavailability associated with that unavailability/cost, and the optimized decision variables are shown in Table 4. In both the cases, the optimized test intervals are decreased for valves and increased for pumps with respect to their initial values. Finally, it is found that important reductions in both unavailability and cost measures have been achieved while all the explicit and implicit constraints are satisfied for the optimized test intervals in both the cases.

2.3 Concluding Remarks on Technical Specification Optimization

Risk-informed decision making ensures safe, economical, and efficient design and operation of nuclear power plants. Test interval optimization, which is one of the important applications of risk-informed approach, has been applied to emergency core cooling system of PHWR. In Sect. 2.2.3, Genetic algorithm has been successfully applied to perform the constrained optimization of test intervals at NPPs, where its capabilities of simplicity, flexibility, easy operation, minimal requirements, and global perspective to find global optimum have been shown. From the case studies it is found that the recommended test strategy is better than the test strategy being followed currently. This methodology provides a framework not only for the mentioned constraints but also other constraints of concern to specific operational scenarios.

3 Risk-Informed In-Service Inspection

Structural Components like piping, welds, fittings etc., are subjected to various loading due to fatigue damage as well as degradation mechanisms present on it. In order to ensure the structural integrity, In-Service Inspection has been taken up at periodic intervals. Some structural components may be very critical, but may not have active high degradation, while others may not be a critical component but have high degradation mechanism. So it has become necessary to perform ISI in a systematic manner consistent with safety level. Since the large number of structural components is present in an NPP, it has become all the more essential to bring out an optimum inspection plan for allocation of inspection resources [9]. Various methodologies developed to achieve this objective are discussed in this section.

Risk-informed in-service inspections programs were initiated by ASME Section XI as an alternative to the current inspection programs. The progression from an implicit risk-informed logic to an explicit risk-informed logic has been seen by many to be a natural progression. A principal difference, however, between the present code and the new risk-informed code, is not only the use of an explicit evaluation of risk but also that this risk is based primarily on the operational details of each specific plant rather than the design analysis. Beginning in late 1988, a multi-disciplined ASME Research Task Force on Risk-Based Inspection Guidelines has been evaluating and integrating these technologies in order to recommend and describe appropriate approaches for establishing risk-informed inspection guidelines. This task force is comprised of members from private industry, government, and academia representing a variety of industries. The NRC, as part of the research effort, applied this technology in pilot studies of inspection requirements for both PWR and BWR plant systems. Later, it requested the ASME Research Task Force to make the risk-informed inspection process consistent with other Probabilistic Safety Assessment (PSA) applications. ASME Section XI formed a Working Group on Implementation of Risk-Based Examination to begin making Code changes based on risk for inspection of passive, pressure boundary components. The first efforts of this group have been to develop Code Cases [10, 11] providing risk-informed selection rules for Class 1, 2, and 3 piping.

The goal of Risk-informed ISI is to allow the use of risk assessment, understanding of component-specific degradation mechanisms, to establish an effective plant integrity management program, which maintains plant safety, while at the same time reducing the burden associated with current ISI requirements. These applications also yield significant safety, worker radiation exposure, and economic benefits. The main advantages of RI-ISI can be summarized as:

1. Decision making based on risk criteria and deterministic information.
2. Better focus on allocating resources to high-safety significant components.
3. Focus on justifying risk increase.
4. In-Service Inspection based on failure modes of components and associated risk.

3.1 RI-ISI Models

There are two independent methods for RI-ISI, viz., ASME/WOG model and EPRI models. Both are discussed in this section.

3.1.1 ASME/WOG Model

The methodology developed by ASME/WOG [12, 13] addresses the quantitative aspect of RI-ISI program, which include:

- Identification of systems and boundaries using information from a plant PSA.
- Ranking of components (piping segments), applying the risk measures to determine the categories that are then reviewed to add deterministic insights in to making final selection of where to focus ISI resources.
- Determination of effective ISI programs that define when and how to appropriately inspect or test the two categories of high-safety significant and low-safety significant components.
- Performing the ISI to verify component reliability and then updating the risk ranking based on inspection and test results.

The first step in Risk-based Inspection is the review of level 1 PSA results of the NPP in concern. The accident sequences, which result in core damage following the occurrences of pre-determined initiating events, are identified. Those basic events, which contribute significantly to the occurrence of the key accident sequence, are identified by applying the appropriate importance measures. These importance measures suggest the importance of systems/components with respect

Table 5 Risk categorization based on importance measures

Risk category	Criterion
Potentially high	RRW > 1.005 and RAW > 2
High	RRW < 1.005 and RRW > 1.001
Low	RRW < 1.001 and RAW < 2

to Core Damage Frequency. Various importance measures, like Fussel-Vesely, Birnbaum Importance, Inspection Importance measure etc., are employed for prioritization, which are discussed in the preceding chapter.

(i) System Prioritization Methodology. There are many importance measures that could be used to rank systems. For example, the Fussel-Vesely (FV) importance measure involves small changes in risk. Importance measures involving larger changes in risk are Birnbaum importance and RAW. Since pipe break probability is a small probability, Birnbaum importance does not reflect the likelihood of failure. A new parameter called inspection importance measure has been developed in order to prioritize the systems for ISI. System level ranking based on Inspection Importance Measure (I^w). Inspection Importance (I^W) of a component is defined as the product of the Birnbaum Importance (I^B) times the failure probability.

$$I^W_{\text{sys}} = I^B_{\text{sys}} \times P_{f_{\text{sys}}} \tag{12}$$

$P_{f_{\text{sys}}}$ System failure probability due to structural integrity failures.

The Inspection Importance is an approximation of the Fussel-Vesely importance of pipe break for the system and has all the useful properties of the Fussel-Vesely importance measure for establishing the inspection priorities.

Birnbaum and Fussell-Vesely importance measures have been suggested by ASME for Risk-Informed In-Service Inspection. In most of the applications, the exact ranking is not important. Guidance and experience for applying importance measures for In-Service Testing/In-Service Inspection is mainly based on expert opinion. A sample categorization is given in Table 5, where RAW refers to Risk Achievement Worth.

(ii) Component (weld) Prioritization Methodology. For the systems selected for more detailed analyses (based on the above prioritization methodology), the most risk-important segments/components should be selected for inspection. Failure Modes and Effects Analysis (FMEA), which is a systematic, logical process for identifying equipment failure modes for a plant, system, or component, has been selected as the methodology for component prioritization. The FMEA inductively determines the effects of such failures will have on the desired operational characteristic of the system being analyzed. The most useful outputs of an FMEA are the assessment of design adequacy of the system to perform its intended function.

The FMEA results can be used to calculate the importance index or relative importance of each weld. This importance index is based on the expected consequence of the failure of weld, as measured as the probability of core damage

Table 6 FMEA sample sheet

(1) Piping section (location)	(2) Failure probability	(3) Failure effect	(4) Recovery action	(5) Core damage probability	(6) Relative importance	(7) Remarks
•	•	•	•	•	•	•
•	•	•	•	•	•	•
•	•	•	•	•	•	•
	•	•	•	•	•	•

resulting from the weld failure. In mathematical terms, the probability of core damage resulting from weld failures is defined as

$$P_{cd} = P_{fi} \times P_{cd|si} \times P_{si|Pf} \times R_i \tag{13}$$

where, P_{cd} is the probability of core damage resulting from weld failure, P_{fi} is the failure probability of weld, $P_{cd|si}$ is the conditional probability of core damage, given system i failure, $P_{si|Pf}$ is the conditional probability of system i failure, given a weld failure, R_i is the probability that operator fails to recover, given system i failure (Table 6).

These rankings also form a basis for determining the inspection category and type of examination required. ASME code case 577 is developed for conducting RI-ISI based on WOG methodology.

3.1.2 EPRI Model

Another methodology has been developed by EPRI. Fleming [14, 15] discusses their methodology (Fig. 5), which analyzes the degradation mechanisms in structures in detail. EPRI methodology blends PSA and deterministic insights.

Risk matrix [16] can be defined as a Decision matrix that is used to categorize the pipe segments into high, medium and low importance, based on degradation mechanism and consequence of its failure (Fig. 6). By examining the service data, a basis has been established for ranking pipe segment rupture potential as high, medium, or low simply by understanding the type of degradation mechanism present (Table 7). Consequence can be quantified through the estimation of Conditional Core Damage Probability (CCDP) .

The matrix defines three broad categories of relative failure potential that are derived from an underlying quantification of pipe rupture frequencies and four categories of relative consequences that are derived from an underlying quantification of conditional probability for a severe core damage accident given a postulated pipe ruptures (Table 8). Different categories are defined which proposed different inspection plans. The bounding values of CCDP and rupture potential are:

The consequence evaluation group is organized into two basic impact groups: (i) Initiating Event and (ii) Loss of Mitigating Ability. In *Initiating Event impact Group*, the event occurs when a pressure boundary failure occurs in an operating

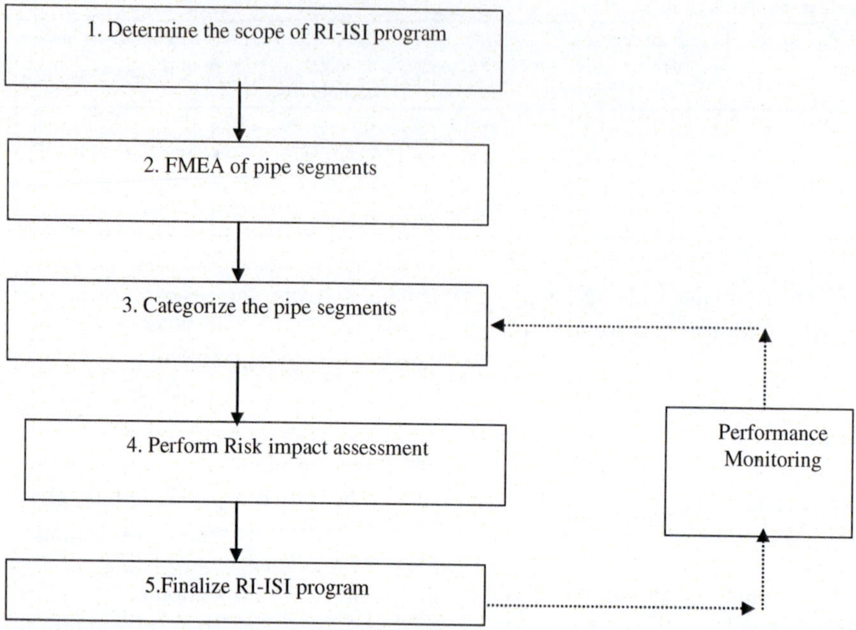

Fig. 5 Flow-chart on RI–ISI program by EPRI

CONSEQUENCE CATEGORY (CCDP)

Likelihood frequency / *Consequence*	None $<10^{-8}$	Low $10^{-8}<$ CCDP $<10^{-6}$	Medium $10^{-6}<$ CCDP $<10^{-4}$	High $>10^{-4}$
High ($>10^{-4}$)	Low 7	Medium 5	High -3 CDF = $10^{-10}-10^{-4}$	High -1 CDF $>10^{-8}$
Medium ($10^{-7}<F<10^{-4}$)	Low 7	Low 6	Medium 5	High - 2 CDF = $10^{-11}-10^{-4}$
Low ($<10^{-7}$) No deg-mech	Low 7	Low 7	Low 7	Medium 4

Fig. 6 Risk matrix

system. This could occur because of loss of fluid (LOCA, Feed water line break) and a loss of system (like service water-cooling). The importance of every initiating event, caused by a pipe failure, needs to be assessed in order to assign it to its appropriate consequence category. CCDP can be directly obtained from the PSA results, by dividing the CDF due to the specific IE by the frequency of that IE. In the *Loss of Mitigating* Ability group, the event describes the pipe failures in safety system. Safety system can be in two configurations, Standby and Demand. While in standby configuration, the failure may not result in an initiating event, but degrades

Table 7 Classification of degradation mechanism

Potential	Degradation mechanism
High	Flow accelerated corrosion, vibration fatigue, water hammer
Medium	Thermal fatigue, corrosion fatigue, stress corrosion cracking, pitting, erosion-corrosion
Low	No degradation mechanism

Table 8 Classification of consequence

	CCDP	Rupture frequency
High	1	1E-4
Medium	1E-4	1E-5
Low	1E-6	1E-6

the mitigating capabilities. After failure is discovered, the plant enters the Allowed Outage Time (AOT). In consequence evaluation, AOT is referred to as exposure time.

$$\text{CCDP}_i = [\text{CDF}_{(\lambda i=1)} - \text{CDF}_{(\text{BASE})}] * T_E \tag{14}$$

where, $\text{CDF}_{(\lambda i\,=\,1)}$ is the CDF given the component failure in a given safety system, $\text{CDF}_{(\text{BASE})}$ is the BASE CDF, λ_i is the Pipe break frequency, T_E is the Exposure Time (Detection time + AOT).

While in demand configuration, the failure occurs when the system/train operation is required by an independent demand. Here, instead of exposure time, time since the last demand is considered, which is the test interval.

$$\text{CCDP}_i = [\text{CDF}_{(\lambda i=1)} - \text{CDF}_{(\text{BASE})}] * T_t \tag{15}$$

where, $\text{CDF}_{(\lambda i\,=\,1)}$ is the CDF given the component failure, $\text{CDF}_{(\text{BASE})}$ is the BASE CDF, λ_i is the Pipe break frequency, T_t is the Mean time between tests or demands.

Measure of Risk due to pipe break:

$\text{CDF}_i = \lambda_i * \text{CCDP}_I$

In order to evaluate the impact of risk from changes in in-service inspection, the change in CDF from both the inspection methodologies has been used as a measure. The model described in Eq. 16 is based on the influence of pipe frequency at a location j due to the inspection program. The change in the risk of core damage at location j that is impacted by the changes in Risk-informed inspection program can be estimated as:

$$\Delta\text{CDF}_j = \left(F_{rj} - F_{ej}\right) * \text{CDF}_j = \left(I_{rj} - I_{ej}\right) * F_{0j} * \text{CCDP}_j \tag{16}$$

where,

$$F_{Aj} = F_{0j} \cdot I_{Aj} \tag{17}$$

$CCDP_j$ is the Conditional Core Damage Probability from pipe rupture at location j. The subscripts "rj" refer to risk-informed approach and "ej" refer to existing strategy.

F_{Aj} =Frequency of pipe rupture at location j subject to inspection strategy A

$$\begin{aligned} F_{0j} &= \text{Frequency of pipe rupture at location } j \text{ subject to no inspection} \\ I_{Aj} &= \text{Inspection effectiveness factor}(0 \; to \; 1) \\ &= \text{This is the probability that the flaw is detected} \\ &= 1 - POD_{Aj} \end{aligned}$$

After the estimation of risk impact or ΔCDF, depending on the acceptable criteria for ΔCDF, the decision shall be made regarding the adoption of inspection strategy. The decision criterion that has been suggested by EPRI is to ensure that the cumulative change in CDF is less than 1E-7/year/system for the employment of the new methodology.

Comparison of RI-ISI Models

The EPRI RI-ISI process includes: selection of RI-ISI program scope, failure modes and effect analysis, risk categorization of pipe elements, selection of inspection locations and examination methods, evaluation of risk impacts of inspection program changes, and final RI-ISI program definition.

After the identification of the critical systems/components, Failure Mode Effect Analysis (FMEA) should be carried out on the basic event. It is essential to identify the prominent failure modes and causes in order to establish the inspection items and guidelines. Risk Matrix is designed with different categories, depending on the CDF values and degradation mechanism for determining the inspection interval. Each segment is assigned the appropriate category depending on its ΔCDF and degradation mechanism.

The EPRI's risk-informed procedure for selecting an ISI program gives a very straightforward approach to the issue. The method introduced in risk-informed fashion combines both the plant-specific PSA information and the deterministic insights in support of the system-specific, detailed ISI program selection. Piping of all systems important to safety are exposed to the selection procedure irrespective of the ASME class (1, 2, 3, or even non-code piping). The selection procedure includes four major steps such as:

- Selection of systems and identification of the evaluation boundaries and functions.
- Failure Mode and Effect Analysis (FMEA) including both consequence evaluation and qualitative degradation mechanism evaluation. These two factors are then used for dividing the systems into pipe segments representing common consequences and degradation mechanisms.
- Risk evaluation is made based on the results of FMEA. The risk matrix is built up on the basis of degradation category (low, medium, high) reflecting the

Table 9 Comparison between WOG and EPRI RI-ISI approaches

STEP	WOG	EPRI
Piping failure probability assessment	Quantitative	Qualitative
Risk evaluation	Classification using RRW	Categorization of segments in three risk regions
Expert panel	Required	Not required
Structural element/ NDE selection	Statistical sampling on target reliability	Significant sampling—25, 10, and 0% from high, medium, and low risk region

potential for large break, and consequence category (low, medium, high) reflecting the core melt potential for limiting break size.

- The division of pipes into segments of various degradation categories is based mainly on qualitative identification of the mechanism, which the pipe segment is exposed to (such as erosion-corrosion, vibration fatigue, water hammer, thermal fatigue, stress corrosion cracking, and others). Consequently, the piping failure data were used to determine the severity and frequency of degradation mechanisms in order to determine the quantitative degradation categories.
- The division of pipes into segments of various consequence categories is based on conditional core damage frequency. High consequence category refers to the conditional core damage frequency class (CCDF) $> 10^{-4}$, medium consequence category to class 10^{-6} <CCDF $< 10^{-4}$, and low consequence category to class CCDF $< 10^{-6}$. The degradation and consequence category pairs determine the risk classes, low, medium, high.
- Finally the pipe segments are divided into two main categories. One contains high and medium risk segments and another category contains low risk segments.

In EPRI's pilot study at least one-fourth (1/4) of the welds in pipe segments of high risk and one tenth (1/10) of welds in pipe segments of medium risk are selected for examination, whereas the welds in pipe segments that fall in the low risk class will continue to be subject to system pressure and leak tests. The examination of specific elements of segments in high and medium risk classes is based on the degradation mechanism, as well as on inspection costs, radiation exposure and accessibility.

The ASME/WOG and EPRI's approaches as well as the NRC's regulatory guide strongly emphasize and recommend that both deterministic and probabilistic engineering insights need to be carefully analyzed and combined for aiding the final decision-making process while selecting the ISI program on piping. A typical approach to combine the information is a panel discussion containing all affecting engineering disciplines. Such a panel discussion is a procedure to reduce the knowledge-based uncertainties which may seriously damage the decision-making process. Table 9 summarizes the comparison between WOG and EPRI RI-ISI approaches.

The NRC's [17] regulatory guide recommends that the potential pipe break probabilities can be estimated by probabilistic fracture mechanics methods.

The related computer codes, complex or simplified, can be used to estimate the piping failures as a function of time. An alternative method is to use expert opinion in conjunction with probabilistic fracture mechanics methods to determine the degradation category of each pipe segment. The degradation categories (low, medium, and high) reflect the potential for large break or rupture.

3.2 ISI & Piping Failure Frequency

Main tasks for RI-ISI revolve around determination of probability of failure and consequence of failure. For quantification of risk in Nuclear Power Plants, Probabilistic Safety Assessment (PSA) models are widely employed, which forms the basis for consequence quantification for RI-ISI. Various methods have been suggested for piping failure parameter estimation like Structural Reliability Analysis (SRA), Service Data Analysis, Expert Opinion, Remaining Life models, etc. The degree to which one relies on one method or another is predicted on the availability of data from service experience, experts or structural reliability or risk models. These aspects were discussed in detail in the preceding section.

SRA employs the use of probabilistic fracture mechanics techniques to calculate the failure probability as a function of time, including the effects of inspection frequency, probability of detection (POD), and degradation mechanism. Through Monte Carlo sampling, the results of tracking a very large number of crack simulations can be used to determine what fraction of cracks will not be detected and repaired before failure results. This methodology provides models for determining the crack growth for different degradation mechanisms also. These models are computationally intensive. The results of these analyses are often driven by uncertainties in defining crack size distribution, stress history, detection probability, and reference flaw size. Some models are available for incorporating ISI as discussed in the preceding section and are not amenable for various issues arising in maintenance activities. In Statistical approach, databases are an important source of information that can support the estimation. Database should comprise the cause of failure, thereby backtracking to the applicable degradation or damage mechanism, which culminated in the pipe failure. There are various problems associated with database ranging from reporting the event to the appropriate root cause analysis of each event reported. Also how far the effect of life management program can be incorporated is still under review.

3.2.1 In-Service Inspection

Nondestructive Testing (NDT), Nondestructive Inspection (NDI), and Nondestructive Evaluation (NDE) denote variations in application of materials evaluation technology that range from process control to the measurement of a material characteristic that is critical to the structural integrity and safe operating life of an engineering system. Some of the important NDT techniques are:

Fig. 7 NDT on piping

- Liquid penetrant inspection.
- Magnetic particle inspection.
- Radiographic inspection (X-ray and gamma ray).
- Electromagnetic inspection.
- Ultrasonic inspection; and
- Thermographic inspection.

Figure 7 shows a picture of inspection on piping. Non-Destructive Testing (NDT) carries an important role in predicting the piping failure frequency. Depending on the technique used the confidence of finding defects varies. If any defect is detected, decision will be taken to undertake repair activity in piping. This will decrease the piping failure frequency and should be accounted for. The efficiency of inspection is quantified through the introduction of the concept of **"Probability of Detection (POD)"**. The **"Probability of Detection (POD)"** concept and methodology have gained widespread acceptance and continuing improvements have enhanced its acceptance as a useful metric for quantifying and assessing NDE capabilities [18]. Since a wide range of NDE methods and procedures are used in "fracture control" of engineering hardware and systems, a large volume of POD data has been generated to validate the capabilities of specific NDE procedures in a multitude of applications. Figure 8 presents a typical POD curve obtained from ultrasound inspection. Sometimes it will generate POD curves for the site equipments. In such cases, models are also developed for determining POD.

Failure parameter of the component gets modified according to the type and frequency of inspection applied on it. Hence, it is essential to account for the frequency of inspection and the type of inspection adopted for a component, while suggesting its failure probability/frequency.

3.2.2 Models for Including ISI Effect on Piping Failure Frequency

Various issues are involved in realistic estimation of probability of failure like incorporating the effects of degradation mechanisms acting on it, repair activities,

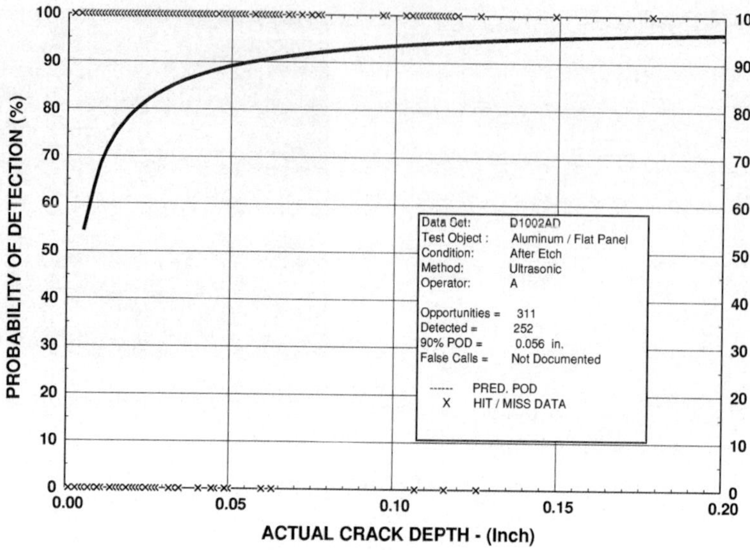

Fig. 8 A typical POD curve obtained from ultrasonic inspection

etc. In the context of RI-ISI [19], the models for piping failure probability estimation need to incorporate the effects of In-Service Inspection frequency, inspection technique involved, etc. The above methods incorporate this information in a manner, which is not amenable for RI-ISI. A suitable model needs to be devised which can be used flexibly to study the effects of inspection interval and techniques. Markov model has been found to be a suitable candidate to study these effects, which can be represented as a state—transition problem.

Piping failure analysis has always been a controversial topic. The unavailability models for active components comprise of failure rate, mission time, and repair and maintenance parameters acting on it. The reliability model of piping systems should meet the following objectives:

- Account for statistical evidence and engineering insights from service experience accumulated through several thousand reactor years of commercial nuclear power plant operating experience.
- Predict the impacts that changes in the in-service inspection program may have on the frequency of pipe ruptures. These changes include adding and removing locations from the inspection program, changing from fixed to randomly selected locations from one inspection interval to the next, and qualitative enhancements to the inspection process that could influence the nondestructive examination (NDE) reliability of a given inspection.
- Account for the full set of pipe failure mechanisms found in the service experience including those due to active degradation mechanisms, severe and normal loading conditions, and combinations of degradation and loading conditions.

- Account for leak before break characteristics of pipe failure modes when appropriate and also account for the possibility to detect and repair a leaking pipe before it degrades to rupture.
- Address uncertainties in the reliability assessment and database development and account for uncertainties in estimating pipe rupture, core damage frequency, and large early release frequencies.
- The models and databases to address the above issues should be provided in forms that can be easily applied by utility personnel in implementing a risk-informed evaluation of the piping inspection program.

During an independent review of the EPRI RI-ISI procedures, an approach to piping reliability assessment was envisioned. This approach makes use of a reliability modeling technique, Markov modeling [20]. A Markov model of a system is defined by assigning two or more discrete states that the system may occupy at any point in time. Transition is permitted from state to state to account for the occurrence of component failures and the possibility that failed components may be repaired. The model is used to develop a set of differential equations, the solution of which is the time dependent probability that the system occupies each state. Other reliability metrics such as system failure rate or hazard rate can also be derived from this model.

In applying the concept to pipes, it was seen that there are natural states that can be assigned to each element of the pipe, such as each weld and each small section of piping material. These states correspond to discreet levels of degradation such as flaw, crack, leak, or rupture as well as the state where the pipe is free of any damage or degradation. The processes that can be modeled in this application of the Markov model include piping degradation either progressively from flaw to leak to rupture, or instantaneously to leak or rupture from any less severe state. The model can also treat the repair processes associated with inspection and detection of critical flaws, detection of leaks, and repair of the damaged pipes prior to occurrence of rupture.

The successful application of the Markov modeling process requires application of the following steps:

1. Development of an appropriate set of states and state-transition possibilities.
2. Definition of the transition rate parameters that dictate the probability of transition from state to state.
3. Development of the differential equations for the Markov model and solution of these equations for the time dependent probability of occupying each state.
4. Development of a hazard rate function to develop the time dependent frequency of pipe ruptures.
5. Development of models for estimating the parameters of the Markov model in terms of observable quantities and reasonable and supportable assumptions. These models include the development of uncertainty distributions for each of the parameters that capture key uncertainties in the degradation processes and in the interpretation of the service experience.

Fig. 9 Three-state Markov model

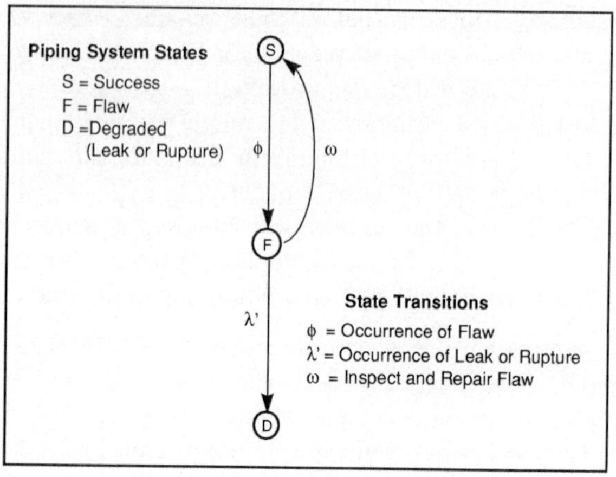

6. Development of a method of integrating the models from different pipe elements and segments into an overall model for a system for application of risk-informed inspection programs.

Discrete State Markov Model for Pipe Failures

The objective of Markov modeling approach is to explicitly model the interactions between degradation mechanisms and the inspection, detection, and repair strategies that can reduce the probability that failure occurs or the failure will progress to rupture. This Markov modeling technique starts with a representation of "piping segment" in a set of discrete and mutually exclusive states. At any instant of time, the system is permitted to change state in accordance with whatever competing processes are appropriate for that plant state. In this application of Markov model the state refers to various degrees of piping system degradation or repairs, i.e., the existence of flaws, leaks, or ruptures. The processes that can create a state change are failure mechanisms operating on the pipe and process of inspecting or detecting flaws and leaks, and repair of damage prior to progression of failure mechanism to rupture.

Three-state Markov model. This model would be applied to a pipe element such as a weld or small section of pipe that is uniquely defined in terms of the presence or absence of degradation mechanisms, loading conditions, and status in the inspection program. The model in Fig. 9 is developed to examine the singular role of the in-service inspection program, which can influence the total failure rate of pipe segments but has little if any impact on the conditional probability that a failure will be a rupture. A limitation of this model is that it does not distinguish between leaks and ruptures, cannot model leak before break, and cannot be used to examine the role of leak detection as a means to reduce pipe rupture frequencies. Another limitation is that leaks and ruptures are only permitted once the system is in the flaw state. This limitation makes the model suitable for degradation type failure mechanisms, but not for severe loading condition related causes such as vibration fatigue or water hammer. These limitations are removed in the next section in which a four-state model is developed and more possibilities are introduced for leaks and rupture transitions from the success state. However, to build up the knowledge about pipe reliability modeling in a step-by-step fashion, this has been found instructive to analyze this more simplified model to understand some basic properties of this approach to reliability modeling such that the necessary details can be built up in an organized fashion.

The relative frequency of pipe ruptures to pipe failures is only a function of the specific failure mechanism that caused the failure as reflected by the "leak before break" characteristic of the failure, and the capability to detect an initially leaking pipe and repair it prior to further degradation to rupture, which in many cases is virtually instantaneous. The model in Fig. 9 will also enable us to determine the time dependent failure frequency of piping systems subject to inspections. Hence, the simplified model in Fig. 9 is adequate to study the impact of changes in the inspection program on the failure frequency of piping systems. As long as changes to the leak detection part of the problem are not affected, one can solve this model for the pipe rupture failure probability and frequency, and use estimates of the conditional probability of pipe ruptures given failures to obtain the corresponding pipe rupture probabilities and frequencies.

Differential Equations and Solution for Markov Model

The differential equations for the model in Fig. 9 are given by:

$$\frac{\mathrm{d}S}{\mathrm{d}t} = -\phi S + \omega F \tag{18}$$

$$\frac{\mathrm{d}F}{\mathrm{d}t} = \phi S - (\lambda' + \omega)F \tag{19}$$

$$\frac{\mathrm{d}D}{\mathrm{d}t} = \lambda' F \tag{20}$$

The left-hand side of each equation represents the rate of change of the probability that the system occupies each state, S for the probability of success, F for the probability of a flaw, and D for the probability of a degraded state, i.e., leak or rupture. The Greek letters are the parameters of the model as defined in Fig. 9. φ is the occurrence rate for flaws, λ' is the occurrence rate for leaks and ruptures given a flaw, and ω is the rate at which flaws are inspected, detected, and repaired. The rate of leaks and ruptures, λ', can be further decomposed by:

$$\lambda' = \lambda_L + \lambda_C, \tag{21}$$

where λ_L is the occurrence rate of leaks given from a flaw state, λ_C is the occurrence rate of ruptures given a flaw state.

Hence, the total pipe failure rate given a flaw used in Fig. 9 corresponds to the sum of the leak and rupture failure rates and the rates are conditional on the existence of a flaw.

The solution of the system of Eqs. 18–20 can be obtained using Laplace transforms or any other suitable technique so long as the boundary conditions are specified. Since for safety-related piping, all are inspected to be free of detectable flaws at the beginning of commercial operation the appropriate boundary conditions are:

$$S\{t = 0\} = 1$$
$$D\{t = 0\} = F\{t = 0\} = 0$$

The time dependent solutions for the state probabilities are given by:

$$D\{t\} = 1 - \frac{1}{(r_1 - r_2)}(r_1 e^{r_2 t} - r_2 e^{r_1 t}) \tag{22}$$

$$F\{t\} = \frac{\phi}{(r_1 - r_2)}(e^{r_1 t} - e^{r_2 t}) \tag{23}$$

$$S\{t\} = 1 - D\{t\} - F\{t\} = \frac{1}{(r_1 - r_2)}[(r_1 + \phi)e^{r_2 t} - (r_2 + A)e^{r_1 t})], \tag{24}$$

where the terms A, r_1, and r_2 are defined according to:

$$A = \phi + \lambda' + \omega \tag{25}$$

$$r_1 = \frac{-A + \sqrt{A^2 - 4\phi\lambda'}}{2} \tag{26}$$

$$r_2 = \frac{-A - \sqrt{A^2 - 4\phi\lambda'}}{2} \tag{27}$$

Hazard Rate for Markov Model

In a PSA model, pipe failures in process systems are normally represented as initiating events. The quantity needed for this case is the initiating event frequency or pipe failure frequency. These initiating event frequencies are normally assumed constant in PSAs. With the Markov model, it is not necessary to make this assumption as whether the failure frequency is constant or not is a byproduct of the particular model. The reliability term needed to represent the pipe failure frequency is the system failure rate or hazard rate, as defined in the following.

To determine the system failure rate or hazard rate we must first determine the system reliability function for this model. Since we are primarily concerned with pipe failures and seek to estimate pipe failure frequencies, we may declare any state except for failure a "success" state, which in this model includes both the success state S and the flaw state F. Using this concept, the reliability function for the Markov model, $r\{t\}$, is given by:

$$r\{t\} = S\{t\} + F\{t\} = 1 - D\{t\} \tag{28}$$

By definition the hazard function and the reliability function are related according to the following equation:

$$h\{t\} = -\frac{1}{r\{t\}}\frac{\mathrm{d}r\{t\}}{\mathrm{d}t} = \frac{1}{(1 - D\{t\})}\frac{\mathrm{d}D\{t\}}{\mathrm{d}t} \tag{29}$$

Applying the solution to the Markov model in Fig. 9, an expression for the hazard function is developed as follows:

$$h\{t\} = \frac{r_1 r_2 (e^{r_1 t} - e^{r_2 t})}{(r_1 e^{r_2 t} - r_2 e^{r_1 t})} \tag{30}$$

Taking the limit of Eq. 28 as $t \rightarrow$ infinity provides us the long-term steady-state hazard rate, h_{SS} as:

$$h_{SS} = -r_1 = \frac{A - \sqrt{A^2 - 4\phi\lambda'}}{2} = \frac{\left(\phi + \lambda' + \omega\right) - \sqrt{(\phi + \lambda' + \omega) - 4\phi\lambda'}}{2} \tag{31}$$

The model in Fig. 9 has now been completely solved for its state probabilities and failure frequencies and is now available for use. Quantification can be completed once the parameter values are estimated for use in specific applications. These equations can be used to compute point estimates of state probabilities and failure frequencies as a function of time, and for use in uncertainty analysis in which uncertainty distributions for each parameter is propagated through the equations in a Monte Carlo Sampling process.

Four-State Markov model. This model consists of four states of pipe segment reflecting the progressive stage of pipe failure mechanism: the state with no flaw, development of flaws or detectable damage, the occurrence of leaks, and occurrence of pipe ruptures. As seen from this model pipe leaks and ruptures are permitted to occur directly from the flaw or leak state. The model accounts for state dependent failure and rupture processes and two repair processes. Once a flaw occurs, there is an opportunity for inspection and repair to account for in-service inspection program that searches for signs of degradation prior to the occurrence of pipe failures. Here the Leak stage L does not indicate actual leak, but represents a stage in which remaining pipe wall thickness is $0.45 \times t$ to $0.2 \times t$ (pipe wall thickness) (Fig. 10).

S is the success (depth of corrosion less than $0.1253t$), F is the flaw (depth of corrosion is $0.1253t$ to $0.453t$), L is the leak stage (depth of corrosion is

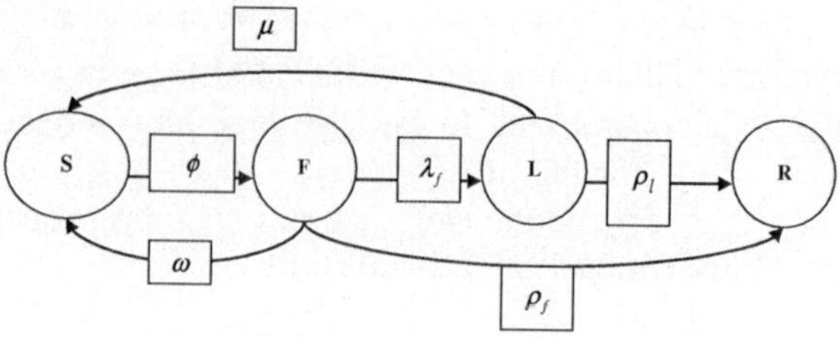

Fig. 10 Markov model for pipe elements with in-service inspection and leak detection

0.453*t* to 0.83*t*), *R* is the rupture (depth of corrosion beyond 0.83*t*), t is the pipe wall thickness

$$\begin{bmatrix} P'_s \\ P'_f \\ P'_l \\ P'_R \end{bmatrix} = \begin{bmatrix} -\phi & \omega & \mu & 0 \\ \phi & -(\omega + \lambda_f + \rho_f) & 0 & 0 \\ 0 & \lambda f & -(\mu + \rho_l) & 0 \\ 0 & \rho_f & \rho_l & 0 \end{bmatrix} \begin{bmatrix} P_s \\ P_f \\ P_l \\ P_R \end{bmatrix} \quad (32)$$

The Markov model diagram describes the failure and inspection processes as discrete state-continuous time problem. The occurrence rates for flaw, leaks, and ruptures are determined from limit state function formulation. The repair rates for flaws and leaks are estimated based on the characteristics of inspection and mean time to repair flaws and leak upon detection. Setting up differential equations for different states and finding the associated time dependent state probabilities can solve the Markov model. These equations are based on the assumption that the probability of transition from one state to another is proportional to transition rates indicated on the diagrams and there is no memory of how current state is arrived at. Assuming the plant life of 40 years, state probabilities are computed at the plant life.

3.3 *Case Study*

The PHWR outlet feeder piping system is taken as a typical case study. There are 306 number of small diameter pipes of diameter ranging from 40 to 70 mm and length 2–22 m that connect coolant channels to the outlet header. The feeder pipe considered in this case study is made of carbon steel A106GrB, with a diameter (*d*) of 70 mm and thickness (*t*) of 6.5 mm. After estimating the degradation rate, it has to be applied in the suitable limit state function to estimate the failure probability.

Assumptions

(i) It has been assumed that Erosion-Corrosion is present in outlet feeder.
(ii) A representative value has been assumed for corrosion rate.
(iii) To estimate the failure probability using FORM, normal distribution has been assumed for all the variables.

Fig. 11 Schematic of primary heat transport system

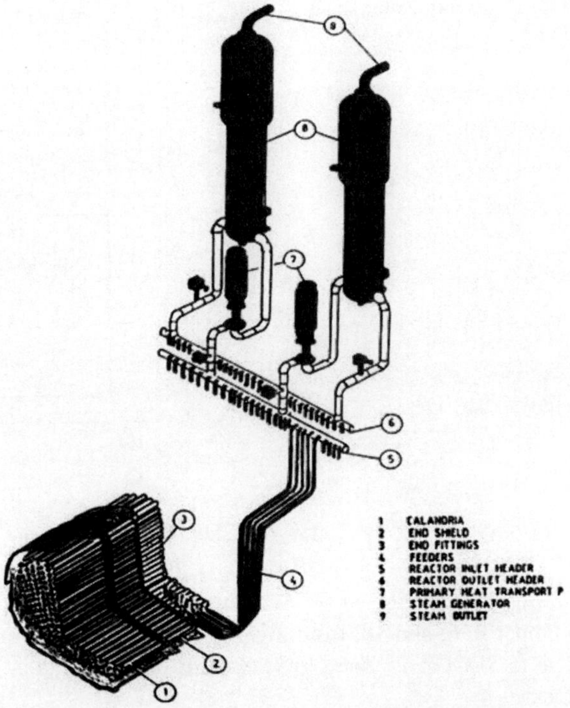

Consequence Analysis of Feeder Failure. The coolant channels are connected via individual feeder pipes to headers at both ends of the reactor. Figure 11 presents the schematic of Primary Heat Transport System, which includes feeder connections. Since feeder failure can result in Small Loss of Coolant Accident (SLOCA), it can be termed as an Initiating Event (IE). From the failure probability obtained from Markov models explained in previous sections, the IE frequency can be estimated using the equation given below:

$$\text{Failure Rate}_{\text{IE,feeder1}} = \frac{\text{Failure Probability}_{\text{IE,feeder1}}}{\text{EOL}} \tag{33}$$

where, EOL is the number of years the plant is licensed (e.g., 40 years).

In the event of feeder failure, Emergency Core Cooling System (ECCS) will be actuated. The ECCS is designed to provide enough coolant to the PHT system and to transport heat from the core to the ultimate heat sink in such a way as to ensure adequate reactor core cooling during all phases of LOCA. Event tree is drawn for this IE and accident sequences are found which can lead to core damage because of this IE. CDF due to the specific IE is estimated by adding the accident sequence frequencies from the IE (Fig. 12).

Conditional Core Damage Probability ($CCDP_i$) for a component failure can be directly obtained from the PSA results, by dividing the CDF due to the specific IE by the frequency of that IE.

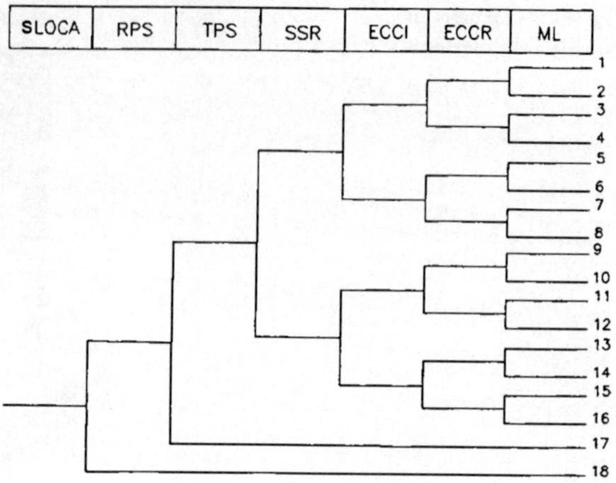

Fig. 12 Event tree for small LOCA

$$\mathrm{CCDP}_i = \frac{\mathrm{CDF}_{\text{due to IE}}}{\mathrm{IE}_{\text{frequency}}} \tag{34}$$

For the case of SLOCA, there are three accident sequences, viz., sequence number 4, 6, and 18, from this IE, which can result in Core Damage. The CCDP due to SLOCA is found to be 8.835E-06, which falls in medium category in risk matrix.

3.3.1 Using Three-State Markov Model

For three-state Markov models, three transition rates are involved as shown in Fig. 9. The fist transition rate φ representing the occurrence of flaw can be found out from limit state function or statistical method. However, in this case study a limit state function has been defined. Success State S represents a situation, in which flaw is less than $0.125 \times t$, and flaw state, F, represents a situation, in which flaw is $0.45 \times t$. ϕ represents transition rate from state S to state F. The limit state function can be defined as

$$G1(d, T) = 0.45 \times t - (d + \text{rate} \times T) \tag{35}$$

d is the undetected flaw $= 0.125 \times t$, T is the time of inspection usually 10 years, Rate is the Erosion-corrosion rate (mm/year).

Corrosion rates can be established either from the operating experience or from models available in the literature.

Table 10 presents mean and variance values for various parameters appearing in the limit state functions.

Next transition rate is defined as Occurrence of degraded state, represented by λ'. Degraded can be referred to as either leak state or rupture state. The equation has already been given in Eq. 21. For parameters like λ', λ_L, and λ_C, we can apply

Table 10 Parameters for failure pressure model with mean and variance

Parameters	Mean values	Variance
Thickness of the pipe (mm)	7	0.148
Outer diameter of the pipe (mm)	72	1.5
Rate of erosion-corrosion (mm/year)	0.051	0.015
Time (year)	40	
Length of defect (mm)	300	

the statistical model like Thomas model. Thomas defined the following relationship between the frequency of catastrophic rupture (λ_C) and frequency of leakage (λ_L);

$$\lambda_C = \lambda_L \cdot 3 \cdot P(C|L) \tag{36}$$

where $P(C|L)$ is the conditional probability of rupture given leakage. $P(C|L)$ has been assumed to be 0.02, considering erosion-corrosion as the dominant degradation mechanism present in the feeder.

λ' can also be found out using a limit state function. Typically, when a piping loses its 80% of wall thickness it is considered to have reached a failed state. So the limit state function can be formulated as

$$G2 = 0.8 \times t - (0.45 \times t + \text{rate} \times T) \tag{37}$$

Third state is the transition from flaw state to success state. This occurs when that particular piping component is subjected to In-Service Inspection. This has been denoted as ω. This parameter in Markov model that accounts for the inspection process can be further defined according to the following model.

$$\omega = \frac{P_I P_{\mathrm{FD}}}{(T_{\mathrm{FI}} + T_R)} \tag{38}$$

where, P_I is the probability that a piping element with a flaw will be inspected per inspection interval. In the case where inspection locations are inspected at random, this parameter is related to the fraction of the pipe segment that is inspected each interval and the capability of the inspection strategy to pinpoint the location of possible flaws in the pipe. When locations for the inspection are fixed, this term is either 0 or 1 depending on whether it is inspected or not. This probability is conditioned on the occurrence of one or more flaws in the segment.

P_{FD} is the probability that a flaw will be detected given this segment is inspected. This parameter is related to the reliability of NDE inspection and is conditional on the location being inspected having an assumed flaw that meets the criteria for repair according to the ASME code. This term is often referred to as the "probability of detection" or POD, T_{FI} is the mean time between inspections for flaws (inspection interval), T_R is the mean time to repair once detected. There is an assumption that any significant flaw that is detected will be repaired.

The software package for structural reliability analysis, STUREL, has been used to estimate the failure probabilities from the limit state functions.

Table 11 Transition rates used in three-state Markov model

Parameters	Values (/year)	Remarks
ϕ	3.812×10^{-4}	G-1
λ'	$\lambda_L = 8.76E\text{-}06$	Thomas model
	$\lambda_C = 1.75E\text{-}07$	
	0.115E-07	G-2 model
ω	0.09	90% POD in 10 years ISI

Table 12 State probabilities

States	State probability (Thomas)	State probability (G)
Success (S)	0.9959	0.9959
Flaw (F)	4.1E-03	4.1E-03
Degraded (D)	1.102E-6	1.375E-9

The solutions are obtained from COMREL module of STUREL which are used to estimate the various transition rates, φ and λ'. Alternatively, λ' has been estimated using Thomas model also. These results are presented in Table 11. These transition rates are applied on the Markov model shown in Fig. 9. Software MKV 3.0 by ISOGRAPH is used for determining the various state probabilities in the Markov model, as shown in Table 12.

The unavailability graph for three-state Markov model, considering the degraded state as unavailable for Thomas model and G function are given in Figs. 13 and 14 respectively. The failure frequencies of the three-state Markov model for Thomas model and G function are depicted in Figs. 15 and 16 respectively.

Degraded state probabilities from Thomas model are found for different POD and ISI interval. Figure 17 shows the degraded state probabilities for different POD. With no repair transition the probability of feeder in degraded state was found to be 2.711E-6. The probability has been found to be increased to twofold from the probability with 10 years of ISI interval and 70% POD detection technique.

Final aim of RI-ISI is to categorizes the components and assign an appropriate inspection category from Risk matrix. The consequence of failure has already been discussed. It falls in medium category in risk matrix. To analyze the impact of different ISI interval and inspection technique on plant risk, the inspection category for these test cases was found out. It has been found that failure frequencies increase by a factor of 100 when Thomas model is used in place of G function. The results and categories obtained after placing them in Risk Matrix are shown in Tables 13 and 14 for Thomas model and G function, respectively. It can be found that it has not made any change in final inspection category, since the failure frequencies obtained from Thomas model and G function fall in the medium range of failure frequency in Risk matrix.

Fig. 13 Unavaila. from Thomas model

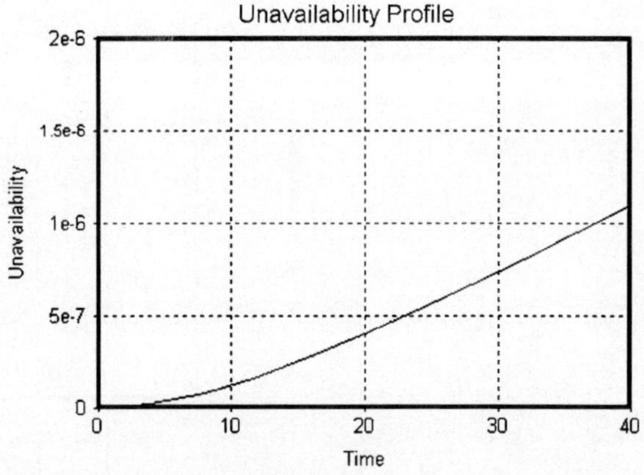

Fig. 14 Unavaila. from G function model

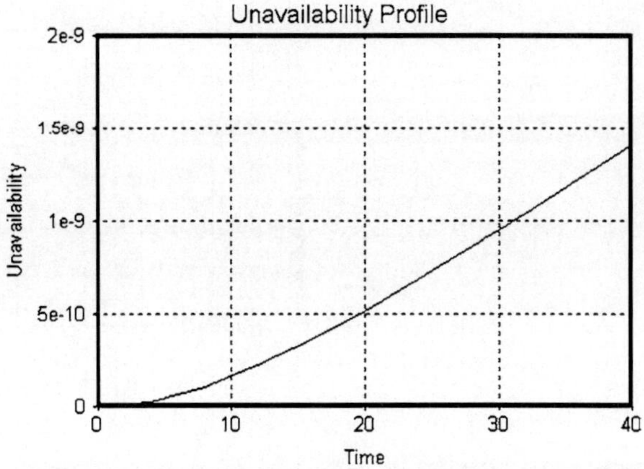

Fig. 15 Failure Freq.—Thomas model

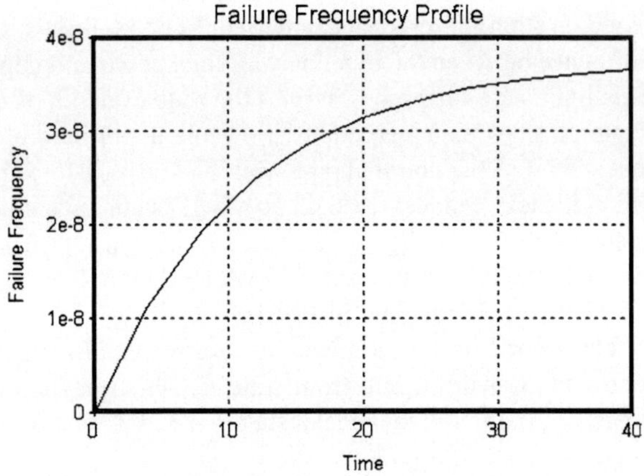

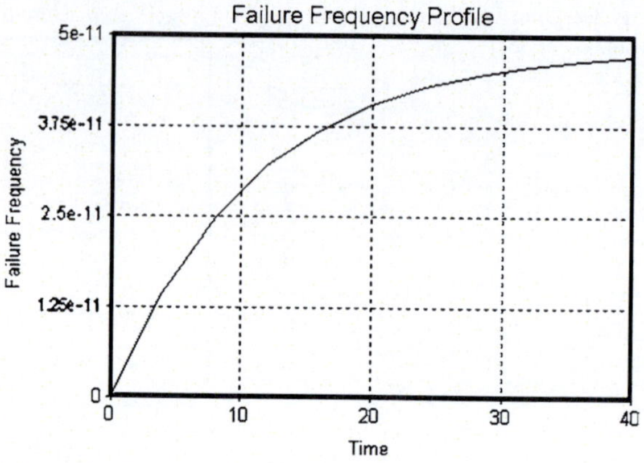

Fig. 16 Failure Freq.—G function model

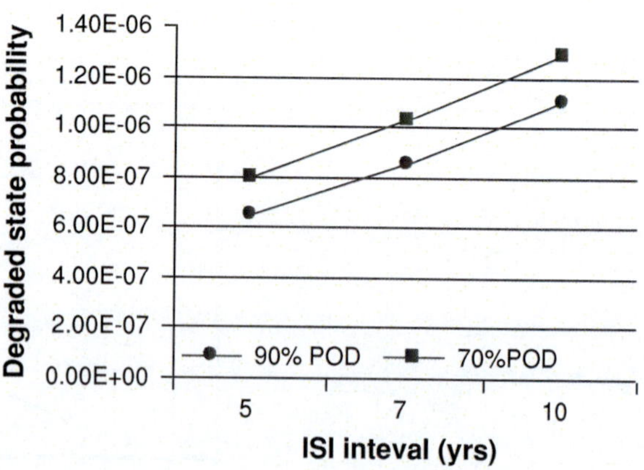

Fig. 17 Impact of inspection and repair strategies on piping failure probability

3.3.2 Using Four-State Markov Model

To determine the different transition rates ϕ, λf, ρ_L, and ρ_f, limit state functions, based on strength resistance, are used. The first limit state function is defined as the difference between the pipeline wall thickness t and depth of corrosion defect [20]. This limit state function describes the state of depth of the corrosion defects with a depth close to their maximum allowable depth before repair could be carried out that is 85% of the nominal pipe wall thickness ($0.45 \times t$). The probability that pipe fall thickness reduces to $0.45 \times t$ will occur at a rate, ϕ, which is defined as occurrence of flaw. So, ϕ represents transition rate from state S, in which flaw is less than $0.125 \times t$, to state F in which flaw is $0.45 \times t$. The limit state function has already been defined in Eq. 35.

The second limit state function is formulated to estimate the transition rate λ_f. λ_f represents transition rate from state F, which has already crossed the detectable range i.e., $0.45 \times t$, to the leak state L, i.e., $0.8 \times t$. The G for this case will be the same as given in Eq. 39.

Table 13 Risk matrix category for Thomas model

ISI interval	90% POD		70% POD	
	Freq (/year)	Category	Freq (/year)	Category
5	1.63E-8	6	1.995E-8	6
7	2.13E-8	6	2.575E-08	6
10	2.55E-8	6	3.225E-8	6

Table 14 Risk matrix category for G model

ISI interval	90% POD		70% POD	
	Freq (/year)	Category	Freq (/year)	Category
5	2.1E-11	6	2.57E-11	6
7	2.73E-11	6	3.3E-011	6
10	3.55E-11	6	4.15E-11	6

There is a probability for the piping reaching directly the rupture state, R, from the flaw state, F, because of encountering the failure pressure in the flaw state. For this case, a different limit state function needs to be formulated. The third limit state function is defined as difference between pipeline failure pressure P_f and pipeline operating pressure P_{op} [20].

$$G3(P_f) = P_f - P_{\mathrm{op}} \tag{39}$$

ω is the parameter in Markov model that accounts for the inspection process and can be further defined according to the following model given in Eq. 38. Another parameter is introduced in four-state Markov model to represent the leak repair. Repair rate

$$\mu = P_{LD}/(T_I + T_R) \tag{40}$$

P_{LD} is the probability that leak in the element will be detected per detection period (Typically assumed as 0.9).

Table 15 presents mean and variance values for various parameters appearing in the limit state functions.

The software package for structural reliability analysis, STUREL, has been used to estimate the failure probabilities from the limit state functions. The solutions are obtained from COMREL module of STUREL, which are used to estimate the various transition rates and are presented in Table 16. These transition rates are applied on the Markov model shown in Fig. 10. Software MKV 3.0 is used for determining the various state probabilities in the Markov model, as shown in Table 17. Modified B31G estimates are considered for ρ_f and ρ_l in the Markov model.

Depending on our definition of failure, state probability of either the leak state or the rupture state can be considered as failure probability of the feeder. The failure frequency of the feeder can be estimated by dividing this probability by the design life of the component, which value can be further employed in RI-ISI for determining its inspection category for In-Service Inspection. The unavailability graph for four-

Table 15 Parameters for failure pressure model with mean and variance

Parameters	Mean values	Variance
Yield strength (MPa)	358	25
Thickness of the pipe (mm)	7	0.148
Ultimate tensile strength (MPa)	455	32
Outer diameter of the pipe (mm)	72	1.5
Rate of erosion-corrosion (mm/year)	0.051	0.015
Load (MPa)	8.7	0.9
Time (year)	40	
Length of defect (mm)	300	

Table 16 Transition rates obtained from COMREL modules

Parameters	Values (/year)	G method
ϕ	3.812×10^{-4}	G-1
λf	2.435×10^{-5}	G-2
ρ_f	0.115×10^{-7}	G-3: modified B31G
ρ_l	1.486×10^{-2}	G-3: modified B31G

Table 17 State probabilities for w = 0.09 and $\mu = 0.084$

States	State probability
Success (S)	0.9956
Flaw (F)	4.362E-03
Leak (L)	9.303E-7
Rupture (R)	3.147E-7

state Markov model, considering the rupture state as unavailable is given in Fig. 18. The failure frequency of the four-state Markov model is depicted in Fig. 19.

Various inspection strategies are tried out changing the inspection interval and detection techniques employed. Figures 20 and 21 present the graphs on the results of these strategies on piping failure probability without and with leak repair respectively.

As per the consequence of failure, it falls in medium category in risk matrix. For different cases of inspection and repair strategies we can find which category the feeder will fall in the Risk matrix. Tables 18 and 19 provide the piping failure frequency, the respective CCDP, and inspection category number from risk matrix for different inspection and repair strategies.

4 Concluding Remarks on Risk-Informed In-Service Inspection

The failure pressure models considered here to define the G function lead to similar failure probabilities for short pipeline service periods. Various parameters are assumed here to be normally distributed, but in actual practice this may not be

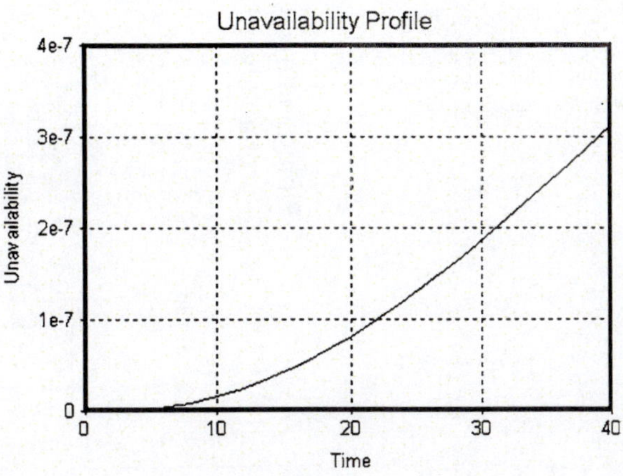

Fig. 18 Unavaila. for four-state model

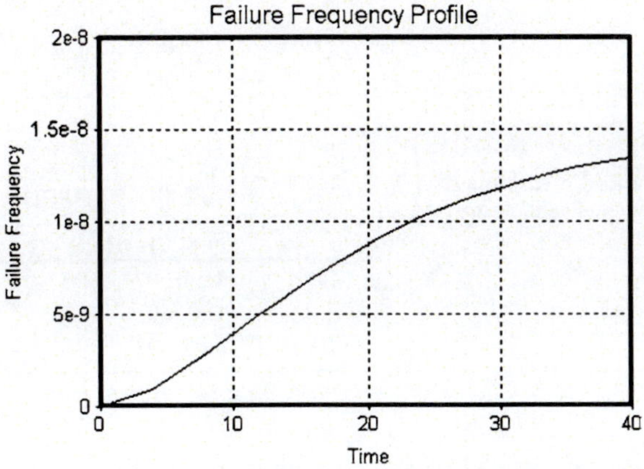

Fig. 19 Failure Freq. for four-state model

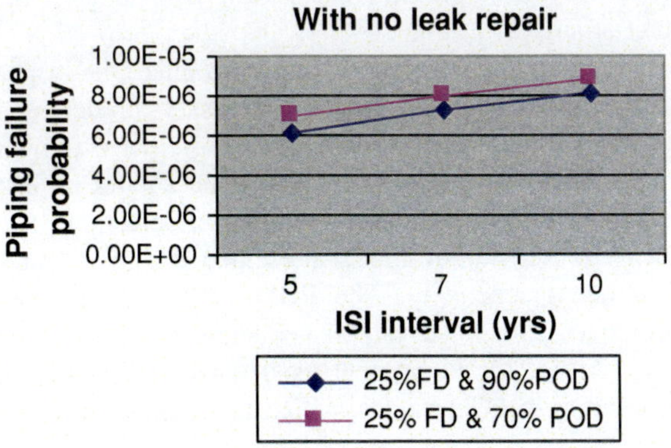

Fig. 20 Impact of inspection and repair strategies on piping failure probability with no leak repair LD leak detection, POD probability of detection, FI flaw inspection

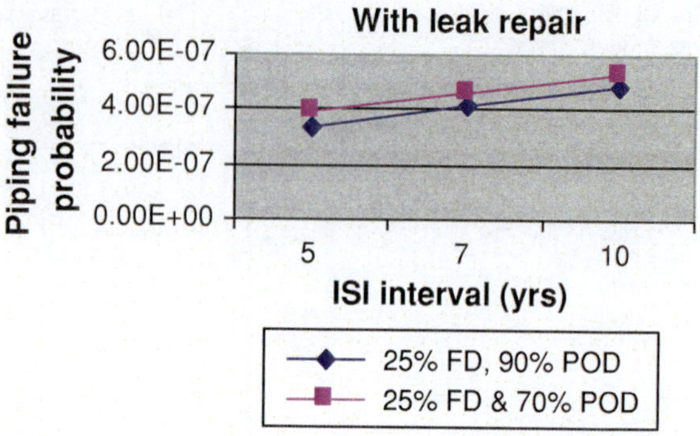

Fig. 21 Impact of inspection and repair strategies on piping failure probability with leak repair

Table 18 Risk matrix category with leak repair

ISI interval	25% FD, 90% POD		25% FD and 70% POD	
	Freq (/year)	Category	Freq (/year)	Category
5	8.45E-9	6	9.85E-9	6
7	1.035E-8	6	1.1625E-08	6
10	1.215E-8	6	1.3325E-8	6

Table 19 Risk matrix category without leak repair

ISI interval	25% FD, 90% POD		25% FD and 70% POD	
	Freq (/year)	Category	Freq (/year)	Category
5	1.54E-07	5	1.74E-07	5
7	1.81E-07	5	1.98E-07	5
10	2.05E-07	5	2.2E-07	5

the case. Instead of applying directly the probabilities obtained from limit state function in RI-ISI evaluation, it is recommended to find the state probabilities using the MARKOV model, since it incorporates the effect of repair and inspection works on the pipeline failure frequency. Markov model also allows formulating a proper inspection program and period depending on the operating condition of the plant at any given time.

The ultimate aim of RI-ISI is to optimize the inspection strategies in terms of risk and cost functions. So it is necessary to address the issues involved in conducting ISI like what should be the optimum frequency of inspection without jeopardizing the risk of the plant, what should be inspection technique adopted which will have maximum probability of detection (POD) of flaw, etc. The terms μ and ω in the Markov model presented in Fig. 10 incorporate ISI frequency and technique respectively. The POD values to be taken for different inspection techniques should be established experimentally taking into consideration, the sensitivity of the equipment used during inspection. There can be a source of uncertainty in POD values, which is assumed to have negligible impact on final failure probability values. It has been seen from Tables 18 and 19 that the changes in inspection and repair strategies can result in change in inspection category. In addition, it gives a direct indication to its effect on plant risk.

References

1. Pranab K. Samanta (1992) Optimisation of technical specifications applications in USA, Lecture 54.4.4. IAEA course: use of PSA in the operation of NPPs
2. Martorell S, Carlos S, Sanchez A, Serradell V (2001) Constrained optimization of test intervals using steady-state genetic algorithms: application to safety systems. Reliab Eng Syst Saf 72:59–74
3. Gopika V, Kushwaha HS, Verma AK, Srividya A (2004) Optimization of ISI interval using genetic algorithms for risk informed in-service inspection. Reliab Eng Syst Saf 86:307–316
4. Vaurio JK (1995) Optimization of test and maintenance intervals based on risk and cost. Reliab Eng Syst Saf 49:23–36
5. Munoz A, Martorell S, Serradell V (1997) Genetic algorithms in optimizing surveillance and maintenance of components. Reliab Eng Syst Saf 57:107–120
6. Vaurio JK (1999) Availability and cost functions for periodically inspected preventively maintained units. Reliab Eng Syst Saf 63:133–140
7. Goldberg DE (1989) Genetic algorithm in search, optimization and machine learning reading. Addison-Wesley, MA
8. Durga Rao K et al (2007) Test interval optimization of safety systems of nuclear power plant using fuzzy-genetic approach. Reliab Eng Syst Saf 92(7):895–901
9. IAEA TECDOC-737, Advances in reliability analysis and probabilistic safety assessment for nuclear power reactors, March 1994
10. ASME Code CASE N-560, Alternative examination requirements for Class1, Category B-J Piping welds
11. ASME Code CASE N-578, Risk informed methods for in-service inspection of pipe welds
12. Balkey et al. Developments on US NRC Approved WOG/ASME research risk informed in-service inspection methodology. A Report
13. Balkey ART et al (1998) ASME risk-based in-service inspection and testing: an outlook for the future. Risk Anal 18:407–421
14. EPRI, USNRC (1999). Risk informed in-service inspection evaluation procedure, TR-112657, July 1999
15. COMED (2000). Risk informed in-service inspection evaluation. Final report, Engineering and Research Inc., July 2000
16. NUREG-1661, Technical elements of risk informed in-service inspection for piping
17. Regulatory Guide 1.178 (1998) An approach for plant specific risk informed decision making: in-service inspection, USNRC, August 1998
18. Rouhan A (2002) Reliable NDT data for risk based inspection for offshore structures. Proceedings of the 3rd European–American workshop on reliability of NDE and demining, Berlin, 2002
19. RIBA PROJECT (2001) Risk informed approach for in-service inspection of nuclear power plant components, EUR 20164 EN, Project Summary, December 2001
20. Fleming KN, Gosselin S, Mitman J (1999) Application of markov models and service data to evaluate the influence of inspection on pipe rupture frequencies. Proceedings of the ASME pressure vessels and piping conference, Boston August 1–5

Risk, Reliability, Safety, and Testing Case Study

Norman Schneidewind

1 Objective

This chapter provides the reader with a case study that illustrates how software risk and reliability analysis can be used to reduce the risk of software failure and improve the overall reliability of the software product. You will learn how sequential testing is used as a process to achieve risk, reliability, and safety goals.

2 Overview of the Principles of Risk-Driven Reliability Model and Test Process

The risk-driven reliability model and testing process borrow concepts from classical sequential testing methodology that is used for hardware, with adaptation to software. Both consumer and producer risk are considered, reflecting the fact the consumer (e.g., customer) and producer (e.g., contractor) have different perspectives concerning what they consider to be tolerable risks of software failure. Similarly, there is also a differentiation based on what the consumer and producer consider to be acceptable reliability. Using the consumer-producer framework, a model and process are developed for executing sequential tests, based on software risk and reliability and model risk and reliability prediction accuracy. Rules are specified for determining at each decision point in testing whether the software and the model prediction accuracy are acceptable. In addition, the test rules serve as stopping criteria for testing (i.e., when it is cost-effective to stop testing).

N. Schneidewind (✉)
Naval Postgraduate School, 1411 Cunningham Road,
Monterey, CA 93943-5219, USA
e-mail: ieeelife@yahoo.com

H. Pham (ed.), *Safety and Risk Modeling and Its Applications*,
Springer Series in Reliability Engineering, DOI: 10.1007/978-0-85729-470-8_13,

Both empirical and predicted quantities are assessed. The test rules are integrated with several levels of criticality of software (i.e., the higher the criticality, the more stringent the tests). Based on lessons learned, the model and process are improved for future applications. The analysis is started by developing a model and process template based on the Poisson distribution of failures. This hypothetical example allows the model and process to be debugged before it is applied to a real application involving the NASA Space Shuttle flight software.

3 Model and Process Basics

This case study is about the development and evaluation of a model and process that uses the risk of software failure to drive test scenarios and reliability predictions. Scenarios involve the comparison of the software's actual outputs, resulting from test scenario execution, with its expected outputs, as documented by a specification [19]. Software actual outputs are empirical values of risk and reliability and the expected outputs are represented by specified threshold values of risk and reliability.

In addition, risk and reliability predictions provide stopping rules for testing. The foundation for these concepts of software testing is based on classical methods addressed to hardware [6], but with significant modifications to tailor the models to software testing and reliability. On the one hand, the classical methods of sequential testing, involving the concepts of consumer and producer risks [6], are very useful for structuring a testing and reliability model. On the other hand, these concepts are lacking in the literature on software testing [3]. Software testing emphasizes techniques such as statement coverage, decision coverage, branch coverage, and data flow coverage [3]. The classical methods are not entirely satisfactory for software because they are based on testing large quantities of homogeneous hardware items. This is not the situation with software because, in many cases, one-of-a kind of software is developed and tested. Thus, the classical methods require modification to be applicable to software.

Another important facet of the risk and reliability process is to evaluate not only the software but the *model* that predicts software risk and reliability, as well. If the model cannot predict accurately, the predictions cannot be used and you must try to validate another model (e.g., Weibull distribution). Thus, in this approach, there is an intimate relationship between software testing and models that provide the predictions for evaluating the outputs of the tests.

One way to analyze the software testing process is to consider the mechanism that drives the number, type, criticality, sequence, and timing of tests. The operational profile—frequency of application functions, weighted by their criticality—is one way [8]. However, the focus of this case study is consumer and producer risk in testing and models for quantifying the risk, with reasonable tradeoffs to balance competing consumer and producer objectives. To achieve this balance is important because on the one hand, the consumer desires highly reliable software at a low cost. On the other hand, the producer desires to deliver software that meets "reasonable" reliability requirements and results in high profit. To make this tradeoff, a balancing act is performed among risk, reliability, test time, acceptance and rejection criteria, and test sequence.

4 Safety Critical Software Considerations

To assist in making informed acceptance decisions, software risk analysis and reliability prediction are integrated to provide a comprehensive approach to implementing test rules designed to reducing risk and increasing reliability. This approach is applicable to all software, and in particular, it is important for certifying safety critical software because achieving improvements in the reliability of software contributes to system safety [5]. In addition, for this type of software, it is critical to have a feedback mechanism during testing to indicate when to continue to test and when to stop testing. Important feedback criteria are level of risk, reliability, and reliability growth. The inspiration for using this feedback mechanism comes from the concept expressed in [1] of using a test manager to monitor the difference between observed reliability and reliability predicted by a model. The difference is fed back into the test process to control the next step in testing. In this case study, the differences between observed and required risk and reliability are used to control the test process.

5 Case Study Approach

The case study approach is to investigate the feasibility of applying the consumer risk—producer risk model of testing to software. This classical method for testing hardware has been used for decades—long before the advent of software. There is no reason why the principles of this approach cannot be applied to software with suitable modifications. The effectiveness of the consumer-producer model, as applied to software, is assessed by documenting the advantages and disadvantages, and the lessons learned.

The analysis begins by developing a risk and reliability model test template that addresses the major issues in consumer and producer risk and reliability. The test template is analogous to the concept described in Stocks and Carrington [15]: a test template framework is a useful concept in specification-based testing (i.e., specification of risk and reliability requirements). The framework can be defined using any model-based specification notation and used to derive tests from model-based specifications (i.e., test sequence and acceptance criteria derived from risk and reliability specifications).

An example from the Poisson probability distribution is used to build the model template. The example is not entirely realistic because the reliability function that is a by-product of this process may not demonstrate reliability growth. In software reliability growth models, growth is possible because faults are removed as they are discovered, and assuming fewer new faults are introduced as old faults are removed, reliability will increase over test time. However, it is better to use a simple probability distribution at the outset to illustrate the model before delving into the analysis of real systems. Later, the NASA Space Shuttle flight software is used to provide a real-world example of applying the model, where reliability growth is part of the modeling process.

The model must be developed carefully and must include risk and reliability objectives. It is important that the model properly maps to the software under test. The method of model construction, when building testing scenarios, is to first build a template for guiding the construction of test sequences. Then in conducting the tests, iterate based on the test results at each stage until either the software is accepted or rejected.

6 Other Reliability Testing Methods

Reliability testing can be conducted at a macro or micro level. The former is used in sequential test scenarios in which the concern is about the big picture of risk, failure occurrence, and how to mitigate the risk to safety by increasing reliability. But in the micro view of testing, the focus is on methods that deal with the specifications, code, and data flow to produce effective fault removal in a cost-efficient manner. Specification-based testing produces test cases based on inputs, outputs, and program states. Code-based testing addresses computation results, predicate coverage, and control flow coverage. In data flow-based testing, test cases are produced to cover the execution space between where variables are declared and where they are used. Yet another method is mutation testing in which mutants of the original code are produced by introducing faults into program statements and observing the resulting execution behavior [4].

Lyu [7] provides a brief description of some of the important white-box testing methods: white-box testing uses the structure of the software to measure the quality of testing. This type of testing includes statement and decision coverage. Statement coverage testing constructs test cases that force each statement or a basic block of code to be executed at least once. Decision coverage constructs test cases that force each decision in the program to be covered at least once. A decision is covered if, during execution, it evaluates to true and in another execution, it evaluates to false [7].

Model-based testing is a technique for generating a suite of test cases from requirements. Testers using this approach concentrate on a data model and generation infrastructure instead of handcrafting individual tests. Several studies have demonstrated how combinatorial test generation techniques allow testers to achieve broad coverage of the input domain with a small number of tests. The authors have conducted several large projects in which they applied these techniques to systems with millions of lines of code [2].

None of these methods is superior to the others in all cases and their effectiveness and efficiency are application dependent. Selected tests at the micro level should be combined with a macro level approach to provide a comprehensive attack on the software risk and reliability problem. In fact, the approach is to do model testing at the micro level (i.e., white-box testing) to provide failure count input to the macro level model (i.e., black-box testing based on top-level specifications). The process does not have to stop there. You can use the two approaches synergistically by feeding black-box testing risk and reliability predictions to white-box testing so that the latter will provide an assessment of likely operational risk and reliability. Then, the white-box strategy would be adjusted to focus testing on the highest risk and lowest reliability software.

7 Risk, Reliability, and Safety Model Development

7.1 Definitions

7.1.1 Risk

According to Software Safety [14], "risk is a function of: the possible frequency of occurrence of an undesired event, the potential severity of resulting consequences, and the uncertainties associated with the frequency and severity". This sounds good but would be difficult to implement because all of the risk factors would not be available in practice. Therefore, you should use a definition to account for, not only the probability of an undesirable event but, in addition, the "severity of resulting consequences", as represented by failure count. Putting these two factors together, we define risk as the expected number of failures (i.e., probability of failure $\times$ failure count).

L_m: risk limit (threshold that risk should not exceed).

Mission critical: an application in which high risk and low reliability would jeopardize the organization's survival.

Safety critical: an application in which high risk and low reliability would jeopardize the safety of the crew and mission.

t_m: Mission duration: length of computer operation, space flight, etc.

"actual" refers to reliability and risk that are computed by using historical data; there is no prediction of the future

"predicted" refers to reliability and risk that are computed by using historical data in order to make forecasts of the future

7.2 Risk Analysis

7.2.1 Actual Risk

In order to have a baseline against which to compute risk prediction errors, start by finding the actual probability of failure for the consumer and producer in Eqs. 1 and 2, respectively:

$P_{ac}(t, r_c)$: actual consumer probability of r_c failures in time

$$t = \frac{r_c(t)}{\sum_{t=1}^{N_c} r_c(t)}, \qquad (1)$$

where N_c is the number of failures that occur on consumer software.

$P_{ap}(t, r_p)$: actual producer probability of r_p failures in time

$$\text{t} = \frac{r_p(t)}{\sum_{t=1}^{N_p} r_p(t)}, \qquad (2)$$

where N_p is the number of failures that occur on producer software.

Then applying the definition of risk (i.e., probability of failure × failure count), compute the *actual* consumer and producer risk in Eqs. 3 and 4, respectively

$$\mu_c(t, r_c) = [p_{ac}(t, r_c)] * r_c \qquad (3)$$

$$\mu_p(t, r_p) = [P_{ap}(t, r_p)] * r_p \qquad (4)$$

7.3 Probability of Failure

Unlike hardware during its operations phase when the time to failure (MTTF) is assumed constant, software has a variable MTTF that is a function of length of time the software has been operating or tested t and the number of failures r that have occurred during this time. Based on these considerations, you can compute the *consumer* MTTF in Eq. 5.

$$m_c(t, r_c) = \left(\frac{t}{\sum_{t=1}^{t} r_c}\right) \qquad (5)$$

Then, it follows that you can compute the *producer* MTTF in Eq. 6:

$$m_p(t, r_p) = \left(\frac{t}{\sum_{t=1}^{t} r_p}\right) \qquad (6)$$

If a Poisson distribution of failure counts is assumed for consumer failures r_c, during time t, with mean time to failure $m_c(t, r_c)$ from Eq. 5, *predict* the probability of failure for the consumer in Eq. 7:

$$P_c(t, r_c) = \left(\frac{t}{m_c}\right)^{r_c} \frac{e^{-\left(\frac{t}{m_c}\right)}}{r_c!} \tag{7}$$

Similar to Eq. 7, if you assume a Poisson distribution of failure counts r_p for the producer, during time t, with mean time to failure $m_p(t, r_p)$, estimated in Eq. 6, you can *predict* the probability of failure for the producer in Eq. 8:

$$p_p(t, r_p) = \left(\frac{t}{m_p}\right)^{r_p} \frac{e^{-\left(\frac{t}{m_p}\right)}}{r_p!} \tag{8}$$

7.3.1 Predicted Risk

Again applying the definition of risk and using Eq. 7, predict the *consumer* risk in Eq. 9:

$$\alpha_c(t, r_c) = [P_c(t, r_c)] * r_c \tag{9}$$

The consumer's risk is not the whole story about risk because you must consider producer's risk in the sequential tests model. As you will recall, the producer wants to minimize the risk of rejecting good software. The producer wants to produce the minimum acceptable software for the consumer, but no more. To do otherwise would result in needless cost for the producer.

Thus, again applying the definition of risk and using Eq. 8, you can compute the *producer* risk in Eq. 10:

$$\beta_p(t, r_p) = [P_p(t, r_p)] * r_p \tag{10}$$

8 Reliability Analysis

8.1 Actual Reliabilities

In order to assess reliability prediction accuracy, it is necessary to compute actual reliabilities so that predicted and actual values can be compared. Therefore, actual consumer reliability $R_{ac}(r, t_c)$ is computed during time t, using the Poisson distribution of failures r_c, as given by Eq. 11:

$$R_{ac}(t, r_c) = 1 - \left(\frac{r_c}{\sum_{t=1}^{N_c} r_c}\right) \tag{11}$$

Similarly, the actual producer reliability $R_{ap}(t, r_p)$ is computed in Eq. 12, using the Poisson distribution of failures r_p:

$$R_{ap}(t, r_p) = 1 - \left(\frac{r_p}{\sum_{t=1}^{N} r_p} \right) \tag{12}$$

8.2 Predicted Reliabilities

Formulate the consumer and producer *predicted*reliabilities by first considering the relationship between the reliabilities $R_c(t)$ and $R_p(t)$, which are the probabilities of survival in the interval t, and the probabilities of failure $P_c(t, r_c)$ and $P_p(t, r_p)$ in the interval t, using Eq. 13.

$$R_c(t) = 1 - P_c(t, r_c), \quad R_p(t) = 1 - P_p(t, r_p) \tag{13}$$

Now obtain $P_c(t, r_c)$ from Eq. 7 and use it in Eq. 14 to predict consumer reliability:

$$R_c(t) = 1 - P_c(t, r_c) = 1 - \left(\frac{t}{m_c}\right)^{r_c} \frac{e^{-\left(\frac{t}{m_c}\right)}}{r_c!} \tag{14}$$

Similarly obtain $P_p(t, r)$ from Eq. 9 and use it in Eq. 15 to predict producer reliability:

$$R_p(t) = 1 - P_p(t, r_p) = 1 - \left(\frac{t}{m_p}\right)^{r_p} \frac{e^{-\left(\frac{t}{m_p}\right)}}{r_p!} \tag{15}$$

Because predictions deal with the highly volatile future they are subject to potentially large prediction errors. Therefore, you should first investigate the prediction error before drawing conclusions about the validity of the prediction results. This is done in the next section.

9 Predictions and Prediction Accuracy

9.1 Risk Prediction

The method of assessing consumer and producer risk prediction accuracy is to see: (1) whether the *mandatory* criteria are satisfied: predicted consumer risk, predicted producer risk, and actual risk are less than the allowable limit; and (2) whether the *desirable* criteria are satisfied: predicted consumer risk and predicted producer risk are less than the actual risk. Mandatory criteria (1) are given in Eqs. 16 and 17 and desirable criteria (2) are given in Eqs. 18 and 19.

$$\text{Mandatory Criteria} \\ \alpha(t,\ r_c),\ \mu_c(t,\ r_c) < L_m(\text{predicted and actual consumer risk} < \text{risk limit}) \tag{16}$$

$$\beta(t, r_p),\ \mu_p(t, r_p) < L_m(\text{predicted and actual producer risk} < \text{risk limit}) \tag{17}$$

$$\text{Desirable Criteria} \\ \alpha(t, r_c) < \mu_c(t, r_c)(\text{predicted consumer risk} < \text{actual consumer risk}) \tag{18}$$

$$\beta(t, r_p) < \mu_p(t, r_p)(\text{predicted producer risk} < \text{actual producer risk}) \tag{19}$$

The rationale for the mandatory criteria is that if the risks exceed the limit, the safety of the software system would be jeopardized.

In the case of the desirable criteria, there would be concern about prediction accuracy if predicted risks exceed actual risks because it would indicate the possibility of large prediction error.

9.2 Risk Prediction Accuracy

Now (16), (17), (18), and (19) are insufficient because, since $\alpha(t, r_c)$ and $\beta(t, r_p)$ are predicted quantities, we need to see whether there is acceptable prediction accuracy with respect to actual consumer risk $\mu_c(t, r_c)$ and actual producer risk $\mu_p(t, r_p)$. You compute the mean square error in Eqs. 20 and 21 for consumer risk and producer risk, respectively.

$$E_{\mathrm{Rc}} = \sum_{t=1}^{T} \frac{[\mu_c(t, r_c) - \alpha(t, r_c)]^2}{T} \tag{20}$$

$$E_{\mathrm{Rp}} \sum_{t=1}^{T} \frac{[\mu_p(t, r_p) - \beta(t, r_p)]^2}{T}, \tag{21}$$

where T is the last time the software is tested.

Next, you test whether the values in Eqs. 20 and 21 satisfy the error thresholds in conditions (22) and (23), respectively, using the mean plus three standard deviations criterion:

$$S_{\mathrm{Rc}} = [\mu_c(t, r_c) - \alpha(t, r_c)]^2 < (E_{\mathrm{Rc}} + 3\sigma) \tag{22}$$

$$S_{\mathrm{Rp}} = [\mu_p(t, r_p) - \beta(t, r_p)]^2 < (E_{\mathrm{Rp}} + 3\sigma) \tag{23}$$

9.3 Reliability Prediction Accuracy

The reliability accuracy test is to compute the mean square error of the difference between (1) actual consumer reliability $R_{ac}(t, r_c)$ and predicted consumer reliability $R_c(t, r_c)$ and (2) between actual producer reliability $R_{ap}(t, r_p)$ and predicted producer reliability $R_p(t, r_p)$. E_{rc} and E_{rp} are the consumer and producer error quantities in Eqs. 24 and 25, respectively.

$$E_{rc} = \sum_{t=1}^{T} \frac{[R_{ac}(t, r_c) - R_c(t, r_c)]^2}{T} \tag{24}$$

$$E_{rp} = \sum_{t=1}^{T} \frac{[R_{ap}(t, r_p) - R_p(t, r_p)]^2}{T} \tag{25}$$

Now, you can test whether the values in Eqs. 24 and 25 satisfy the error threshold conditions (26) and (27), based on the mean plus three standard deviations criterion, respectively:

$$S_{rc} = [R_{ac}(t, r_c) - R_c(t, r_c)]^2 lt; (E_{rc} + 3\sigma) \tag{26}$$

$$S_{rp} = [R_{ap}(t, r_p) - R_p(t, r_p)]^2 lt; (E_{rp} + 3\sigma) \tag{27}$$

10 Tradeoff between Consumer's Risk and Producer's Risk

As mentioned in the Overview, it is desirable to balance the conflicting objectives of minimizing both consumer's risk and producer's risk. To do this, use the difference $[\alpha(t, r_c) - \beta(t, r_p)]$ between predicted consumer and producer risks to analyze whether balance has been achieved. You examine the minimum of this quantity, noting how close it is to zero (degree of balance), when it occurs in test time (test resources necessary to achieve balance), and the failure counts at this test time (reliability of the software when balance achieved).

11 Test Rules

One of the most difficult aspects of testing is to answer the question: "when to stop testing?" Myers suggests to stop testing when a given number of faults have been discovered and corrected [10]. While this approach is *indirectly* related to reliability, and is certainly better than stopping when we run out of money and time, it is better to use criteria that are *directly* related to risk and reliability. With this approach, you can key the stopping rules to achieving acceptable levels of risk and reliability. This concept is embodied in the test rules below.

Test rules should also include the criticality of the software being tested (see Fig. 1, Part 2). This factor is mentioned in [12], where the authors state: "Many commercial products are not fully prepared for use in high assurance situations. In spite of the criticality of these applications, there currently exists a dearth of software assurance techniques to assess the robustness of both the application and the operating system under strenuous conditions. The testing practices that ordinary commercial products undergo are not thorough enough to guarantee reliability. High assurance applications require software components that can function correctly even when faced with improper usage or stressful environmental conditions". Our aim is to guarantee reliability by using a model and test schema that requires the software to pass several reliability and risk checks before it can be certified by imposing the most stringent test conditions in acceptance tests for mission critical and safety critical software.

Based on the roadmap in Fig. 1 and the mandatory and desirable risk criteria previously formulated, you specify the rules for the *software* and *model* accept–reject decisions. Accept *software* if the *software* rules evaluate to "true". Accept *model* if the *model* rules evaluate to "true". Mandatory rules are designed to ensure that there are no unacceptable risks in the operation of the software, whereas desirable rules are designed to ensure reasonable prediction accuracy. To be certified as safe to deploy, consumer and producer software must pass all parts of two sequential tests. If there is a failure to pass any part of the first test, the software is given a second chance to pass the two tests, after faults are removed (see Fig. 1, Part 2).

Risk

Mandatory for Software

Risks of failure of consumer and producer software must be less than risk limit.

a. $(t, r_c) < L_m$ (predicted consumer risk < risk limit)
b. $(t, r_p) < L_m$ (predicted producer risk < risk limit)
c. $\mu_c(t, r_c) < L_m$, $\mu_p(t, r_p) < L_m$ (actual consumer and producer risk < risk limit)

Mandatory for Prediction Model

Consumer and producer risk prediction model errors must be less than error limits.

d. $[\mu_c(t, r_c) - \alpha(t, r_c)]^2 < (E_{Rc} + 3\sigma)$ (consumer error < consumer error limit)
e. $[\mu_p(t, r_p) - \beta(t, r_p)]^2 < (E_{Rp} + 3\sigma)$ (producer error < producer error limit)

Desirable for Software

Desire predicted risks to be less than actual risks.

f. $\alpha(t, r_c) < \mu(t, r_c)$
g. $\beta(t, r_p) < \mu(t, r_p)$ (predicted producer risk < actual producer risk)

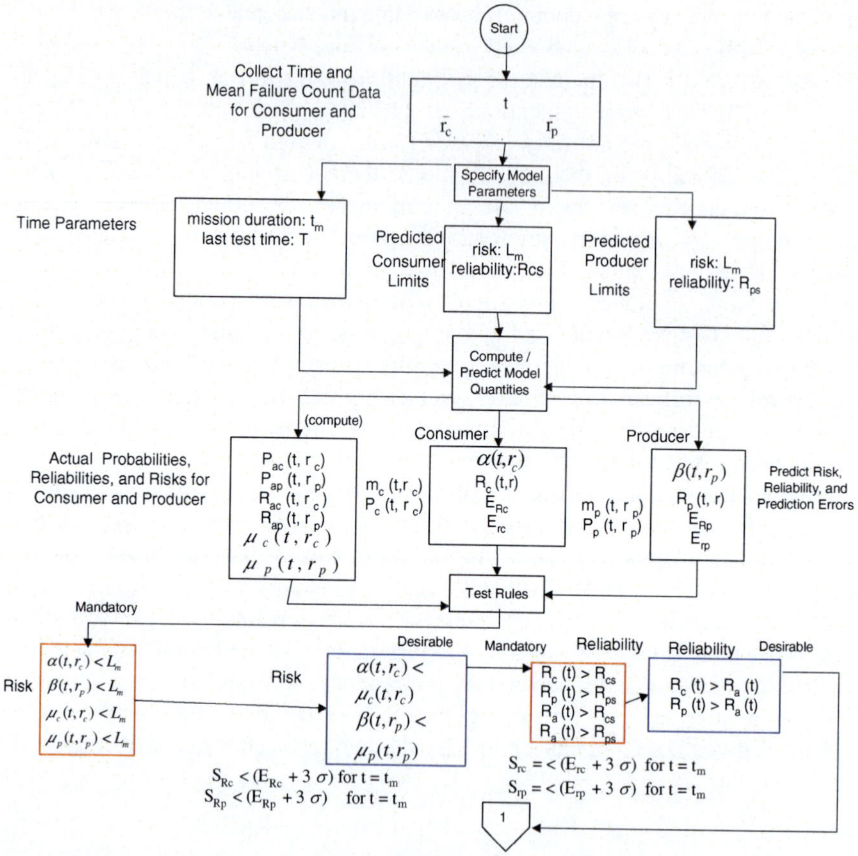

Fig. 1 Risk-driven testing and reliability process (Part 1), **b** Risk-driven testing and reliability process (Part 2)

Desire consumer to have lower risk than producer.

h. $\alpha(t, r_c) = \beta(t, r_p)$ (predicted consumer risk < predicted producer risk)

Reliability

It is important that *both* risk and reliability satisfy stringent specifications. Therefore, in addition to the risk test rules that were just developed, the reliability test rules are developed below.

Mandatory for Software

Predicted and actual reliabilities must be greater than specified reliabilities

a. $R_c(t) > R_{cs}$ (predicted consumer reliability > specified consumer reliability)
b. $R_p(t) > R_{ps}$ (predicted producer reliability > specified producer reliability)
c. $R_a(t) > R_{cs}$ (actual reliability > specified consumer reliability)

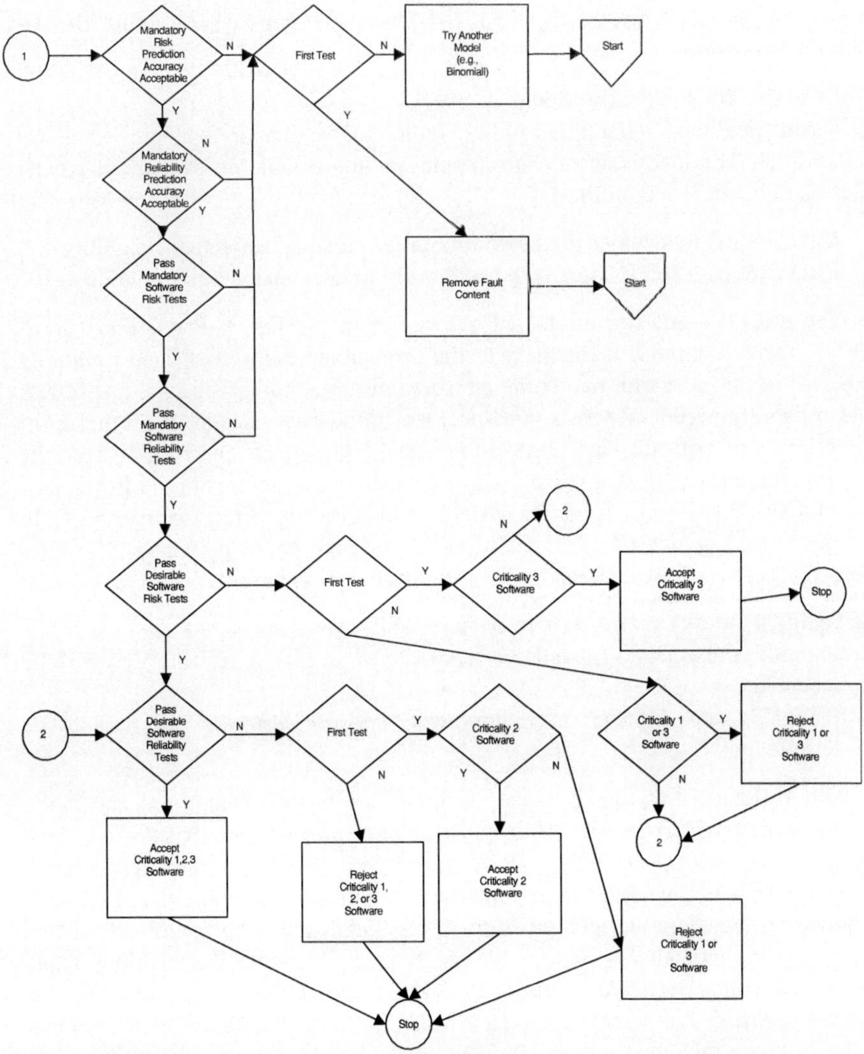

Fig. 1 (continued)

d. $R_a(t) > R_{ps}$ (actual reliability > specified producer reliability)

Mandatory for Prediction Model

Consumer and producer reliability prediction errors must be less than error limits.

e. $[R_{ac}(t, r_c) - R_c(t, r_c)]^2 < (E_{rc} + 3\sigma)$ (consumer prediction error must be less than error limit)

f. $[R_{\mathrm{ap}}(t, r_p) - R_p(t, r_p)]^2 < (E_{\mathrm{rp}} + 3\sigma)$ (producer prediction error must be less than error limit)

Desirable for Prediction Model

Desire predicted reliabilities of the future to be greater than present actual reliabilities. The idea is an *attempt* to achieve future reliabilities that are greater than the historical reliabilities.

g. $R_c(t) > R_a(t)$ (consumer predicted reliability greater than actual reliability)
h. $R_p(t) > R_a(t)$ (producer predicted reliability greater than actual reliability)

The test rules and definitions are put in context by Fig. 1, Part 1 and Part 2. These figures comprise a roadmap to the sequential testing, risk, and reliability process. In this scenario, two complete software tests, comprising the use of risk and reliability acceptance tests, are specified. In addition, acceptance criteria for three levels of criticality are used in the accept/reject decision process: from the most to the least critical: Criticality 1 (e.g., mission critical—Shuttle flight software), Criticality 2 (e.g., operating system), and Criticality 3 (e.g., spreadsheet). In the scenario, any one of the three types of software could be tested. The relationships between test rules and criticality are the following:

Criticality 1: must pass *all* tests to be accepted,
Criticality 2: must pass *all* mandatory tests and *all desirable* reliability tests to be accepted,
Criticality 3: must pass *all* mandatory tests and *all desirable risk* tests to be accepted,

Note that according to Fig. 1, Part 2, the testing process cannot proceed to the *desirable* risk and reliability tests until all *mandatory* tests are satisfied, both prediction accuracy and software tests. In addition, in Part 2, passing desirable reliability tests is considered more important than passing desirable risk tests. This is based on the logic that compared to risk, reliability can be quantified, understood, and applied, whereas, setting the risk limit, for example, is subjective. Thus, passing *desirable reliability tests* is associated with Criticality 2 software and passing *desirable risk tests* is associated with Criticality 3 software. Note also that if the software fails the first test, another model could be considered, such as the Binomial (see Fig. 1, Part 2).

12 Example Problem, Using Poisson Distribution

In the two tests in Fig. 1, Part 2, use the Poisson distribution to predict consumer and producer risk and reliability. The Poisson distribution is used to illustrate the risk and reliability analysis process because it is a typical distribution that is used in reliability modeling. Using this distribution produces increasing reliability with test and operational time. This means that the software is run, faults that cause failures are removed, and reliability improves, as a consequence. A typical example is that a personal computer runs for a while, and then multiple failures occur, and are cleared with a reboot. Some researchers call this time "soak time" [8]. The reboot allows errant application code to be cleansed (i.e., faults are removed) such that the software can operate until the next incident occurs. In contrast, in our second example, involving safety critical software like the Shuttle flight software, such a scenario could not be tolerated. This point will be elaborated when the second example is introduced.

The following parameters are specified in the example problem (note that for illustrative purposes, the units of the quantities are immaterial):

$$t : \text{time} = \text{testplus operational time}(1, \ldots, t_N) = 1, \ldots, 28$$

Note: to make the plots easy to read, risk and reliability quantities are plotted against "test time" in the figures. This "test time" includes test time and operational time.

t_m: = *desired* mission duration = 8
N: number of failure count intervals = 28
L_m: risk limit = 0.500000

It is reasonable to ask on what basis the risk limit is chosen. Admittedly, it is somewhat subjective, but it is based on the following consideration: risk is: (probability of r failures) $\times$ (r failure count). For this illustrative software, the assumption is made that the probability is 0.5 for $r = 1$ failure, or $L_m = 0.5$. The reason for six decimal places in the risk limit is that risks can be so close to zero that they must be computed to six places to be compared with the limit.

R_{cs}: specified *minimum* consumer reliability =0.9000
R_{ps}: specified *minimum* producer reliability, where $R_{ps} < R_{cs} = 0.8500$

The choice of reliability thresholds is, again, a bit subjective but the important point is that the relationship must be $R_{ps} \leq R_{cs}$, reflecting the desire to force greater reliability from the consumer's perspective than from the producer's.

For the Second Test, in order to illustrate the benefit of fault reduction and consequent failure reduction, mean failure count is reduced by approximately 50%. This action results in reducing consumer and producer risk and increasing consumer and producer reliability. Thus, the mean of Poisson distributed failure counts $\bar{r}_c$ for the consumer was reduced from 1.11 to 0.57. In addition, the mean of Poisson distributed failure counts $\bar{r}_p$ for the producer was reduced from 1.39 to 0.61.

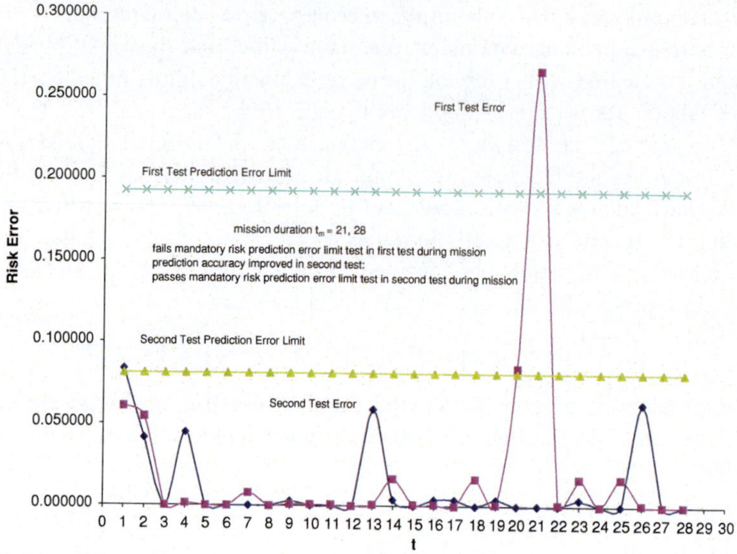

Fig. 2 First test and second tests: consumer risk prediction error: (actual − predicted)^2 versus test time t

12.1 Error Analysis

12.1.1 Risk Predictions

As seen in Fig. 2, consumer risk passes its mandatory risk prediction accuracy test for the Second Test, but fails the First Test *during the mission*. In addition, as Fig. 3 shows, producer risk passes its mandatory risk accuracy test for both the First Test and the Second Test *during the mission*. This result would not be comforting for the consumer because, despite the good results achieved by the producer, faults would have to be removed from the consumer software before it is acceptable.

12.1.2 Reliability Predictions

Figures 4 and 5 tell us that both consumer and producer reliability prediction error passes their mandatory tests for First and Second Tests *during the mission*. At this point, considering the combination of risk prediction and reliability error results would lead us to conclude that it is too risky to release the software, and that more testing is necessary to remove more faults.

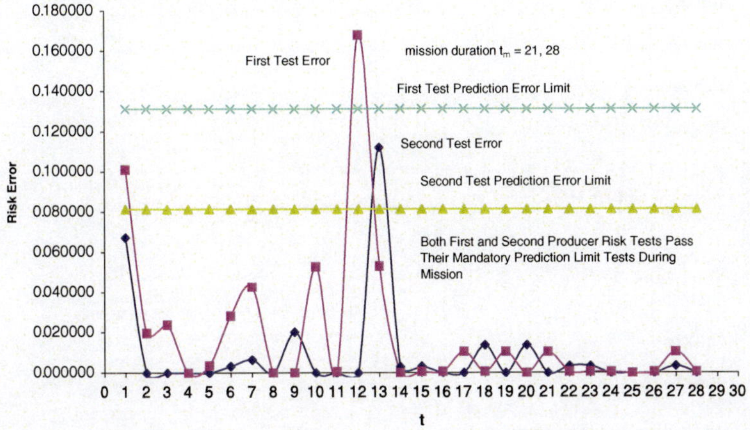

Fig. 3 First test and second tests: producer risk prediction error: (actual − predicted)^2 versus test time t

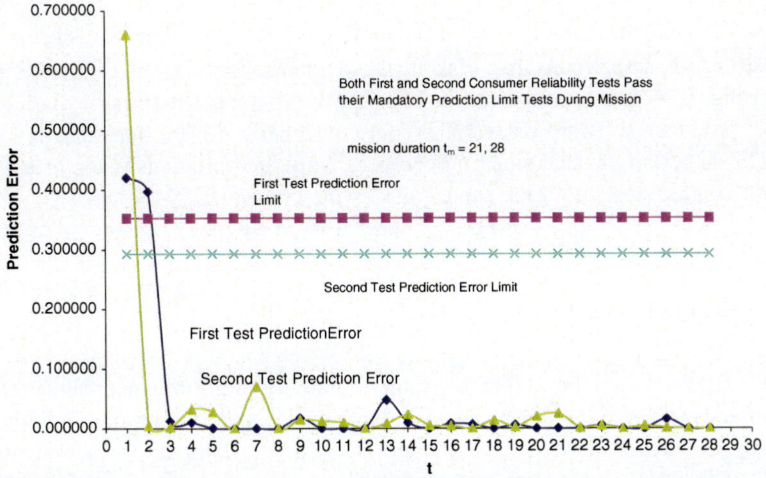

Fig. 4 First test and second tests: consumer reliability prediction error: $(R_{ac}(t) - R_c(t))^2$ versus test time t

12.2 Risk Analysis

Note that risk functions, being rather subjective, can have variations that obscure the underlying patterns [11]. The implication of this situation is that you can expect some variation in the prediction of risk over test time.

Figures 6 and 7 address consumer risk and producer risk, respectively. You see that, unfortunately, just one case of not meeting the criterion causes the mandatory

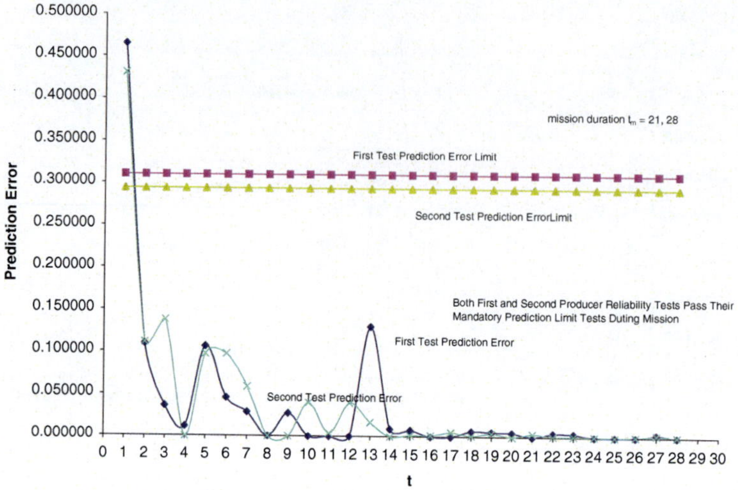

Fig. 5 First and second tests: producer reliability prediction error: $((R_{\mathrm{ap}}(t) - R_p(t))^2$ versus test time t

risk test to fail: actual consumer risk $\mu_c(t, r_c)$ for the First Test exceeds the risk limit in Fig. 6. All tests are passed in Fig. 7. A consistent pattern is beginning to emerge: producer software passes tests, but consumer software does not pass all tests. This is not an unusual situation because, as implied previously, the producer does not conduct tests with the same rigor as the consumer.

12.3 Reliability Analysis

Figure 8 tells us that the actual consumer reliability fails the mandatory test, although the predicted reliability passes it. Note that it is more important for actual reliability to pass because it is based on actual failure data, as opposed to predictions, which may or may not be accurate. Thus, consumer software continues to have problems. Figure 9 is more encouraging because all reliabilities pass the second test. The scenario of this outcome per Fig. 1, (Part 2) would be fault removal from the consumer software, repeat and pass the first test, and pass the second test.

12.4 Risk Tradeoff Analysis

Risk tradeoff analysis is also a part of the model roadmap. It provides additional insight for making the accept/reject decision based on the relationship between

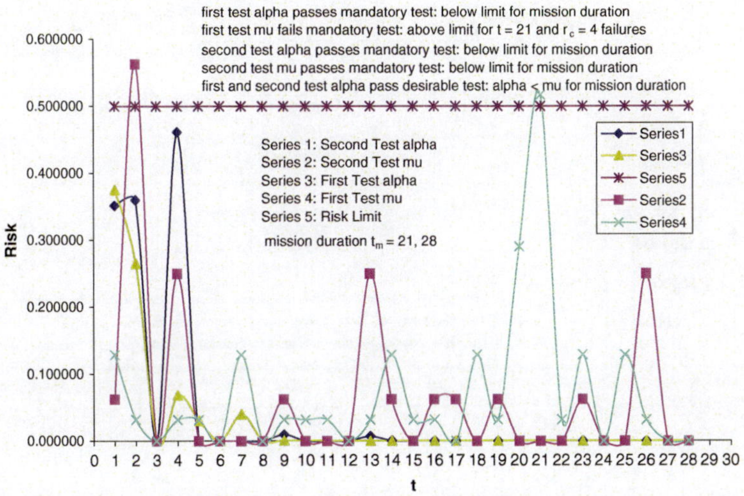

Fig. 6 First and second tests: predicted consumer risk (alpha) and actual consumer risk (mu) versus test time t

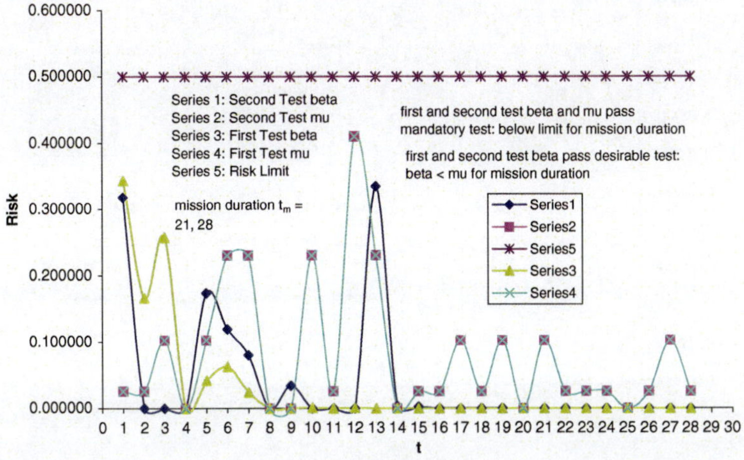

Fig. 7 First and second tests: predicted producer risk (beta) and actual producer risk (mu) versus test time t

consumer and producer risk. One objective of Fig. 10 is to determine the amount of test time required for consumer risk to equal producer risk—this is a good balance point. Another objective is to determine whether (consumer risk—producer risk) is predicted to occur during the mission. For example, while equality is obtained during the second test at $t = 8$, with zero failures, it is not obtained until $t = 21$ during the first test, with zero failures. In other words, the requirements are

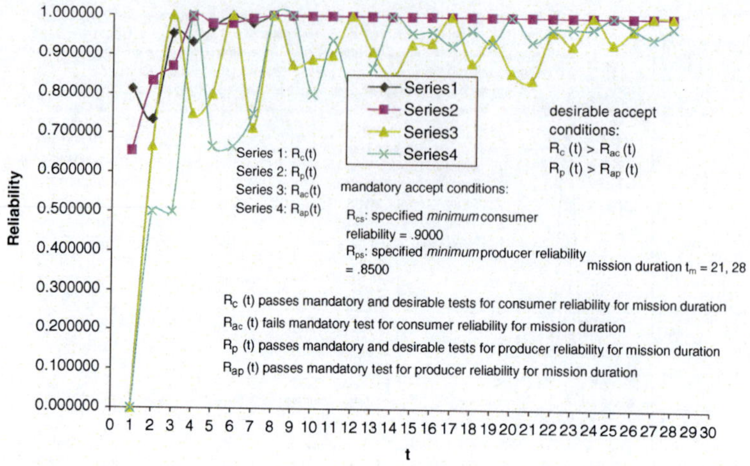

Fig. 8 First test: predicted consumer reliability $R_c(t)$, predicted producer reliability $R_p(t)$, actual consumer reliability $R_{ac}(t)$, and actual producer reliability $R_{ap}(t)$ versus test time t

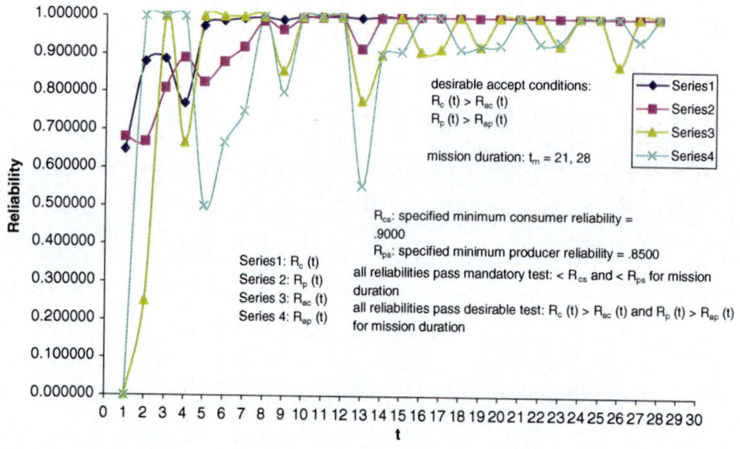

Fig. 9 Second test: predicted consumer reliability ($R_c(t)$), predicted producer reliability ($R_p(t)$), actual consumer reliability ($R_{ac}(t)$), and actual producer reliability ($R_{ap}(t)$) versus test time t

not completed during test time ($t = 1, 20$); they are not completed until the beginning of the mission at $t = 21$. Thus, the decision would be to reject the software until it can satisfy the second test.

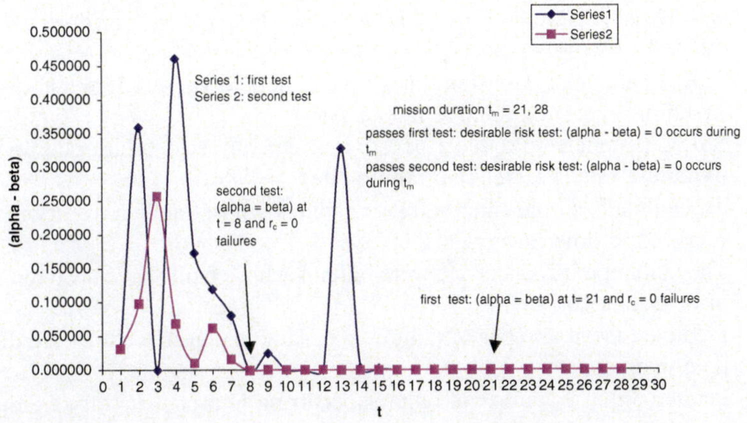

Fig. 10 First and second tests: (predicted consumer risk–predicted producer risk) (alpha–beta) versus test time t

13 NASA Space Shuttle Application

Now use the roadmap in Fig. 1 and apply it to the Shuttle flight software, using the Schneidewind Software Reliability Model (SSRM) [13]. Any software reliability growth model (srgm) would suffice for this purpose. An assumption of srgm's is that reliability will increase with time, as faults are removed as they are discovered. Thus, a sufficiently long test time is required to: (1) collect sufficient failure data in order to estimate model parameters and (2) allow reliability growth to take place (e.g., reliability reaches an acceptable level). Once (1) and (2) have been accomplished, you can *predict* the reliability of the software for the specified mission duration t_m.

The first step is to provide definitions of model quantities:

13.1 Definitions and Assumption

Only the definitions that are specific to the Shuttle are given here. Previous definitions are not repeated.

t_{cp} Test time or operating time when predicted consumer reliability, $R_c(t, r_c)$, = predicted producer reliability, $R_p(t, r_p)$

s_c SSRM parameter: Consumer starting interval for using observed failure data in parameter estimation

s_p SSRM parameter: Producer starting interval for using observed failure data in parameter estimation

a_c SSRM parameter: Consumer failure rate at the beginning of time interval s_c

a_p	SSRM parameter: Producer failure rate at the beginning of time interval s_p
b_c	SSRM parameter: Consumer negative of derivative of failure rate divided by failure rate (i.e., relative failure rate)
b_p	SSRM parameter: Producer negative of derivative of failure rate divided by failure rate (i.e., relative failure rate)
$r_c(t)$	The number of consumer software failures whose faults have been removed in time interval t
$r_p(t)$	The number of producer software failures whose faults have been removed in time interval t
$m_c(t)$	Predicted mean number of failures occurring in consumer software during time interval t
$m_p(t)$	Predicted mean number of failures occurring in producer software during time interval t
N_c	Total number of failures that occur in consumer software over all time intervals
N_p	Total number of failures that occur in producer software over all time intervals

Typically, in Shuttle software, there is a one-to-one relationship between faults and failures. Thus, this assumption is made in the analysis.

13.2 Test Rules

Since the Shuttle uses a reliability growth model, a modified set of test rules is called for to capture this important characteristic. To compute reliability growth quantitatively, Jeff Tian suggests that reliability growth measure *purification level*, the ratio between the number of faults (failures) removed during testing over the total faults (failures) at the beginning of testing. In this analysis it is convenient to define the purification level as equal to actual reliability in Eqs. 28 and 29. Tian states that the purification level captures overall reliability growth and testing effectiveness [16]. The objective of using purification level is to produce tests that have high testability (i.e., use tests that will cause failures to be detected and faults to be exposed and removed).

Reliability growth is related to the principle of *testability*, as described by Voas and Kassab [18]: “Software testability is a characteristic that either suggests how easy software is to test, how well the tests are able to interact with the code to detect defects, or some combination of the two”. The authors suggest it is useful to employ the perspective that software testability is a measure of how good test cases will be at making defects detectable. In addition, Voas states [17]: “A program is said to have high testability if it tends to expose faults during random black-box testing, producing failures for most of the inputs that execute a fault. A program has low testability if it tends to protect faults from detection during random black-box testing, producing correct output for most inputs that execute a fault". We embody these concepts in the test rules below that are designed to expose faults and failures (testability) that will result in reliability growth.

13.3 Purification Levels

The actual purification level (i.e., actual reliability) for consumer and producer, using failure counts, is computed in Eqs. 28 and 29, respectively:

$$\rho_c = 1 - \left(\frac{r_c(t)}{\sum_{t=1}^{N_c} r_c(t)}\right) \tag{28}$$

$$\rho_{\mathrm{p}} = 1 - \left(\frac{r_p(t)}{\sum_{t=1}^{N_p} r_p(t)}\right) \tag{29}$$

Since it is important to assess the validity of the prediction system, by comparing the predictions with the actual purification levels, predict the purification levels. You do this by using Eqs. 30 and 31 for the consumer and producer, respectively.

$$\rho_{\mathrm{cp}} = 1 - \left(\frac{m_c(t)}{\sum_{t=1}^{N_c} m_c(t)}\right) \tag{30}$$

$$\rho_{\mathrm{pp}} = 1 - \left(\frac{m_p(t)}{\sum_{t=1}^{N_p} m_p(t)}\right) \tag{31}$$

The reason Eqs. 30 and 31 are predicted quantities, as opposed to Eqs. 28 and 29, is that the former include predicted mean failures $m_c(t)$ and $m_p(t)$, whereas the latter include observed failure counts $r_c(t)$ and $r_p(t)$.

13.4 Test Rules

For the Shuttle, all previous test rules apply, with the following additions dealing with reliability growth and purification level:

Mandatory for Software Reliability Growth Model
Consumer Reliability, predicted during times t_i and t_{i+1} : $R_c(t_{i+1}) > R_c(t_i)$ for all i

Producer Reliability, predicted during times t_i and t_{i+1} : $R_p(t_{i+1}) > R_p(t_i)$ for all i

Desirable for Purification Level

Predicted Consumer Purification Level > Actual Consumer Purity Level : $\rho_{cp} > \rho_c$

Predicted Producer Purification Level > Actual Producer Purity Level : $\rho_{pp} > \rho_p$

Reliability growth and purification level tests have been added because, for safety critical systems like the Shuttle, it is important to demonstrate reliability growth, as contributing to the safety of the crew and mission. As pointed out by [9], it may be necessary for an organization to demonstrate the reliability of its product "as delivered". For example, there could be a test where the consumer "buys off" the product from the producer. If this is the case for safety critical software, the test model and schema must enforce a high standard of reliability (and risk) before the product is accepted.

The Shuttle test rules, based on modifying the original roadmap with reliability growth and purification level criteria, are shown in Fig. 11, Parts 1 and 2.

13.5 Risk Analysis

In the case of the Shuttle, the consumer and producer risk equations are developed, giving effect to the way that failure data is generated. There are several streams of failure data available for a given software release (i.e., operational increment (OI)): one from the producer (contractor), another from the customer (NASA), and another from the Shuttle simulator. One failure stream is used for the consumer and another for the producer. The logic of this is that the producer tests the software generating one stream, provides the software to the customer, and the customer tests the software generating another stream. The consumer attempts to increase the reliability over that delivered by the producer by continuing to test and remove faults. Thus, the next step is to formulate the probability of failures, assuming a Poisson distribution of failures, occurring at time t, using Eqs. 32 and 33 for the consumer and producer, respectively. These equations will be used in the formulation of consumer and producer risk.

$$P_c(r_c) = \left(m(t)^{r_c} e^{-m(t)}\right)/r_c! \quad (32)$$

$$P_p(r_p) = \left(m(t)^{r_p} e^{-m(t)}\right)/r_p! \quad (33)$$

In order to estimate the *mean number offailuresm*(t) in Eqs. 32 and 33, use Eq. 34: From SSRM [13]:

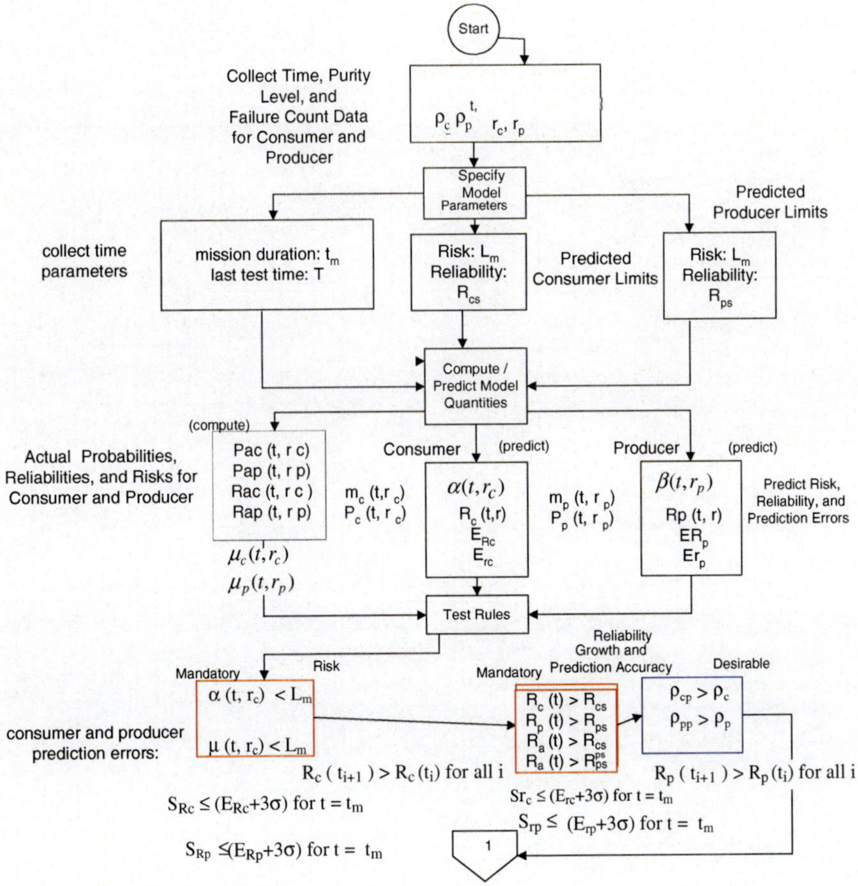

Fig. 11 **a** Shuttle risk-driven testing and reliability process (Part 1), **b** Shuttle risk-driven testing and reliability process (Part 2)

$$m(t) = \frac{a}{b}\left(e^{-b(t-s)} - e^{-b(t-s+1)}\right) \tag{34}$$

Recalling that risk is the (probability of an undesirable event times the consequences of the event), use Eqs. 32 and 33 to develop the following equivalences, based on the Poisson distribution:

$$\text{Consumer risk} = \alpha(t, r_c) = P_c(r_c) * r_c = \left[\left(m_c(t)^r e^{-m_c(t)}\right)/r_c!\right] * r_c \tag{35}$$

$$\text{Producer risk} = \beta(t, r_p) = P_p(r_p) * r_p = \left[\left(m_p(t)^r e^{-m_p(t)}\right)/r_p!\right] * r_p \tag{36}$$

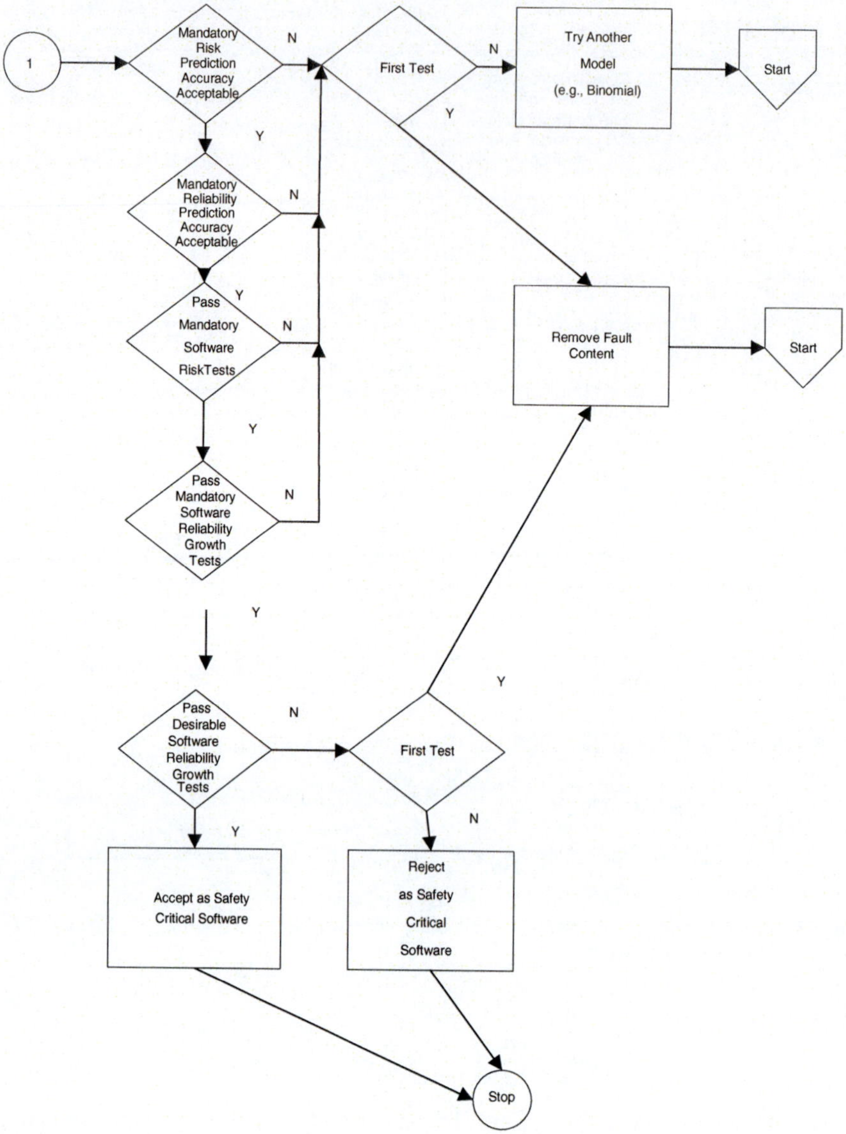

Fig. 11 (continued)

With respect to actual risk, since it is based on empirical failure counts, use Eqs. 37 and 38, respectively, for actual consumer probability, $P_{ac}(t, r_c)$, and actual producer probability $P_{ap}(t, r_p)$.

$P_{\mathrm{ac}}(t, r_c)$: actual consumer probability of r_c failures in test time

$$t = \frac{r_c(t)}{\sum_{t=1}^{N_c} r_c(t)} \tag{37}$$

$P_{ap}(t, r_p)$: actual producer probability of r_p failures in test time

$$t = \frac{r_p(t)}{\sum_{t=1}^{N_p} r_p(t)} \tag{38}$$

This then leads to the equations for actual consumer risk and actual producer risk in Eqs. 39 and 40, respectively.

$$\mu_c(t, r_c) = P_{ac}(t, r_c) * r_c \tag{39}$$

$$\mu_p(t, r_p) = p_{ap}(t, r_p) * r_p \tag{40}$$

13.6 Reliability Analysis

Using SSRM [13], the general form of consumer and producer reliability at time t is given by Eq. 41:

$$R(t) = e^{-[a(e^{-(b(t-s+1))})]} \tag{41}$$

The reliability at time t_{cp} when consumer reliability is equal to producer reliability can be found by equating $R_c(t)$ to $R_p(t)$, using Eq. 41, and solving for $t = t_{cp}$. This is of interest because when $t > t_{cp}$, it is desirable for $R_c(t) > R_p(t)$, meaning that at this value of test time, the consumer has achieved a reliability greater than that delivered by the producer. The solution is found in Eq. 42:

$$t_{cp} = \frac{\left[\frac{\log(a_p)}{\log(a_c)} - b_c(s_c - 1) + b_p(s_p - 1)\right]}{(b_p - b_c)} \tag{42}$$

For the purpose of comparing predicted with actual values, the actual reliability is computed as follows for the consumer and producer reliability, in Eqs. 43 and 44, respectively:

$$R_{ac}(t, r_c) = 1 - \left(\frac{r_c}{\sum_{t=1}^{N_c} r_c}\right) \tag{43}$$

$$R_{ap}(t, r_p) = 1 - \left(\frac{r_p}{\sum_{t=1}^{N_p} r_p}\right) \tag{44}$$

14 NASA Space Shuttle Application

The following parameters are specified, understanding that the failure data observed during tests is used for estimating model parameters. Then the fitted model is used to make forecasts for the prediction range.

t: consumer test time = 1,...,25; consumer prediction range = 26,...,45
t: producer test time = 1,...,36; producer prediction range = 37,...,45
t_m: = *desired* mission duration = 8 (45–37)

Test time can be different for consumer and producer because each chooses to test a different amount of time, dependent on their risk and reliability objectives. For example, the producer may have more resources than the consumer to do testing and, therefore, test for a longer time, and, in addition, is getting paid by the consumer to do testing. This difference leads to different prediction times, given the end of the mission: $t = 45$. Of course, the mission duration must be the same for consumer and producer. A mission duration of 8 days is typical for the Shuttle.

$$\bar{r_c} : \text{mean of consumer failure distribution} = 0.2400 \text{ failures}$$

$$\bar{r_p} : \text{mean of producer failure distribution} = 0.1818 \text{ failures}$$

Since r_c is needed in the computation of consumer risk for the prediction range $t = 26,\ldots,45$, and there are no historical values available for this range, Eq. 34 is used to compute their mean values. Likewise, this equation is used to predict r_p for producer risk in the prediction range $t = 37,\ldots, 45$.

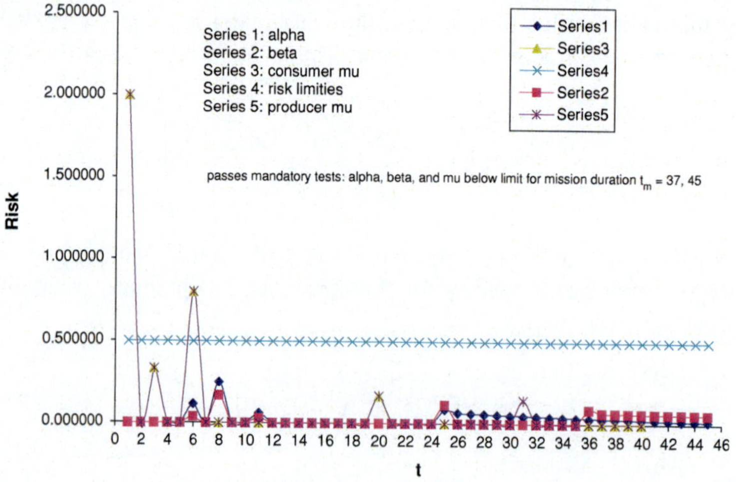

Fig. 12 Shuttle test: predicted consumer risk (alpha), predicted producer risk (beta), and actual risk (mu) versus time t

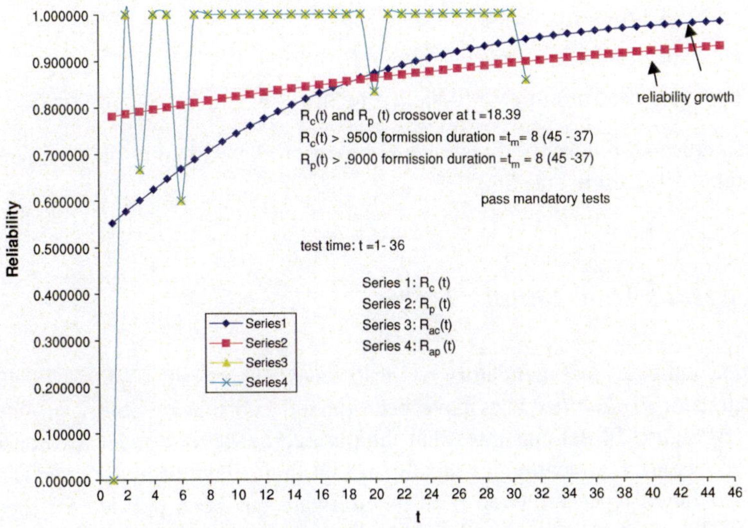

Fig. 13 Shuttle first test: consumer reliability $R_c(t)$, producer reliability $R_p(t)$, actual consumer reliability $R_{ac}(t)$, and actual producer reliability R_{ap} versus time t

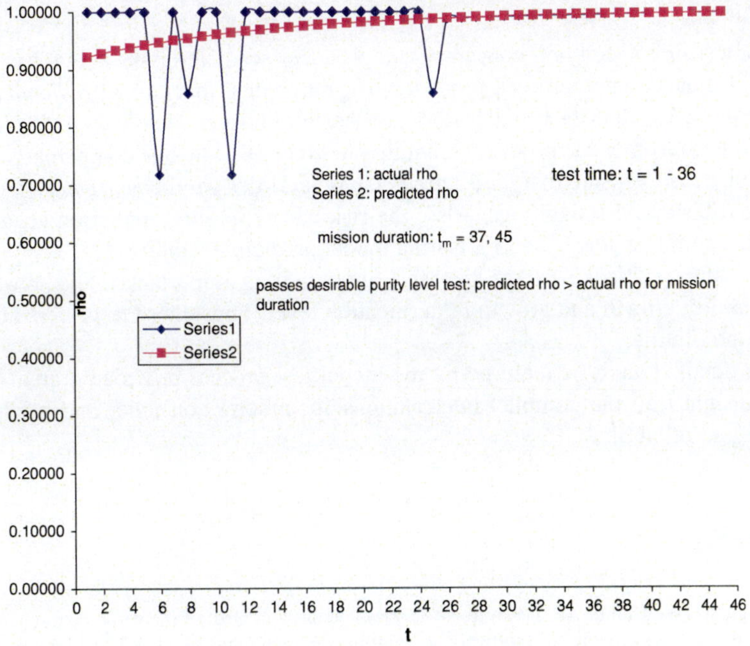

Fig. 14 Shuttle first test: consumer purity level (rho) versus time t

L_m : risk limit = 0.500000

R_{cs} : specified minimum consumer reliability = 0.9500

R_{ps} : specified minimum producer reliability, where $R_{ps} < R_{cs} = 0.9000$

The choice of reliability thresholds is based on the criticality of the mission to the safety of the crew and Shuttle.

14.1 Results from Shuttle Tests

Figure 12 indicates that mandatory *risk* tests have been passed. Figure 13 indicates that mandatory *reliability* tests have been passed and that reliability growth has occurred. Figure 14 demonstrates that the purification level test for the consumer has been passed; the meaning of this test is that faults have been removed (actual) or are predicted to be removed (i.e., the software has been purified for the consumer). The producer test was also passed, but its plot is not shown because the result is almost identical to the consumer test. The final outcome is that this *safety critical* software would be accepted.

15 Summary

1. The test rule specifying consumer and producer predicted risks being less than their actual counterparts is a good idea. The rationale is that *future predicted* risks should be *less* than the actual historical risks. Furthermore, it is also a good idea to require consumer and producer reliability to exceed their actual counterparts. The rationale is to ensure that *future predicted* reliabilities would be *greater* than the actual historical reliabilities. Also, the risk and reliability prediction accuracy rules should be adequate for ensuring model predictive validity.
2. For safety critical systems like the Shuttle, it is important to demonstrate reliability growth and growth in purification level. Thus, these tests were added for the Shuttle.
3. The detailed analysis required by the model test process provided a great deal of insight into the complex interrelationships among consumer and producer risk and reliability.

References

1. Cangussu JW, Mathur AP, DeCarlo RA (2001) Feedback control of the software test process through measurements of software reliability. Proceedings of the 12th international symposium on software reliability engineering, 2001. ISSRE 2001, 27–30 Nov, pp. 232–241

2. Dalal SR, Jain A, Karunanithi N, Leaton JM, Lott CM, Patton GC, and Horowitz BM. Model-based testing in practice. In: Proceedings of 21st international conference on software engineering (ICSE-99), May 16–22, 1999, Los Angeles, California, USA, pp. 285-294
3. Horgan JR, Mathur AP (1996) Software testing and reliability, Handbook of software reliability engineering. In: Lyu MM (eds) Computer Society Press
4. Jursto N, Moreno AM, Vegas S, Solari M (2006) In search of what we experimentally know about unit testing. IEEE Softw 23(6):72–79
5. Keller T, Schneidewind NF (1997) A successful application of software reliability engineering for the NASA space shuttle. Software reliability engineering case studies, International symposium on software reliability engineering, Nov 3, Albuquerque, Nov 4, pp 71–82
6. Loyd DK, Lipow M (1962) Reliability: management, methods, and mathematics. Prentice-Hall, Inc
7. Lyu MR (1998) An integrated approach to achieving high software reliability. Proceedings of the aerospace conference, vol 4, pp 123–136, 03/21/1998–03/28/1998, Snowmass at Aspen, CO
8. Musa JD (1999) Software reliability engineering: more reliable software, faster and cheaper, 2nd edn. Authorhouse, Bloomington
9. Musa JD, Iannino A, Okumoto K (1987) Software reliability: measurement, prediction application. McGraw-Hill, USA
10 Myers GJ (1979) The art of software testing. Wiley, New York
11. Nelson CR (1973) Applied time series analysis for managerial forecasting. Holden-Day, Inc, San Francisco
12. Schmid M, Ghosh A, Hill F (2000) Techniques for evaluating the robustness of windows NT software. DARPA information survivability conference and exposition, January Hilton Head
13. Schneidewind NF (1997) Reliability modeling for safety critical software. IEEE Trans Reliab 46(1):88–98
14. Software Safety (1997) NASA Technical Standard, NASA-STD-8719.13A, Sep 15
15. Stocks PA, Carrington DA (1993) Test templates: a specification-based testing framework. Proceedings of the 15th international conference on software engineering, May 17–21, 1993, Baltimore, pp 405–414
16. Tian J (1995) Integrating time domain and input domain analyses of software reliability using tree-based models. IEEE Trans Softw Eng 21(12):945–958
17. Voas J (1991) Factors that affect software testability. Proceedings of the 9th Pacific Northwest Software. Quality Conf, pp 235–247, Oct 1991, Pacific Northwest Software Quality Conference, Inc, Portland
18. Voas J, Kassab L (1999) Using assertions to make untestable software more testable. Softw Qual Prof J 1(4), Sep 1999
19. Whittaker JA (2000) What is software testing? And why is it so hard? IEEE Softw 17(1):70–79

Human Grasp Prediction and Analysis

Tim Marler, Ross Johnson, Faisal Goussous, Chris Murphy, Steve Beck and Karim Abdel-Malek

1 Introduction

Given that one of the critical motivations for using virtual humans is to simulate the interaction between humans and products, and given that using one's hands are a primary means for interaction, then simulating human hands is arguably one of the most important elements of digital human modeling (DHM). Consequently, there is much research and development in this area, ranging from basic model development to detailed simulations of specific joints and tendons. However, when considering hand simulation and analysis within the context of a complete high-level DHM, the culmination of hand-related capabilities is grasping prediction. Thus, the focus of this chapter is on postural simulation and analysis capabilities of the overall hand as a component of a complete high-level DHM, with an eye toward grasping prediction. Within this context, the fundamental necessary elements one must consider when modeling the hand are highlighted. The intent is to provide general guidelines for creating computational models of hands and to present novel modeling and simulation techniques.

As with any model, the appropriate fidelity is the most important element. In this context, *model* refers to the underlying computational structure of the hand, separate from simulation and analysis that the model may be used for. All simulation and analysis capabilities depend on the fundamental mathematical model. Analyses are only as accurate as the model with which they are conducted. Despite the expanse of work with 2-D hand models, for practical interaction with current virtual environments, the model should be in 3-D. Furthermore, the structure of the model must be easily altered to facilitate various types of analyses and studies that

T. Marler (✉) · R. Johnson · F. Goussous · C. Murphy · S. Beck · K. Abdel-Malek
Center for Computer Aided Design, University of Iowa,
111 Engineering Research Facility, Iowa City, Iowa 52242, USA
e-mail: tmarler@engineering.uiowa.edu

H. Pham (ed.), *Safety and Risk Modeling and Its Applications*,
Springer Series in Reliability Engineering, DOI: 10.1007/978-0-85729-470-8_14,

incorporate various anthropometric cross-sections. The primary advantage of using virtual models is the ability to simulate and rerun experiments or tests under different conditions, easily and quickly. This advantage is not fully recognized unless the human model can be altered to represent different humans.

Given a sound underlying model (i.e., a skeletal system that incorporates the proper number of joints), the simulation capabilities must be predictive. Computational models of a human should be able to predict an outcome based on a set of input parameters that may change. In this way, models can be used to formulate and test hypotheses; they can be used to answer questions. Although a human model can provide a tool for evaluating products, using a model that cannot be varied in order to test hypotheses concerning the human itself, defeats a substantial purpose of actually developing models in the first place. However, all components of a human model should allow for direct user manipulation, to refine and/or alter predicted outcomes if necessary.

In accordance with providing predictive capabilities, one of the most critical and helpful forms of prediction is posture (for the fingers) prediction with associated feedback and analysis. Posture prediction capabilities should then culminate in grasping prediction, which is a key element of hand functionality that must be incorporated in any DHM.

With most segments/components of the human body, it is insufficient to model them independently; interdependencies must be considered. This is especially true with the hand, which is a complex extension of the body. Furthermore, this close relationship between hand modeling and overall human modeling is poignant in the context of reach-related tasks. Thus, when developing hand-related modeling and simulation capabilities, the natural progression is to focus eventually on the hand, arm, and body as a complete system.

One final critical element in modeling the hand, especially when striving to predict human behavior or actions, is cognitive modeling. Much of one's tendencies when reaching for or grasping objects depend on one's personal history, state of mind, knowledge, and decision-making processes. However, few researchers in the modeling-and-simulation arena have made the connection between whole-body performance, hand-modeling capabilities, and cognitive modeling. Thus, the focus of this work is on predictive modeling and simulation capabilities for the hand, with extension to the whole body, from a biomechanical perspective.

These critical components of hand modeling and simulation are demonstrated in the context of ongoing work at the University of Iowa's Virtual Soldier Research (VSR) Program. Nonetheless, this work serves a platform for a broad discussion regarding the most advantageous direction for research and development surrounding the human hand as a constituent of DHM in general.

Because of the applicability to the field of robotics, the problem of hand modeling and grasp synthesis has a long history and has received much attention both in and out of the DHM arena. Many of the techniques used to solve the robotic grasping problem can be readily transferred to the virtual world with only slight modifications. Other algorithms are specifically tailored to virtual humans.

Here, we classify and summarize the primary works according to the method used to achieve a grasp, based on a brief review provided by Goussous et al. [1].

Rule-based techniques classify the part of the object to be grasped as one of several previously stored shape primitives, such as a sphere, pyramid, cube, or cylinder. These systems then contain rules for grasping each of these primitives [2–6]. More recently, Xue et al. [7] have also incorporated into their shape primitives semantic information related to grasping. Simplified information about the task or physical properties of the object is supplied by the user and embedded in the shape description. A primary shortcoming of all these rule-based approaches is that they are not suitable for automatic grasping of arbitrary objects, because the decision about which primitive to use is either left to the user or embedded in the object model during the design stage. Even if we assume that the latter problem has been solved, there are still many objects that cannot be intuitively classified as one of the few geometrical primitives.

Researchers have attempted to produce intelligent grasping systems that can learn from previous successful grasps and adapt them to new objects. There are many examples of the use of neural networks for grasping in the literature [8–11]. Many of these approaches attempt to learn and control low-level grasping behaviors, such as finger joint angles or wrist orientation. These actions can be more efficiently determined using inverse kinematics. Another drawback is that most of this work has been theoretical and has been applied only to specific types of objects with well-known geometry.

Pelossof et al. [12] use support vector machines (SVM) to associate a successful grasp with a given object. The objects are modeled as superquadrics, which are not general enough to describe any arbitrary object. In addition, this work focused on robotic grippers rather than humanoid grasps.

Much of the grasping literature is concerned with optimization-based methods for determining appropriate grasping postures. Most of these techniques use the quality metric proposed by Ferrari and Canny [13], or variations thereof, as an objective function in an optimization problem. Some of these techniques assume the availability of a closed-form description of the object surface [14–17]. Other techniques generate a large number of arbitrary grasps and then use the quality metric to rank the grasps and pick the most appropriate one [4, 18–20]. Some authors use genetic algorithms and developed customized quality measures to be used as fitness functions in these algorithms [21–23]. However, most methods that rely on optimizing a grasp quality function are not ideal for use with virtual humans for the following reasons:

1. Most of these methods assume the existence of a closed-form description of the object surface. Such descriptions are not readily available in 3-D virtual environments where objects are typically modeled with large sets of polygons.
2. Calculation of grasp quality measures can be computationally expensive, thus inhibiting any real-time application to grasp synthesis. This is especially true when genetic algorithms are used.

3. These approaches are restricted by the number of contact points on the object surface. This is not applicable to power grasps where the surface of the palm is in contact with the object at many points.

As an alternative to the optimization-based techniques, data-driven grasping techniques exploit the idea that data obtained offline about grasps can be used to synthesize similar grasps online. For example, ElKoura and Singh [24] use a database of human grasps to enhance results obtained from an inverse kinematics algorithm. Ehrenmann et al. [25] utilize a data glove to record grasping actions that are later used to teach a robot manipulation task in similar environments. Bohg and Kragic [26] use a database of objects with labeled grasping points to train a machine learning algorithm for use on new objects. Li and Pollard [16] use a database of grasps obtained through motion capture data and try to adapt these grasps to novel objects through the use of shape-matching algorithms. Similarly, Miyata et al. [27] rely on motion capture data to select starting hand poses for grasp posture generation. Aleotti and Caselli [28] use virtual reality to program their system to grasp by demonstration. These systems generally simulate natural grasps, because the database itself will contain precise grasps that were carefully generated from actual human postures. However, it can prove challenging to adapt the grasps in a database to new objects and situations in a virtual environment.

In summary, there are several deficiencies in the state of the art that prevent one from directly applying any of the aforementioned methods to the problem of virtual human grasping. These include:

1. None of the methods consider the effect of the upper body on the feasibility of the grasp. They consider the grasp complete when the positions and orientations of the fingers and wrist are calculated.
2. Robotic grasping does not necessarily apply to human hands, which are unique in their anatomy and complexity.
3. Few methods are able to address the problem of grasping arbitrarily shaped objects.

The methods outlined in this chapter respond to these deficiencies and culminate in new approaches for predicting human grasps.

One of the most critical aspects of human modeling and simulation is the hand, which often acts as the primary interface with one's environment. In turn, one of the most critical and complex elements of modeling the hand is grasping. In the context of digital human modeling (DHM), the hand cannot be considered independently. Rather, it must be considered as an integral element of the complete human body, especially with respect to predicting human performance. Thus, the intent of this chapter is to present novel modeling and simulation capabilities related to the hands with a focus on performance prediction and analysis, and with culmination in new methods for grasping prediction. In addition, general and critical aspects for consideration when working with hand models are highlighted. Although much work has been completed in the robotics arena, the human hand

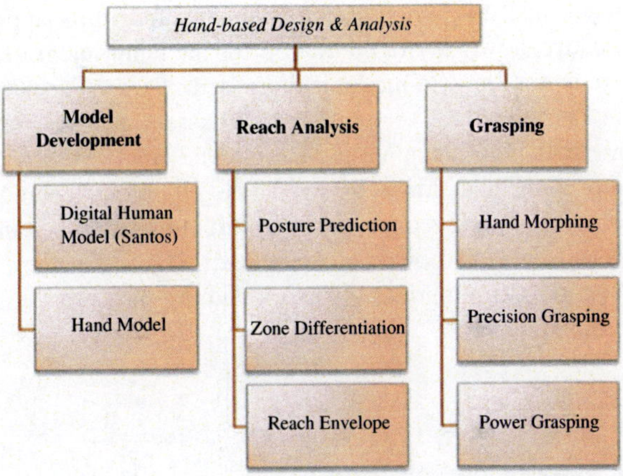

Fig. 1 Overview of hand capabilities

tends to be relatively complex, so the degree to which developments with robotics may be leveraged is limited. Furthermore, little work has been completed with grasping arbitrarily shaped objects while considering the effects of the complete body. Thus, following the development of a high-fidelity hand model integrated with the Santos human model, we propose a suite of predictive reach-analysis capabilities. We then propose a series of tools for grasping prediction, including new techniques for morphing between a library of standard grasps, optimization-based precision-grasp prediction that leverages collision detection for finger wrapping, and an extensive algorithm for power-grasp prediction based on shape matching and integration with whole-body posture prediction. With respect to the underlying hand model, we find that variable anthropometry and using a 3D modeling are critical. With respect to grasping, optimization-based predictive capabilities are necessary such that a user can study what drives human performance. Furthermore, we find that system integration between the body and the hand is another critical component. Given a sound model and predictive capabilities, a variety of analyses tools are necessary in order to use the human model for design and analyses of products and processes. The newly developed capabilities show promising and practical results.

1.1 Overview of the Chapter

This chapter is divided into three primary sections, as shown in Fig. 1. Section 1 addresses modeling. The underlying skeletal model of the hand and the skin model that provides the realistic appearance of the hand are both presented. These models are discussed as components of the Santos DHM, which is summarized as well. Section 2 focuses on basic predictive capabilities and feedback. Optimization-based posture prediction for the fingers is introduced. Then, two new methods for reach analysis are presented that leverage this approach to posture prediction. Finally, how posture prediction for the hand integrates with posture prediction for the overall human model is discussed. Section 3 continues to build on the material in the preceding sections and discusses methods for predicting grasping. Three methods for grasping prediction and analysis are discussed: interactive shape morphing that draws on avatar development techniques from Sect. 1, precision grasping that leverages posture prediction discussed in Sect. 2 along with collision detection, and power grasping that involves a new shape-matching algorithm. Section 4 summarizes this work and presents broader issues of hand modeling and simulation.

2 Model Development

The hand-modeling and simulation capabilities discussed in this chapter do not operate independently; they are one component of a complete digital human model called Santos. Santos houses a variety of components and capabilities, ranging from dynamic motion prediction to muscle models to physiological models [29–31]. One of the things that makes Santos unique is his ability to serve as a platform for multi-scale modeling, which involves integrating various type of models and predictive capabilities. Given the capabilities of the hand model that relate to the whole body, the discussion of Santos focuses on the skin and skeletal models, and on posture prediction. Subsequently, a high-fidelity hand model is presented as an extension of Santos. In developing the model, which is then used for simulation, the focus must be on functional fidelity and on visual realism. Furthermore, any human model must have variable anthropometry, thus able to represent humans of different sizes.

2.1 Santos

In the past a typical human model mesh was created by hand using various software packages. This process was truly a black art, and for years a realistic human model was considered the holy grail of computer graphics. Even today a realistic human model is difficult to achieve through traditional modeling methods and can require weeks to months of effort and expense. The advent of 3-D scanning technology has significantly reduced the time if not the expense necessary to develop a human model. While the best source of data for overall anatomic accuracy, a raw body scan is not suitable for use as an avatar as it may contain many more polygons than are necessary to describe the form. There may also be areas of the body that were parallel to the scan line or obscured by another body part, and therefore no surface points were created to represent them (Fig. 2).

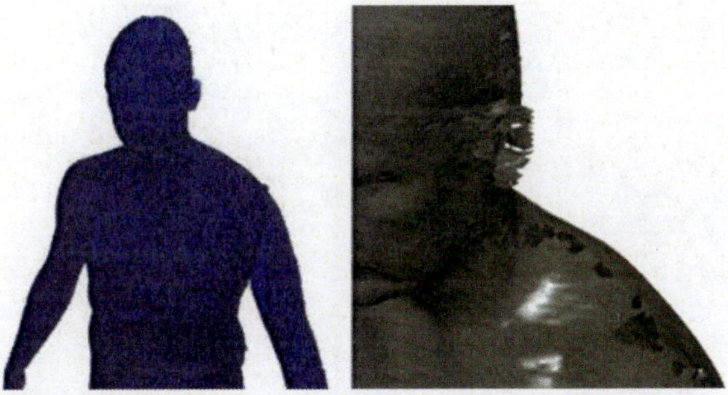

Fig. 2 The dense wire mesh of a scan with an example of the holes scans often produce

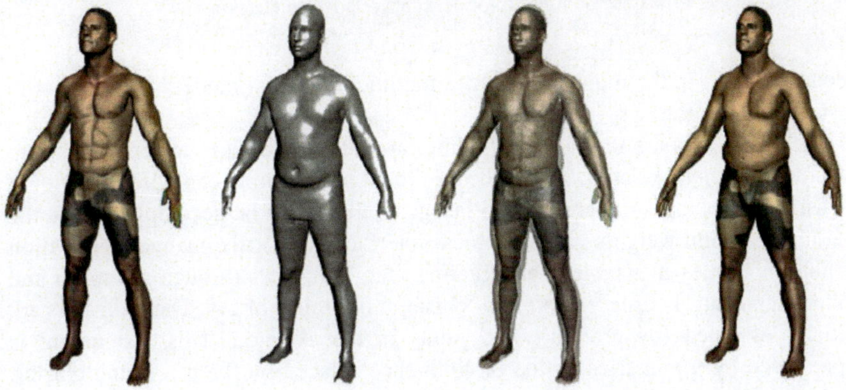

Fig. 3 Process of morphing a pre-existing avatar into a scan template, to form a new avatar

This would create potentially large holes in the geometry effectively omitting anatomic data.

Body scans are instead used as 3-D templates for desired body types, and a single, highly refined 3-D mesh, optimized for use in a real-time environment, is modified to reflect specific body (or hand) types, as shown in Fig. 3. These differences in overall shape or type, not just anthropometric dimensions, constitute the morphology of a body or body segment.

Once a 3-D model of a desired body type has been developed, a hierarchical grouping structure of local coordinate axes (or kinematic skeleton) must be imposed on the model to enable the mesh geometry to move in a way that is recognizably human. A highly accurate 3-D computer model of a human skeleton (Fig. 4) is used as visual confirmation to (a) ensure that the kinematic skeleton is accurately mapped within the 3-D body type model and (b) ensure that the movement of the kinematic skeletal joints is as biomechanically accurate as possible. Care must be taken in the placement of kinematic joints to ensure that the skin will deform in a realistic manner and that the various parts of the body pivot about the points at which they should. Ultimately, the joints in the skeleton are represented with a computational model, and in general, the number of degrees of

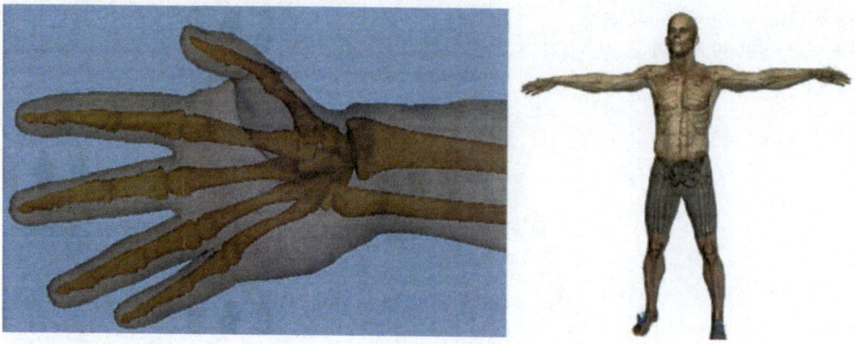

Fig. 4 3-D model of a skeleton used to locate joints relative to the skin

freedom (DOF) in the model constitutes the fidelity of that model, with respect to posture and motion.

Once the joints are in place, a skinning algorithm is used to attach the skin to the computational skeleton. The procedure for skin weighting can differ depending on which modeling software is used but generally it can be accomplished by the 'painting' of skin weights via a graphical interface, and/or the manual application of number values in a table. The skinning algorithm goes through the mesh and divides the control vertices (cvs) into 'clusters' or groups of cvs. These clusters are initially assigned (by proximity) to joints in the skeleton. This assignment is represented by a numerical value or 'influence' placed on the cvs in the cluster. That value controls a cv's movement around the joint or joints that influence it. A cluster can be assigned to multiple joints in varying degrees, allowing for smooth morphing and stretching of skinned surfaces. Although the surface has been skinned and the skin has the ability to deform smoothly, that does not mean that it will deform in an anatomically correct fashion, as shown in Fig. 5. Skin weighting is required before the skin will rotate, stretch, and pull around the joints realistically.

Once a visual skeletal and skin system are constructed, simulating human posture, whether it be for an arm or a hand, depends largely on how the skeleton is modeled. One way to view a skeleton is as a kinematic system, or series of links with each pair of links connected by one or more revolute joints. Therefore, a complete human body can be modeled as several kinematic chains, formed by series of links and revolute joints, as shown in Fig. 6.

The links that connect the joints essentially represent skeletal components. Because the dimensions of these links provide input to the model prior to running a simulation, they can be changed on the fly and can thus enable variable anthropometry. q_i is a *joint angle* and represents the rotation of a single revolute joint. There is one joint angle for each degree of freedom (DOF). $q = [q_i, \ldots, q_n]^T \in R^n$ is the vector of joint angles in an n-DOF model and represents a specific posture. Each skeletal joint is modeled using one, two, or three kinematic revolute joints. $\mathbf{x}(\mathbf{q}) \in R^3$ is the position vector in Cartesian space that describes the location of

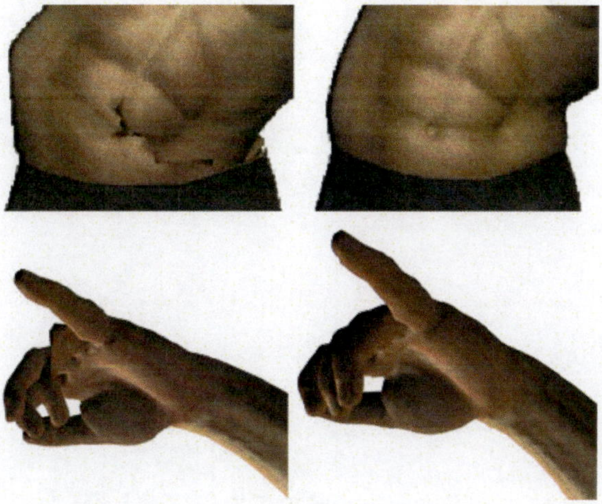

Fig. 5 Skin mesh before and after skin weighting

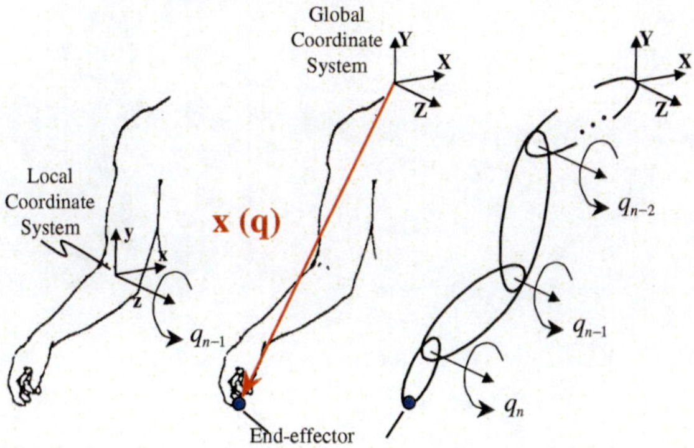

Fig. 6 A kinematic chain of joints

the end-effector as a function of the joint angles, with respect to the global coordinate system. For a given set of joint angles $\mathbf{q}$, $\mathbf{x}(\mathbf{q})$ is determined using the Denavit-Hartenberg (DH)-method [32].

Using the DH-method, $\mathbf{x}(\mathbf{q})$ is expressed in terms of a series of transformations ${}^{i-1}\mathbf{T}_i$ and is calculated as follows:

$$\mathbf{x}(\mathbf{q}) = \left(\prod_{i=1}^{n} {}^{i-1}\mathbf{T}_i \right) \mathbf{x}_n \tag{1}$$

where $\mathbf{x}_n$ is the position of the end-effector with respect to the nth frame and n is the number of DOFs. Note that the rotational displacement $\boldsymbol{q}_i$ changes the value of $\boldsymbol{\theta}_i$. The link lengths between each of the joints are variable and can be set based on anthropometric data, thus representing various population variations. The kinematic system for Santos, to which hands are added, is shown in Fig. 7. In Fig. 7, each cylinder (a revolute joint) represents a degree of freedom that articulates about the indicated z-axis.

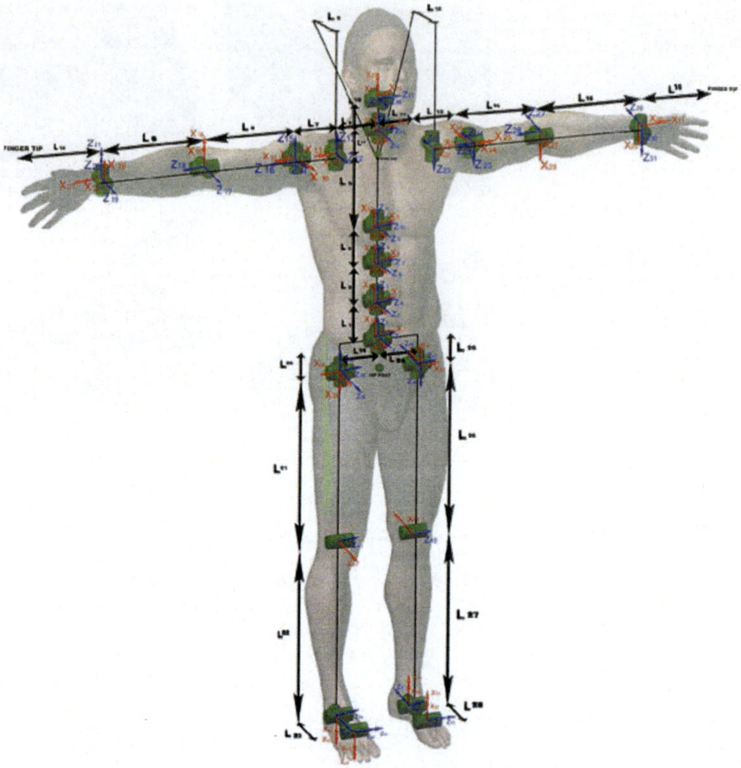

Fig. 7 Kinematic system for Santos

2.2 *Hand Model*

Using the methods described above, we have created a 25 DOF model (not including three DOFs for the wrist) of the human hand. This model is a variation on the work by Pena-Pitarch et al. [33]. A right-handed version of the model is shown in Fig. 8. q_1 through q_3 represent the three wrist DOFs, including pronation/supination, and q_4 through q_{28} represent joints in the thumb and four fingers. The pointer and middle finger each have 4 DOFs; the thumb has 5 DOFs; and the ring and pinky finger have 6 DOFs apiece. Note that although the human hand does have approximately 25 DOFs (the precise number is disputable), the range of motion (ROM) for q_6 (metacarpo-phalangeal joint in the thumb) is minimal. This is true for q_{17} and q_{23} as well.

As with the overall human model, the anthropometry for the hand is variable, as shown in Fig. 9. Each link length can be altered on the fly, or the overall size of the hand can be altered to represent different anthropometric cross-sections. In addition, altering the position or orientation of the DOFs simply entails altering DH-parameters and is thus relatively easy from a developmental perspective.

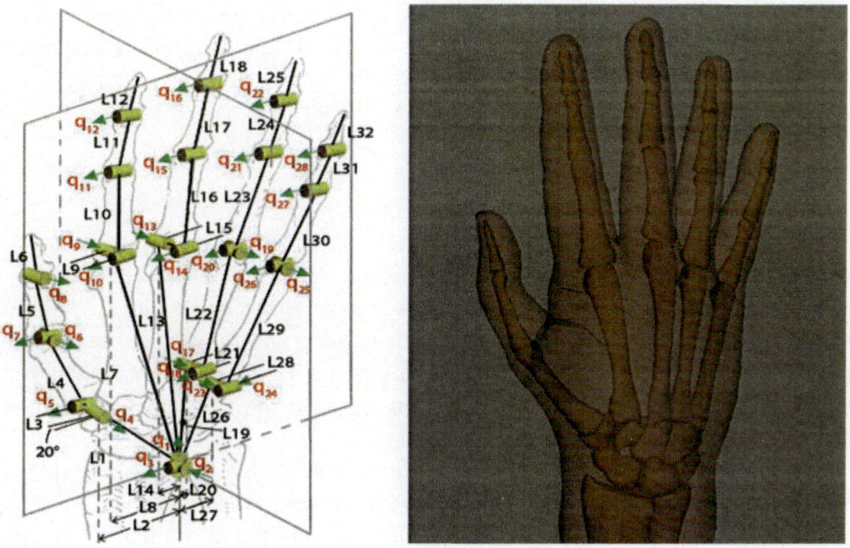

Fig. 8 Kinematic system for the hand

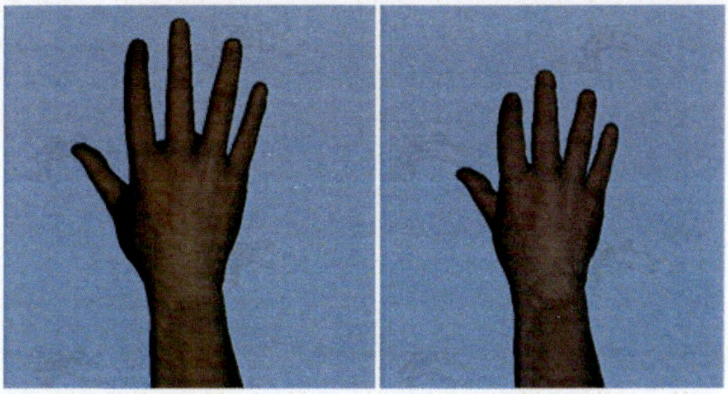

Fig. 9 Variable anthropometry for the hand

3 Posture Prediction and Reach Analysis

Using the same optimization-based technique for posture prediction that was developed for the Santos model [34–36], this section discusses posture prediction for the hand, which will then be utilized in the development of new grasping-prediction techniques. When simulating human behavior, a key criterion is the ability to predict behavior from a mathematical model, rather than reproduce prerecorded data. Such predictive capabilities allow one to test various hypotheses by altering the mathematical model, with fewer restrictions than a data-based model, which may be overly limited with respect to the scenarios that can be simulated. Novel methods for reach analysis, which leverage the optimization-based method for prediction, are also summarized. These satisfy the necessity for feedback, which is another critical element of any useful simulation capability.

Finally, the ability to link hand-based predictive capabilities with the complete body is presented, thus satisfying the necessity to link the hand to the body and demonstrate interconnectivity of various models.

3.1 Prediction

The posture of the human model is determined by solving an optimization problem, for which the design variables are $\mathbf{q} = [q_1, \ldots, q_n]^T \in R^n$ measured in units of radians. The first constraint, called the distance constraint, requires an end-effector (anywhere on the body of hand) to contact a target point. In addition, each joint angle is constrained to lie within predetermined limits. q_i^U represents the upper limit, and q_i^L represents the lower limit. Together, these limits define the ROM for each DOF. These limits are derived from anthropometric data.

The objective functions for the optimization problem are human performance measures, such as joint displacement. This performance measure is proportional to the deviation from the neutral position, which is selected as a relatively comfortable posture. q_i^N is the neutral position of a joint. Because some joints articulate more readily than others, a weight w_i is introduced to stress the relative stiffness of a joint. The final joint displacement is given as follows:

$$f_{\text{Joint Displacement}}(\mathbf{q}) = \sum_{i=1}^{n} w_i \left(q_i - q_i^N\right)^2 \tag{2}$$

With this objective function, the predicted posture generally gravitates toward the neutral position. In addition, a second performance measure, effort, is defined as follows:

$$f_{\text{Effort}}(\mathbf{q}) = \sum_{i=1}^{n} w_i \left(q_i - q_i^I\right)^2 \tag{3}$$

where q_i^I is the initial position (prior to posture prediction) of a joint. With this objective function, the predicted postures tend to gravitate toward one's starting posture.

The optimum posture for the system shown in Fig. 7 is determined by solving the following optimization problem:

$$\begin{aligned}
&\text{Find: } \mathbf{q} \in R^{\text{DOF}} \\
&\text{to minimize: Performance Measure}(\mathbf{q}) \\
&\text{subject to: distance} = \left\| \mathbf{x}(\mathbf{q})^{\text{end-effector}} - \mathbf{x}^{\text{target point}} \right\| \le \varepsilon \\
&q_i^L \le q_i \le q_i^U; \quad i = 1, 2, \ldots, \text{DOF}
\end{aligned} \tag{4}$$

where ε is a small positive number that approximates zero. The objective function(s) in (4) models what drives the posture, while the constraints represent the boundary conditions of what is being modeled. For instance, the contact between bones and cartilage that is actually responsible for restricted ROM is highly complex. Thus, rather than model the internal contact problem for every joint, high-level joint limits is simply imposed as constraints that must be met. (4) is solved using the software SNOPT [37], which uses a sequential quadratic programming algorithm. Note that performance measures can be used as objective functions in (4) or simply as analysis tools that are evaluated at specified postures (sets of joint angles).

Posture prediction for the fingers of a hand uses this same optimization-based approach, with similar objective functions and constraints. In addition, coupling constraints that enforce the relative motion for the finger joints, self-avoidance, and integration between the hand and the complete body are included and are discussed as follows.

The coupling constraints represent the fact that not all joints in the hand are independent. For example, the top knuckle and middle knuckle cannot be moved separately, but instead both rotate during a finger flexion movement. These intra-finger coupling constraints are represented with linear equations based on the literature [5] and are included in (4) as follows:

$$\begin{aligned} &q_i - {}^2\!/_3 q_{i-1} \leq \varepsilon; \quad i = 12, 16, 22, 28 \\ &q_7 - 2q_5 + 60^\circ \leq \varepsilon \\ &q_8 - 1.4q_6 \leq \varepsilon \end{aligned} \tag{5}$$

Although some work suggests that the relationship shown in the first constraint is not completely linear [38], this approach is used as an adequate approximation.

Additional constraints in the problem formulation also prevent the hand from intersecting other geometry and from intersecting itself. This approach to obstacle avoidance and collision avoidance is based on the work by Johnson et al. [39, 40]. Essentially, the avatar (hand) and geometry in the virtual environment are represented with sphere-based surrogate geometry. The spheres used to represent the hand and/or various pieces of geometry are then incorporated in constraints that prevent the intersection of adjacent spheres. These constraints require that the distance between two spheres is greater than the sum of their respective radii, as follows:

$$\begin{aligned} &\text{Position}(O) \cdot \text{Position}(B, \mathbf{q}) - (\text{radius}(O) - \text{radius}(B))^2 \geq \varepsilon; \\ &\quad \text{for all obstacle spheres O and body spheres } B \end{aligned} \tag{6}$$

$$\begin{aligned} &\text{Position}(B_1, \mathbf{q}) \cdot \text{Position}(B_2, \mathbf{q}) - (\text{radius}(B_1) - \text{radius}(B_2))^2 \geq \varepsilon; \\ &\quad \text{for all pairs of body spheres } B_1 \text{ and } B_2 \end{aligned} \tag{7}$$

Results for hand self-avoidance are shown in Fig. 10.

One of the most critical elements in accurately simulating hand posture and motion is integration with the complete body. To work only with a disembodied hand is inaccurate and extremely limiting, especially if one is interested in the

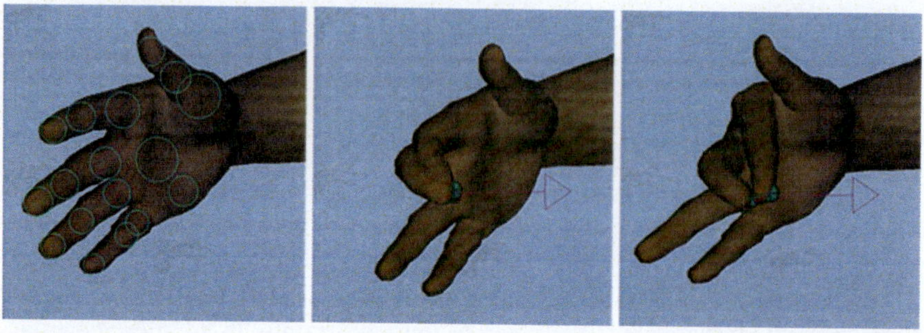

Fig. 10 Sphere-based surrogate geometry (*left*), hand-posture prediction without self-avoidance (*middle*), and with self-avoidance (*right*)

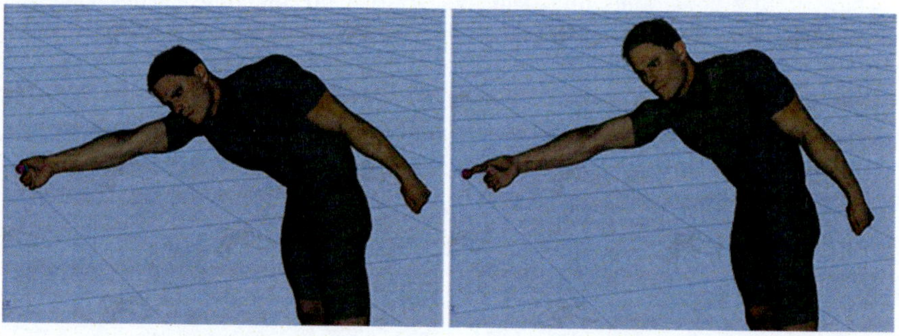

Fig. 11 Whole-body posture prediction without hand (*left*) and with hand (*right*)

global position and orientation of the hand. Thus, the computational skeletons shown in Figs. 7 and 8 are combined. The formulation in (4) is versatile enough to allow for additional degrees of freedom. The result is a human model with 107 degrees of freedom when considering the two hands. Results when predicting posture while using a combined hand-body joint-displacement performance-measure are shown in Fig. 11.

3.2 Analysis

Digital human modeling is only as useful as the feedback it provides. Certainly, visual cause and effect, given changes in problem parameters, is critical. That is, feedback in the form of predicted postures and/or motion provides a primary tool for analysis. However, additional numerical feedback is necessary for quantifying results and developing metrics for evaluation. The optimization-based approach described above is especially useful in this respect, as it provides numerical output with every posture, in the form of performance-measure values and joint angles.

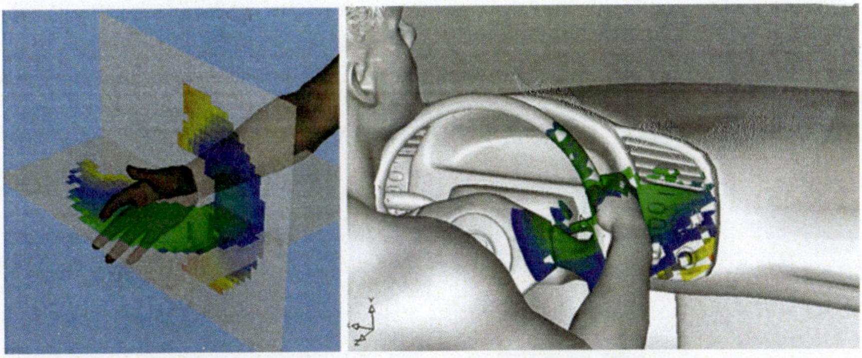

Fig. 12 Zone differentiation for product design (*left*) and product analysis (*right*)

However, because this approach is extremely fast, running in real time, it is possible to output performance measure values for millions of potential target points surrounding the body or just the hand. This ancillary tool is called *zone differentiation* and is based on the work presented by Yang et al. [41, 42]. As shown in Fig. 12, the relative values of a performance measure (hand-based joint displacement in this case) at different target points are output and color coded, for many different target points. The color green, for instance, indicates target points for which the predicted hand posture has a relatively low value for joint displacement. This tool can be used either for product analysis, where existing geometry is colored, or for product design, where a potential package space is evaluated.

Another way to visualize the reachable space of a particular point on a hand is with a *reach envelope*. Unlike zone differentiation, the reach envelope does not further distinguish reachable areas based on performance-measure values, and this allows a reach envelope to be generated much faster than a zone. In fact, the reach envelope calculation does not depend at all on posture prediction. Instead, each DOF is essentially swept through its complete range of motion, and any points that the end-effector contacts are marked as reachable. The algorithm for determining this reach envelope is outlined as follows:

1. Find the chain of DOFs from the end-effector's parent joint to the root joint (the desired base of the reach envelope).

For each of these DOFs, calculate $\Delta\theta_i$, which is the angle through which each DOF is articulated when it is swept through its range of motion. This value depends on the resolution of the reach envelope and the maximum distance of the end-effector to the joint.

2. Set the angle for each DOF on the chain to its lower limit q_i^L.

Start with the outermost DOF (the one furthest from the root joint), and increment its angle by $\Delta\theta_i$ until it reaches the upper joint limit q_i^U. At each

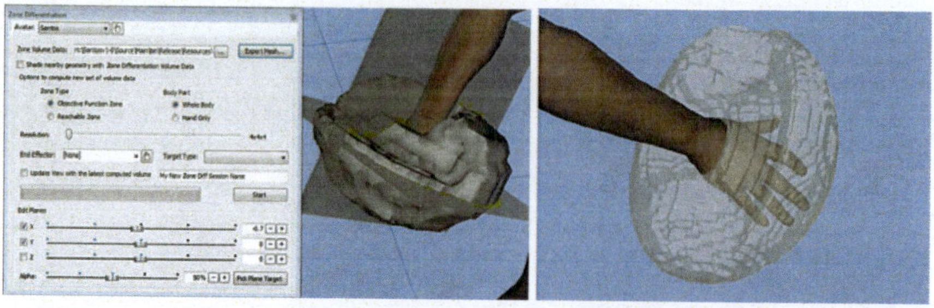

Fig. 13 Reach envelopes for *right* pointer finger, including wrist

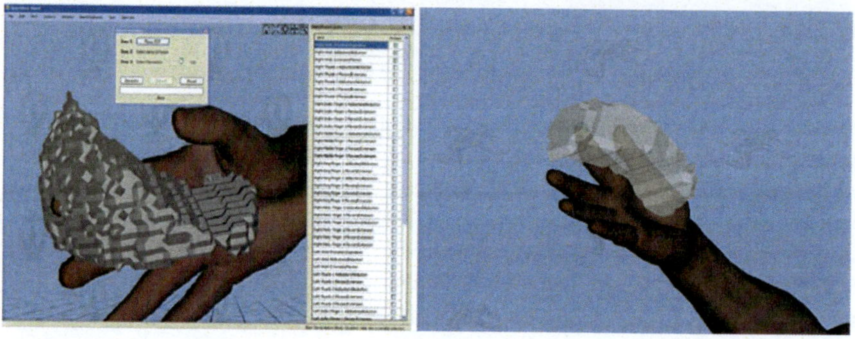

Fig. 14 Reach envelopes for *right* pointer finger with frozen/restricted wrist

increment, transform the local end-effector position into the parent DOF's reference frame, and store the position.

3. Repeat Step 4 for each DOF continuing up the chain to the root joint, but instead of just transforming the one local end-effector position, transform the entire set of local, reachable positions calculated during the previous DOF's sweep.
4. Once all the points have been transformed to the root frame, they can be pigeonholed into the grid. That is, each cell can be marked as either reachable or non-reachable depending on whether it contains any of the generated reachable points.
5. Finally, marching cubes [43] can be used to generate a mesh (3-D surface) that surrounds the reachable cells of the grid.

This process can be further complicated when joint-coupling constraints must be respected. To increase speed, we have implemented an intermediate pigeonholing technique which is used with every DOF along the chain. Essentially, each point's position is rounded to some small degree, and then duplicate points are discarded. This process effectively combines points that are very close together, and thus reduces the total number of necessary transformations. Examples of the reach envelope are shown in Figs. 13 and 14.

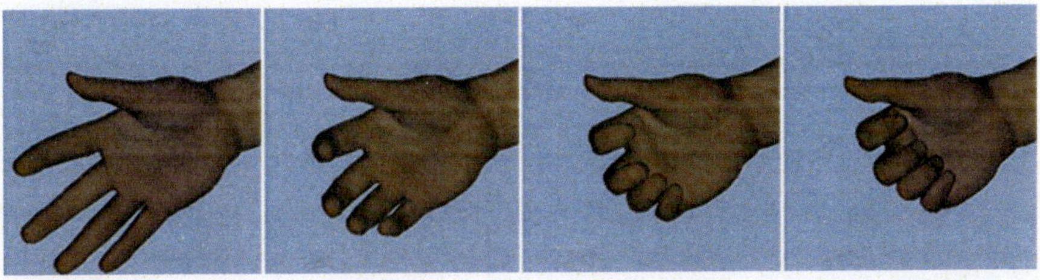

Fig. 15 Various stages of morphing a grasp between an open hand and a closed hand with abducted thumb

4 Grasping Analysis

Because of the influence on grasping of the overall body posture, one's history, and one's immediate decisions, predicting grasps accurately for a general case while integrating with a larger whole-body model is especially difficult. Consequently, a few steps can be taken to offset this complexity, and these are seen in various forms in much of the literature. First, predictive capabilities are made more task-specific. In this case, we present two grasping-prediction approaches: one for precision grasps and one for power grasps, according to the original taxonomy of Napier [44]. Secondly, the user is involved, thus providing a substitute for cognition, which typically dictates the grasping strategy. In this case, we allow for manual manipulation of the hand as well as the ability to morph between common pre-defined grasps. In addition, during the grasp-prediction process, the user has the option to make choices regarding the preferred results.

4.1 Morphing

With respect to manual manipulation of the hand, the user is able to articulate any and all DOFs in Fig. 8, for what is often called single-finger control. Concurrently, the joint limits that define the ROM for each DOF are enforced. Regardless of the extent to which one can predict human behavior, the option for user interaction is paramount as far as software use is concerned.

In addition to being able to articulate each joint, it is possible to morph between any two pre-defined or user-specified grasps or hand postures. A library of existing grasps is provided based on the work of Cutkosky [45] and Feix et al. [46]. The various grasps or postures are defined by a set of joint angles (one for each DOF). The user can then select two postures. Morphing is then conducted by linearly interpolating between the two specified joint angles for each DOF in the hand. The result is a series of infinitely many postures, as represented in Fig. 15.

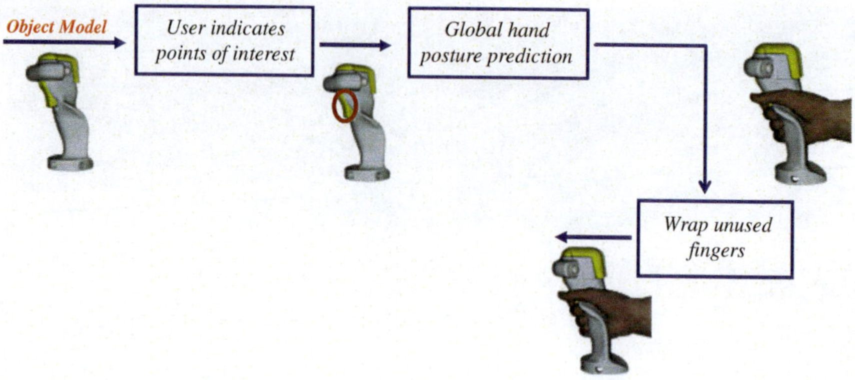

Fig. 16 Algorithm for precision grasping

4.2 Precision Grasping

The general approach to precision grasping entails predicting hand posture while including the global DOFs (three rotations and three translations) for the wrist as deign variables in (4). These new degrees of freedom may be included in the hand performance measure, which is used as the objective function in (4). Concurrently, user-specified points of interest are used as target points for corresponding user-specified end-effectors on the hand. Any number of target points (and corresponding end-effectors) may be specified. Any fingers that are not governed by specified targets are automatically wrapped around the object being grasped. That is, they are articulated until a collision detection algorithm indicated contact between the finger(s) and the object. This process is summarized in Fig. 16.

The novelty of this approach is the inclusion of the global DOFs of the hand with posture prediction, which inherently specify the position and orientation of the hand relative to the object being grasped. However, determining just how these DOFs factor in the objective function can be difficult and often degrades to an unscientific process of trial-and-error. This potential difficulty is resolved by integrating the hand with the whole-body posture prediction, as demonstrated in Fig. 10. Then, the global DOFs for the hand are represented by the position and orientation of the wrist, which are inherently determined by the predicted body posture.

4.3 Power Grasping

Figures 17 and 18 summarize the new methodology used to synthesize a grasp. This work is based on and extends the method proposed by Goussous et al. [1], which contends that power grasps are governed in large part by the shape of

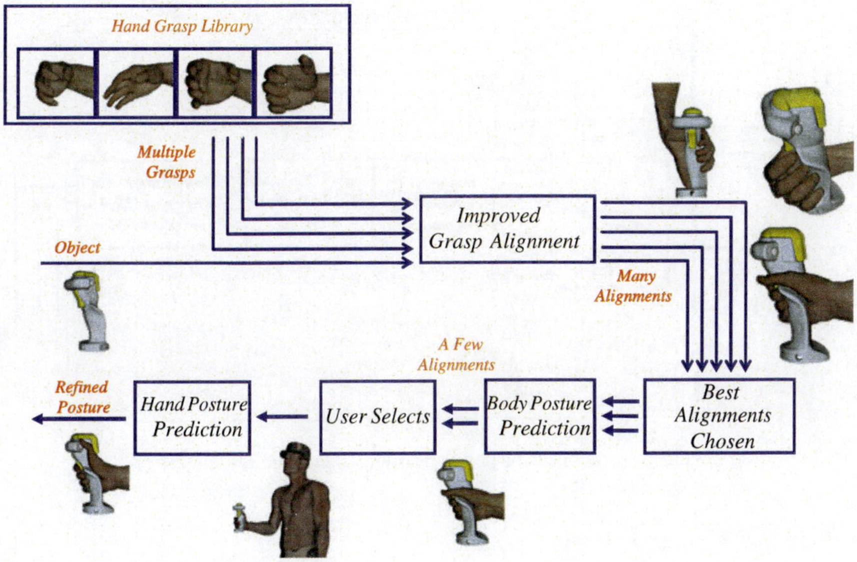

Fig. 17 Algorithm for power grasping

the object being grasped. With this approach, information about the object shape/geometry is first sampled in the virtual world. This information describes the polygonal composition of the object mesh. The program then iterates through a collection of stored hand grasp shapes (pre-defined grasp postures) that are frequently used in grasping, and for each shape the grasp alignment module calculates several possible hand alignments (position and orientation of the pre-defined grasp posture). Alignments that result in object collisions are discarded, and the best of the remaining alignments are input to an upper-body posture prediction module. This module calculates the proper whole-body joint angles that result in the hand being in the given position and orientation required by the grasp alignment, subject to joint limits. Grasp alignments that result in infeasible posture solutions are discarded, leaving the user with a small collection of acceptable hand shapes and alignments to choose from. The user can be involved in this process, and thus provide a cognitive element, by either making choices about the final outcome or by population the library of stored hand shapes, which essentially represents the user's history and knowledge of grasps.

Following is a detailed explanation of each of the components in Figs. 17 and 18. In general, the input for the overall algorithm is a library of hand grasp shapes. The output is a single hand shape with a specified orientation and position. The input for the grasp alignment component is again the library of hand shapes, and the output is subset of hand grasp shapes as well as a set of alignments (position and orientation) for each hand grasp shape.

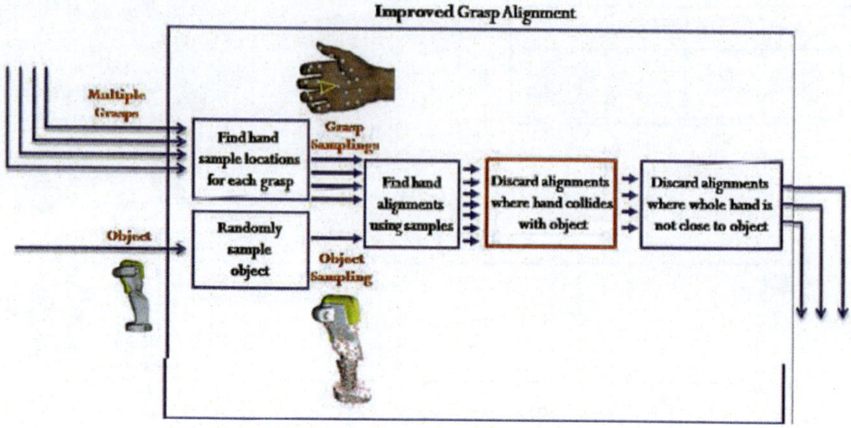

Fig. 18 Grasp alignment component of power grasping algorithm

4.4 Hand Grasp Library

The software application contains a default library of common hand grasp shapes [46]. New shapes can be created by manually manipulating the joints of the virtual hand or by using hand-posture prediction. For more accurate results, the grasp predictor can consider all of the hand grasps in the library, or for faster results, the user can specify a subset of these grasps to pass on to the next phase. Each selected grasp is passed to the grasp alignment phase, along with the object to be grasped.

4.5 Grasp Alignment

The grasp alignment phase attempts to find the most appropriate alignment for the most appropriate hand grasp shape. It does this by considering each hand grasp shape and finding the alignment for each grasp that produces the closest match between the hand and the object. Then, the *closeness* scores for the candidate alignment/grasp combinations are compared, and the best overall alignments and grasps are passed on to the next phase. Each stage in the grasp alignment phase, as shown in Fig. 18, is described in more detail below.

4.5.1 Hand Sampling

The hand samples are Cartesian points on the surface of the hand that serve to represent the contact surface in a simple manner. Unlike the object samples, these are chosen manually, and are defined by a given vector offset from a particular hand or finger joint. When assigning the samples, locations are selected

Fig. 19 Sampling Example

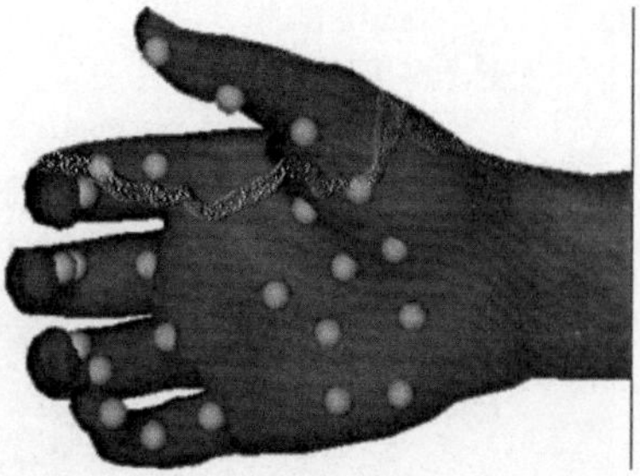

empirically in a way that captures the shape of the palm and fingers as precisely as possible. Special attention is paid to the deformable parts of the palm, which play a prominent role in power grasping. Twenty-four sample points for the hand are used for the results shown here. Figure 19 shows the hand samples that are used throughout the experiments. For each hand grasp shape passed to the alignment phase, the positions of the hand sample points are calculated relative to the global frame (wrist) given their local sample positions and the joint angles.

4.5.2 Object Sampling

We define an object sample point as a Cartesian point in 3-D (x, y, z) on the surface of an object. The sampling algorithm presented by Osada et al. [47] is used and allows the user to specify the number of sample points to be considered. The algorithm then chooses a polygon in the mesh to be sampled, at random with probability proportional to its area. Thus, there is a higher chance of sampling a large polygons than a small one, and the sample points are not concentrated in sharp parts of the object (where mesh polygons would be smaller), but instead are evenly spread out. After a polygon is chosen for sampling, the location of the sample point on the triangle is randomly chosen. This process is then repeated a number of times equal to the number of desired sample points.

4.5.3 Sample Alignment

The sample alignment stage takes as its input sample points for all of the selected hand grasp shapes from the library, and one set of object sample points. For each candidate hand shape, it is necessary to calculate the best hand position and orientation that most closely matches the hand sample points to the object samples. This is done with a random sample consensus algorithm [16, 48, 49]. A triplet of points on the hand is picked and designated as the control frame (Fig. 20).

Then, triangles formed by the object sample points are tested for their similarity to the control frame. Each possible triple of points on the object's surface is checked. For a triangle to be considered similar, the lengths of the corresponding sides must be within a user-specified threshold ε_d. For each triangle that matches

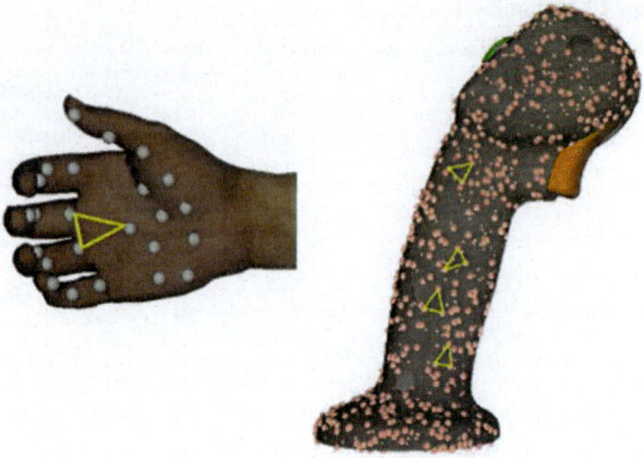

Fig. 20 The control frame on the hand and the partial frames on a joystick, shown in *yellow*

the control frame, a 4 × 4 transformation matrix is calculated that transforms the control frame to the partial frame.

4.5.4 Collision Detection

Next, the transformation matrix is used to transform the hand collision spheres (Fig. 10) to the object space. The hand spheres are then checked for intersection with any of the object spheres. If a particular alignment results in collisions, the alignment is discarded.

4.5.5 Closeness Calculation

Next, the hand samples are transformed by this matrix so that the object samples and hand samples are in the same space. The distance from each transformed hand sample to its nearest neighbor on the object surface is calculated, and transformations that result in samples violating the distance threshold are discarded.

The alignments are then sorted by their closeness score, which is the sum of the distances from the hand samples to their nearest object sample. The alignments with lowest closeness and that obey the distance threshold and the collision constraints are passed on to the next stage.

4.6 Integration with Body Posture Prediction

As with posture prediction and precision grasp prediction, it is important to integrate the hand with the rest of the DHM. However, because the grasp alignment process is relatively complex, the whole-body posture-prediction component is decoupled. That is, whole-body posture is predicted and evaluated

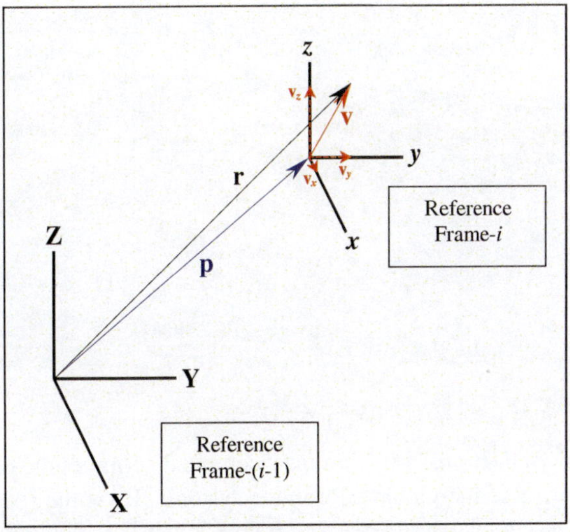

Fig. 21 Reference frame transformation

independently, based on input from the hand model. The input takes the form of the hand location and orientation. The location and orientation from the grasp alignment stage are essentially modeled as constraints for the hand in Fig. 7. The location of the wrist is used as a target point for the wrist joint-center on the body. The orientation of the hand is easily incorporated as an additional constraint in (4). The process of developing orientation constraints by leveraging the advantages of the DH-method is described as follows, based on the work of Goussous et al. [1].

With a specified set of parameters that characterizes how the local coordinate systems in Fig. 7 are related, a single transformation matrix in the DH-method transforms the coordinates of a point in reference frame-i to coordinates in terms of frame-($i-1$). The matrix can be decomposed as follows:

$$\mathbf{T} = \begin{bmatrix} \mathbf{R}_{3\times 3} & \mathbf{p}_{3\times 1} \\ \mathbf{0}_{1\times 3} & 0 \end{bmatrix} \quad (8)$$

The matrix R is responsible for the rotation of frame-i with respect to frame-($i-1$), and the vector p represents the translation of frame-i with respect to frame-($i-1$), as shown in Fig. 21. Thus, $\mathbf{r} = \mathbf{p} + \mathbf{R}_v$.

The columns of R represent the direction of the axes for frame-i, in terms of frame-($i-1$). The first column represents the direction in which the x-axis points; the second column represents the y-axis; and the third column represents the z-axis. By constraining portions of R, we can constrain the orientation of frame-i. Specifically, we can dictate in which direction (in terms of the global coordinate system) each axis of a local coordinate system points. This is done by specifying values for each column of R. Considering that the axes are orthogonal, only two axes can be constrained at a time.

Each transformation matrix for the Santos model ${}^{i-1}\mathbf{T}_i$ includes an independent rotation matrix ${}^{i-1}\mathbf{R}_i$. The cumulative transformation matrix ${}^{0}\mathbf{T}_n$ is the product of

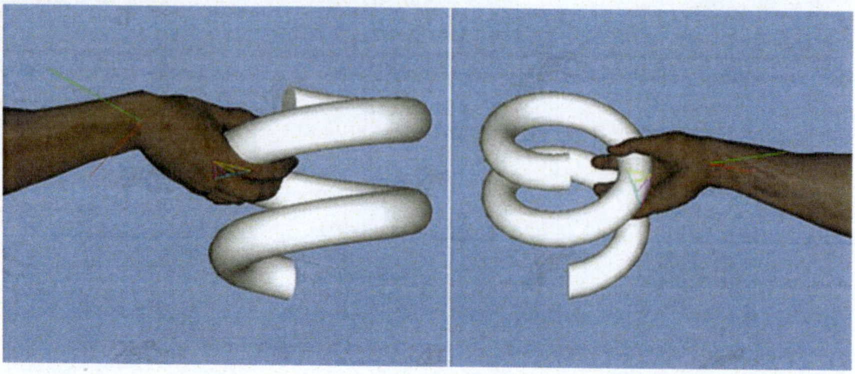

Fig. 22 Coil grasping task

all of the transformation matrices and determines the position of the end-effector in terms of the global coordinate system. ${}^{0}\mathbf{R}_n$ is the rotation matrix incorporated in this cumulative matrix, and it determines the orientation of the nth reference frame with respect to the global reference frame. Thus, to constrain the orientation of the hand, which is associated with the nth reference frame, we constrain the components of ${}^{0}\mathbf{R}_n$.

To write an independent constraint for each component of ${}^{0}\mathbf{R}_n$ can be cumbersome. Instead, we combine the components into a single inequality constraint as follows:

$$\left\{\left[{}^{0}R_n(1,1)-l_{11}\right]^2+\left[{}^{0}R_n(2,1)-l_{21}\right]^2+\left[{}^{0}R_n(3,1)-l_{31}\right]^2\right\} \\ +\left\{\left[{}^{0}R_n(1,2)-l_{12}\right]^2+\left[{}^{0}R_n(2,2)-l_{22}\right]^2+\left[{}^{0}R_n(3,2)-l_{32}\right]^2\right\}\leq\gamma \quad (9)$$

where γ is a small positive number approximating zero. For this study, $\gamma = 1 \times 10^{-8}$. This constraint is incorporated in the formulation in (4). l_{ij} indicates the constrained value for ${}^{0}R_n(i,j)$. The first three terms in (9) relate to the x-axis, while the second three terms relate to the y-axis. It is possible to simplify (9) and only constrain the orientation of one axis.

4.7 Results

The power-grasp process described above is demonstrated on an academic case and a more practical application. Figure 22 shows a grasp prediction result for a coil object. All of the hand shapes from the library were considered. The coil was positioned about chest high and within arm's reach of the avatar. The best overall alignment and grasp produced by the grasping algorithm are shown. This means that the particular hand shape in the alignment shown produced the best combination of closeness to the object and upper-body posture comfort (a discomfort

Fig. 23 Handlebar grasping task

performance measure is minimized in (4)) of all the various alignments and hand shapes tested by the algorithm. This example ran in under 10 s using an Intel Core Duo running at 3.0 GHz.

Figure 23 illustrates a second example, where Santos predicts an appropriate grasp for a motorcycle handlebar. Typically, with most currently available DHM tools, one would simulate a grasp for a case like this by manually manipulating the hand or drawing on a stored grasp posture. Such a process is possible with Santos as well. However, by using the newly developed grasp prediction capability, this process becomes relatively easy. Figure 23 shows the avatar's hand comfortably wrapped around the handlebar grip. The grasp prediction was quickly able to find a comfortable hand shape and a corresponding upper-body posture.

5 Conclusions

This chapter has presented a three pronged approach to hand simulation and analysis: model development, reach analysis, and grasping prediction. Each of these sections builds on the previous, with grasping prediction as a pinnacle of hand-based predictive capabilities. Although each aspect involves new research and development, this chapter also highlights critical elements that must be considered with any effort to model and/or simulate the hand. First, the most important aspect is the fidelity of the fundamental model and the freedom to alter the underlying model. Secondly, to be most effective, a virtual model should be predictive, able to predict performance based on sound mathematical models. Finally, in the context of predictive DHMs, the influence of the body on the hand is too great to disregard. Any model of the human hand should be integrated with a model of the human body.

A key element that is not addressed directly with this work is cognitive modeling. Grasping can depend heavily on decision-making and on past experiences. This potential dependence is mitigated to some extent by allowing for user input, both with regard to reach analysis and grasping. Although body posture and shape matching are key elements of grasping, some form of cognition needs to be involved. In the absence of an actual cognitive modeling engine, we allow the user to interact (use his/her inherent cognitive model) by either altering the database or by choosing from a set of returned results. Modifying the database is comparable to modeling one's memory, whereas interacting with algorithm is comparable to modeling decision-making abilities. Nonetheless, integration of grasping prediction with cognitive modeling is a fertile area for future work.

Another important area for future work is validation. With respect to posture prediction, the kinematic chains that form various fingers are not particularly redundant. That is, there is less potential variability in finger posture for a given target point than there is, for instance, with the arm. Nonetheless, as a matter of thorough scientific practice, all predictive capabilities must be validated objectively. To date, the work presented in this chapter has only been validated subjectively. With respect to grasping, validation becomes more critical, especially when considering a range of potential grasping strategies for more complex objects. Finally, although the methods described in this chapter constitute new and interesting developments in and of themselves, they also provide a platform for further studying how and why people behave the way they do. Development of modeling and simulation capabilities is simple a means to an end, the end typically being scientific study, education, and problem solving. One of the unique aspects of optimization-based prediction is the ability to study which performance measure(s) governs human behavior most accurately, and the methods presented in this chapter facilitate such studies well.

References

1. Goussous F, Marler T, Abdel-Malek K (2009) A new methodology for human grasp prediction. IEEE Trans Syst Man Cybern Part A Syst Humans 39(2):369–380
2. Carenzi F, Gorce P, Burnod Y, Maier M (2005) Using generic neural networks in the control and prediction of grasp postures. The European symposium on artificial neural networks, Bruges
3. Sanso RM, Thalmann D (1994) A hand control and automatic grasping system for synthetic actors. Proceedings of EUROGRAPHICS'94, vol 13, pp C168–C177
4. Miller A, Knoop S, Christensen H, Allen P (2003) Automatic grasp planning using shape primitives. IEEE Int Conf Robot Autom, ICRA'03, vol 2, pp 1824–1829
5. Rijpkema H, Girard M (1991) Computer animation of knowledge-based grasping. Proc ACM SIGGRAPH'91 25(4):339–348
6. Tomovic R, Bekey G, Karplus W (1987) A strategy for grasp synthesis with multifingered robot hands. Proc IEEE Int Conf Robot Autom 4:83–89
7. Xue Z, Kasper A, Zoellner JM, Dillmann R (2009) An automatic grasp planning system for service robots. Proceedings of the ICAR 2009 14th international conference on advanced robotics, Munich
8. Gorce P, Rezzoug N (2005) Grasping posture learning with noisy sensing information for a large scale of multifingered robotic systems. J Robot Syst 12:711–724
9. Jagannathan S, Galan G (2004) Adaptive critic neural network based object grasping control using a three finger gripper. IEEE Trans Neural Netw 15(2):395–407

10. Moussa M (2004) Combining expert neural networks using reinforcement feedback for learning primitive grasping behavior. IEEE Trans Neural Netw 15(3):629–638
11. Taha Z, Brown R, Wright D (1997) Modeling and simulation of the hand grasping using neural networks. Med Eng Phys 19(6):536–538
12. Pelossof R, Miller A, Allen P, Jebara T (2004) An SVM learning approach to robotic grasping. Proc IEEE Int Conf Robot Autom 4:3512–3518
13. Ferrari C, Canny J (1992) Planning optimal grasps. Proceedings of the 1992 IEEE international conference on robotics and automation, Nice
14. Katada Y, Svinin M, Matsumura Y, Ohkura K, Ueda K (2001) Optimization of stable grasps by evolutionary programming. Proceedings of the 32nd international symposium on robotics, pp 1503–1508
15. Kim B, Yi B, Oh S, Suh I (2004) Non-dimensionalized performance indices based optimal grasping for multi-fingered robot hands. Mechatronics 14(3):255–280
16. Li Y, Pollard N (2005) A shape matching algorithm for synthesizing humanlike enveloping grasps. IEEE-RAS international conference on humanoid robots (Humanoids 2005)
17. Liu G, Xu J, Wang X, Li Z (2004) On quality functions for grasp synthesis, fixture planning and coordinated manipulation. IEEE Trans Autom Sci Eng 1(2):146–162
18. Borst C, Fischer M, Hirzinger G (1999) A fast and robust grasp planner for arbitrary 3-D objects. Proceedings of the IEEE international conference on robotics and automation, Detroit, May
19. Hester, R., Cetin, M., Kapoor, C., and Tesar, D. (1999), "A criteria-based approach to grasp synthesis", *Proceedings of the 1999 IEEE International Conference on Robotics and Automation*, Detroit, Michigan
20. Toth E (1999) Stable object grasping with dexterous hand in three dimensions. Periodica Polytechnica SER EL. ENG 43(3):207–214
21. Berenson D, Kuffner J, Choset H (2008) An optimization approach to planning for mobile manipulation. Proceedings of the IEEE international conference on robotics and automation, Pasadena, May
22. Fernandez J, Walker I (1998) Biologically inspired robot grasping using genetic programming. Proceedings of the 1998 IEEE international conference on robotics and automation, Leuven
23. Globisch R (2005) Automated grasping for articulated structures using evolutionary learning algorithms. Master's thesis, University of Johannesburg, South Africa, April 2005
24. ElKoura G, Singh K (2003) Handrix: animating the human hand. In: ACM SIGGRAPH/ Eurographics symposium on computer animation, 2003
25. Ehrenmann M, Rogalla O, Zollner R, Dillmann R (2001) Teaching service robots complex tasks: programming by demonstration for workshop and household environments. Proceedings of the 2001 international conference on field and service robots, vol 1, Helsinki, pp 397–402
26. Bohg J, Kragic D (2009) Learning grasping points with shape context. Robot Auton Syst 58(4):362–377
27. Miyata N, Kouchi M, Mochimaru M (2006) Posture estimation for screening design alternatives by DhaibaHand—cell phone operation. Proceedings of the SAE 2006 digital human modeling for design and engineering conference, 2006-01-2327
28. Aleotti J, Caselli S (2006) Grasp recognition in virtual reality for robot pregrasp planning by demonstration. Proceedings of the 2006 IEEE international conference on robotics and automation, Orlando
29. Abdel-Malek K, Arora J, Yang J, Marler T, Beck S, Kim J, Swan C, Frey-Law L, Kim J, Bhatt R, Mathai A, Murphy C, Rahmatalla S, Patrick A, Obusek J (2009) A physics-based digital human model. Int J Veh Des 51(3/4):324–340
30. Marler T, Arora J, Beck S, Lu J, Mathai A, Patrick A, Swan C (2008) Computational approaches in DHM. In: Duffy VG (ed) Handbook of digital human modeling for human factors, ergonomics. Taylor and Francis Press, London
31. Yang J, Kim JH, Abdel-Malek K, Marler T, Beck S, Kopp GR (2007) A new digital human environment and assessment of vehicle interior design. Comput-Aided Des 39:548–558

32. Denavit J, Hartenberg RS (1955) A kinematic notation for lower-pair mechanisms based on matrices. J Appl Mech 22:215–221
33. Pena-Pitarch E, Yang J, Abdel-Malek K (2005) SANTOS^{TM} Hand: a 25 degree-of-freedom model. SAE International, Iowa City, June 14–16, 2005-01-2727 DHM
34. Liu Q, Marler T, Yang J, Kim J, Harrison C (2009) Posture prediction with external loads—a pilot study. SAE Int J Passeng Cars Mech Syst 2(1):1014–1023
35. Marler RT, Arora JS, Yang J, Kim H–J, Abdel-Malek K (2009) Use of multi-objective optimization for digital human posture prediction. Eng Optim 41(10):295–943
36. Marler T, Yang J, Rahmatalla S, Abdel-Malek K, Harrison C (2007) Validation methodology development for predicted posture. SAE digital human modeling conference, June, Seattle, Society of Automotive Engineers, Warrendale
37. Gill P, Murray W, Saunders A (2002) SNOPT: an SQP algorithm for large-scale constrained optimization. SIAM J Optim 12(4):979–1006
38. Rmstrong TJ, Chaffin DB (1978) An investigation of the relationship between displacements of the finger and wrist joints and the extrinsic finger flexor tendons. Biomechanics 11:119–128
39. Johnson R, Smith BL, Penmatsa R, Marler T, Abdel-Malek K (2009) Real-time obstacle avoidance for posture prediction. SAE digital human modeling conference, June, Goteborg, Society of Automotive Engineers, Warrendale
40. Johnson R, Fruehan C, Schikore M, Marler T, Abdel-Malek K (2010) New developments with collision avoidance for posture prediction. 3rd international conference on applied human factors and ergonomics, July, Miami
41. Yang J, Verma U, Penmatsa R, Marler T, Beck S, Rahmatalla S, Abdel-Malek K, Harrison C (2008) Development of a zone differentiation tool for visualization of postural comfort. *SAE 2008 World Congress*, April, Detroit, Society of Automotive Engineers, Warrendale
42. Yang J, Verma U, Marler T, Beck S, Rahmatalla S, Harrison C (2009) Workspace zone differentiation tool for visualization of seated postural comfort. Int J Ind Ergon 39:267–276
43. Lorensen W, Cline H (1987) Marching cubes: a high resolution 3-D surface construction algorithm. Comput Graph (SIGRAPH 87 Proc) 21(4):163–170
44. Napier J (1956) The prehensile movements of the human hand. J Bone Joint Surg 38B(4): 902–913
45. Cutkosky MR (1989) On grasp choice, grasp models, and the design of hands for manufacturing tasks. IEEE Trans Robot Autom 5(3):269–279
46. Feix T, Pawlik R, Schmiedmayer H, Romero J, Kragic D (2009) The generation of a comprehensive grasp taxonomy. In: Robotics, science and systems conference: workshop on understanding the human hand for advancing robotic manipulation, June
47. Osada R, Funkhouser T, Chazelle B, Dobkin D (2002) Shape distributions. ACM Trans Graph 21(4):807–832
48. Chen C, Hung Y, Cheng J (1997) RANSAC-based DARCES: a new approach to fast automatic registration of partially overlapping range images. Technical Report, Institute of Information Science, Academia Sinica, TR-IIS-97-019
49. Fischler M, Bolles R (1981) Random sample consensus: a paradigm for model fitting with applications to image analysis and cartography. Commun ACM 24(6):381–395

About the Editor

Hoang Pham is currently Professor and Chair of the Department of Industrial and Systems Engineering at Rutgers University. Before joining Rutgers university, he was a senior engineering specialist at the Boeing Company, Seattle, and the Idaho National Engineering Laboratory, Idaho Falls. He received the B.S. degree in mathematics, B.S. degree in computer science, both with high honors, from Northeastern Illinois University, Chicago, the M.S. degree in statistics from the University of Illinois, Urbana-Champaign, and the M.S. and Ph.D. degrees in industrial engineering from the State University of New York at Buffalo. His research interests include software reliability, system reliability modeling, maintenance and system ability modeling.

He is the author or coauthor of six books, and has published more than 110 journal articles, 50 conference papers, and edited more than 10 books including the *Handbook of Reliability Engineering* and *Springer Handbook of Engineering Statistics*. He is editor-in-chief of the *International Journal of Reliability, Quality and Safety Engineering*, associate editor of the *IEEE Transaction on Systems, Man and Cybernetics*, and an editorial board member of a dozen journals. He is also the editor of Springer Series in Reliability Engineering and World Scientific Series of Industrial and Systems Engineering. He has been conference chair and program chair of over 30 international conferences and workshops. He has received numerous awards including the 2009 IEEE Reliability Society Engineer of the Year Award. He is a fellow of the IEEE and fellow of IIE.

H. Pham (ed.), *Safety and Risk Modeling and Its Applications*,
Springer Series in Reliability Engineering, DOI: 10.1007/978-0-85729-470-8,

Index

H. Pham (ed.), *Safety and Risk Modeling and Its Applications*,
Springer Series in Reliability Engineering, DOI: 10.1007/978-0-85729-470-8,